Neuroradiology Board's Favorites

Samer Hoz

Asmaa H. AL-Sharee · Mustafa Ismail
Ali A. Dolachee · Osman Elamin
Oday Atallah · Maliya Delawan

Editors

Neuroradiology Board's Favorites

100 MRI-Based Pathology-Proven Cases
Supplied with 170 MCQs

 Springer

Editors
Samer Hoz
Department of Neurosurgery
University of Pittsburgh Medical Center
Pittsburgh, PA, USA

Mustafa Ismail
Department of Neurosurgery
Neurosurgery Teaching Hospital
Baghdad, Iraq

Osman Elamin
Department of Neurosurgery
Jordan Hospital and Medical Center
Amman, Jordan

Maliya Delawan
College of Medicine
Gulf Medical University
Al Jurf, Ajman, United Arab Emirates

Asmaa H. AL-Sharee
Department of Neuroradiology
Neurosurgery Teaching Hospital
Baghdad, Iraq

Ali A. Dolachee
Department of Surgery
University of Baghdad
Baghdad, Iraq

Oday Atallah
Department of Neurosurgery
Hannover Medical School
Hannover, Niedersachsen, Germany

ISBN 978-3-031-64263-0   ISBN 978-3-031-64261-6   (eBook)
https://doi.org/10.1007/978-3-031-64261-6

This Springer imprint is published by the registered company Springer Nature Switzerland AG
The registered company address is: Gewerbestrasse 11, 6330 Cham, Switzerland

If disposing of this product, please recycle the paper.

*To my lovely family: Sawsan, Saad, Arwa,
Farah, Ward, Samhar, Anoona, Sanaa, Faris,
and Luay. To the GYL team, Prof. Andaluz,
Prof. Prestigiacomo, Kathleen Smith, and
Paolo Palmisciano.*
Samer S. Hoz

*To my Dad (May Allah bless his soul) and to
my Mom and to my sister and brother and my
lovely family.*
Asmaa H. AL-Sharee

To my mom Khafia who I adore and love.
Mustafa Ismail

*To my lovely family which supports me in
this long way, neurosurgery.*
Ali A. Dolachee

*To my exceptional parents, Fatima and
Alamin, my guiding lights. To my rock, my
beloved wife Seroona, your support knows no
bounds. To my cherished sisters, Samah and
Suhaila, my endless love.*
Osman Elamin

*To my homeland, Palestine, as well as to
Emad, Khuloud, Aseel, and Lara, my
cherished family, for being the heartbeat of
my creative journey.*
Oday Atallah

Foreword 1

Like most empirical sciences, medicine builds a new layer of information daily. Hence, the task of embracing the entirety of the canon is impossible for any individual. Fortunately, the internet is a suitable depository of a wealth of information. Thus, the formidable task ahead for the medical educators is to transform the immense volume of data into knowledge and knowledge into understanding.

Many outstanding neuroradiology textbooks are a good compass towards the sources of information. This text ventures one step forward. The authors and editors shaped the chapters so that the reader glides from data into knowledge and, with self-reflection, the devoted reader achieves understanding of clinical neuroradiology.

Those who care for individual patients know very well that no two identical representations of any pathology exist. This fact is true in any branch of medicine and more so in clinical neurosciences. For that reason, the excellent clinician aims to comprehend the mechanism of the disease and how neuroimaging tools represent it. Sometimes, the busy neurologist, neurosurgeon, or neuroradiologist is contended by a single sentence summarizing the diagnosis and hurries into writing a report or setting a course of action. But, also, there are those less in number but by no means few whose gaze searches for curves, for shades, and even for the condition of structures centimeters away from the region of interest. With its trove of explanatory paragraphs for each correct answer, this book is for the curious among the many representatives of clinical neurosciences.

We should not be fooled by the apparent practical intention behind its conception. Besides the treasure trove of tools for succeeding in the board exam, the reader will discover in this text the three steps of the epistemological arch; those who want information will find plenty of clinical pearls, those who wish knowledge will also find them in the rich explanatory paragraphs, and those who want to understand the subject will reach that stage once they read *Neuroradiology Board's Favorites* from cover to cover.

I will not delay your trip into the intellectual adventure of reading this book.

Jorge A. Lazareff
Emeritus, Department of Neurosurgery,
David Geffen School of Medicine,
University of California Los Angeles, UCLA,
Los Angeles, CA, USA

Foreword 2

The best books are those which are clear and illustrative. That is especially true in books targeting students. This book, *Neuroradiology Board's Favorites*, is an excellent example of such a book. Young neurosurgeons preparing for professional examinations will find it the most valuable aid. It covers the main neuroradiology topics in the brain and spine based on MRI images of proven pathology followed by related MCQs.

Writing such a book is always challenging as the authors went through hundreds of cases to select the well-documented 100 cases.

Being a neurosurgeon and teacher, one feels so proud of these rising stars, young neurosurgeons, and neuroradiologists to author this board favorite book.

We are confident that the book will be of great help not only to trainees but also to qualified neurosurgeons refreshing their knowledge.

A. Hadi Al Khalili
Emeritus, Department of Neurosurgery,
Baghdad University,
Baghdad, Iraq

Key Features

- *The Neuroradiology Board's Favorites* is a case-based review book focused on the imaging of pathology-proven lesions related to the central nervous system.
- The mission of this book is to help readers revise the core concepts and maintain knowledge of neuroradiology from radiology, neurology, and neurosurgical perspectives.
- This study companion has a unique design and is structured in two sections, totaling *15 chapters*. It includes *100 clinical scenarios and imaging*, supplemented by more than *150 multiple-choice questions* in a convenient format suitable for self-study.
- The multiple-choice questions are attached to each case scenario in a style that mirrors the format adopted by the majority of local, regional, and international board examinations. The answers and explanations appear immediately below the questions to enhance readability.
- This book is an adjunct to existing texts and does not intend to be the primary source of information; it aims to help readers identify their relevant strengths and weaknesses in the area and is based on the most up-to-date best practice evidence.
- This book is targeted at students of radiology, neurology, neurosurgery, and neurosciences. In addition, it can be beneficial for residents, fellows, and young attendings to prepare for exams or practice.

Preface

Dear Reader,

We are delighted to present to you *Neuroradiology Board's Favorites*, a case-based review book focused on the imaging of pathology-proven lesions related to the central nervous system. The field of neuroradiology, given its wide spectrum of conditions, stands as a challenging discipline to master. Becoming proficient in neuroradiology requires dedication, a keen eye for detail, and the ability to connect clinical information with radiological findings.

Neuroradiology Board's Favorites is meticulously crafted to help students, residents, fellows, as well as junior attending physicians and surgeons excel in this field. Within these chapters, you will encounter a rich collection of 100 clinical scenarios with accompanying imaging studies, providing a comprehensive overview of the subject matter.

This book employs a systematic approach to foster a deep understanding of central nervous system pathologies and enhance diagnostic capabilities. For each case, you will find detailed imaging descriptions and explanatory sections highlighting characteristic features associated with each presented disease condition, along with over 150 multiple-choice questions to solidify essential knowledge. In addition, each case is supplemented with clinical information related to the specific disease condition, effectively bridging the gap between theory and practice.

As you delve into the chapters, we sincerely hope that *Neuroradiology Board's Favorites* becomes a valuable companion in your journey to excellence in neuroradiology.

Contents

Contributors

Rokaya H. Abdalridha College of Medicine, Babylon University, Babylon, Iraq

Alkawthar M. Abdulsada Azerbaijan Medical University, Baku, Azerbaijan

Fatimah O. Ahmed College of Medicine, Al-Mustansiriyah University, Baghdad, Iraq

Awfa Aktham Department of Neurosurgery, Tokyo General Hospital, Nakano, Japan

Teeba A. Al-Ageely College of Medicine, University of Baghdad, Baghdad, Iraq

Zainab K. A. Al-Araji College of Medicine, Al-Nahrain University, Baghdad, Iraq

Zinah A. Alaraji College of Medicine, Al-Nahrain University, Baghdad, Iraq

Ali Q. Al-Asady College of Medicine, University of Baghdad, Baghdad, Iraq

Sajjad G. Al-Badri College of Medicine, University of Baghdad, Baghdad, Iraq

Sama Albairmani College of Medicine, University of Al-Iraqia, Baghdad, Iraq

Saja A. Albanaa College of Medicine, University of Baghdad, Baghdad, Iraq

Sadeem A. Albulaihed College of Medicine, Alfaisal University, Riyadh, Saudi Arabia

Usama AlDallal School of Medicine, Royal College of Surgeons In Ireland—Medical University of Bahrain, Busaiteen, Bahrain

Linah Alduraibi Sulaiman Al Rajhi University, Qassim, Saudi Arabia

Mostafa H. Algabri College of Medicine, University of Baghdad, Baghdad, Iraq

Hagar A. Algburi College of Medicine, University of Baghdad, Baghdad, Iraq

Mohammed E. Al-Hamadani College of Medicine, University of Baghdad, Baghdad, Iraq

Younus M. Al-Khazaali College of Medicine, Al-Nahrain University, Baghdad, Iraq

Fatimah K. Al-Kishawi Al-Kindy College of Medicine, University of Baghdad, Baghdad, Iraq

Wafa M. Almuallim College of Medicine, Al-Rayan Colleges, Al-Madinah Al-Monawarah, Saudi Arabia

Muntadher H. Almufadhal College of Medicine, University of Baghdad, Baghdad, Iraq

Mahmood H. AlObaidy College of Medicine, Al-Nahrain University, Baghdad, Iraq

Asmaa H. AL-Sharee Department of Neuroradiology, Neurosurgery Teaching Hospital, Baghdad, Iraq

Khalid M. Alshuqayfi Faculty of Medicine, King Abdulaziz University, Jeddah, Saudi Arabia

Mahmood F. Alzaidy College of Medicine, University of Baghdad, Baghdad, Iraq

Oday Atallah Hannover Medical School, Hannover, Germany

Ameer M. Aynona College of Medicine, University of Babylon, Babylon, Iraq

Ahmed Y. Azzam October 6 University Faculty of Medicine, Giza, Egypt

Muslim M. Badr Department of Neuroradiology, Neurosurgery Teaching Hospital, Baghdad, Iraq

Mahmood S. Bilal College of Medicine, Al-Mustansiriyah University, Baghdad, Iraq

Maliya Delawan College of Medicine, Gulf Medical University, Ajman, United Arab Emirates

Ali A. Dolachee Department of Surgery, Al-Kindy College of Medicine, University of Baghdad, Baghdad, Iraq

Osman Elamin Neurosurgery Department, Jordan Hospital and Medical Center, Amman, Jordan

Toka Elboraay Faculty of Medicine, Zagazig University, Elsharqia, Egypt

Minaam Farooq King Edward Medical University, Mayo Hospital Lahore, Lahore, Pakistan

Hussein M. Hasan College of Medicine, University of Baghdad, Baghdad, Iraq

Samer S. Hoz University of Pittsburgh Medical Center (UPMC), Pittsburgh, PA, USA

Abbas F. A. Hussein College of Medicine, Babylon University, Babylon, Iraq

Osama S. Idris College of Medicine, University of Mansoura, Mansoura, Egypt

Khadija J. Ismael College of Medicine, Diyala University, Diyala, Iraq

Mustafa Ismail College of Medicine, University of Baghdad, Baghdad, Iraq

Neurosurgery Teaching Hospital, Baghdad, Iraq

Hasan M. Jabbar College of Medicine, University of Misan, Misan, Iraq

Sajjad N. Majeed College of Medicine, Al-Mustansiriyah University, Baghdad, Iraq

Sara A. Mohammad College of Medicine, University of Baghdad, Baghdad, Iraq

Tabarek F. Mohammed College of Medicine, University of Baghdad, Baghdad, Iraq

Ahmed Muthana College of Medicine, University of Baghdad, Baghdad, Iraq

Ali M. Neamah College of Medicine, University of Baghdad, Baghdad, Iraq

Zainab Q. Saadi College of Medicine, University of Baghdad, Baghdad, Iraq

Saleh A. Saleh College of Medicine, University of Baghdad, Baghdad, Iraq

Haneen A. Salih Department of Molecular and Medical Biotechnology, College of Biotechnology, Al-Nahrain University, Baghdad, Iraq

Maen Saris Dar Al-Shifaa Hospital, Kuwait, Kuwait

Abbreviations

3D	Three dimensional
A	Anterior cerebral artery segments (A1–A5)
ABC	Aneurysmal bone cysts
ACA	Anterior cerebral artery
ADC	Apparent diffusion coefficient (MRI)
ADEM	Acute disseminated encephalomyelitis
AIDS	Acquired immunodeficiency syndrome
AVM	Arteriovenous malformation
C	Cervical nerve roots (C1–C8)
C+	Contrast
CCVMs	Cerebral cavernous venous malformations
Cho	Choline
CISS	Constructive interference in steady state (MRI)
CN	Cranial nerve
CNS	Central nervous system
CPA	Cerebellopontine angle
Cr	Creatine
CRS	Caudal regression syndrome
CSF	Cerebrospinal fluid
CT	Computed tomography
CTA	Computed tomography angiography
D	Dorsal nerve roots (D1–D12)
DAI	Diffuse axonal injury
DNETs	Dysembryoplastic neuroepithelial tumors
DRIVE	Driven equilibrium radiofrequency reset pulse (MRI)
DSA	Digital subtraction angiography
DVA	Deep venous anomaly
DWI	Diffusion-weighted imaging (MRI)
EPI	Echo planner imaging
FASI	Focal areas of signal intensity
FCD	Focal cortical dysplasia
FDG	F-fluorodeoxyglucose
FFE	Fast field echo (MRI)
FIESTA	Fast imaging employing steady-state acquisition

FLAIR	Fluid-attenuated inversion recovery
FSE	Fast spin echo (MRI)
FXTAS	Fragile X-associated tremor/ataxia syndrome
GBM	Glioblastoma multiforme
GCT	Giant cell tumor
(Gd)	Gadolinium
GdDTPA	Gadolinium-diethyle-netriamine penta-acetic acid
GRE	Gradiant recalled echo (MRI)
HIE	Hypoxic-ischemic encephalopathy
IAC	Internal auditory canal
ICA	Internal carotid artery
ICP	Intracranial pressure
IIH	Idiopathic intracranial hypertension
IV	Intravenous
L	Lumbar nerves root (L1–L5)
LTM	Lateral thoracic meningocele
M:F	Male to female ratio
MCA	Middle cerebral artery
MCP	Middle cerebellar peduncles
MELAS	Mitochondrial encephalomyopathy with lactic acidosis and stroke-like episodes
MOG	Myelin oligodendrocyte glycoprotein
MOGAD	Myelin oligodendrocyte glycoprotein antibody-associated disease
MPR	Multiplaner reformation (CT)
MRA	MR angiography
MRI	Magnetic resonance imaging
MRS	MR spectrometry
MS	Multiple sclerosis
NAA	N-acetylaspartate
NF	Neurofibromatosis (NF1, NF2)
PD	Proton density (MRI)
PET	Positron emission tomography
PXA	Pleomorphic xanthoastrocytomas
rCBV	Relative cerebral blood volume
S	Sacral nerve roots (S1–S5)
SB	Syringobulbia
SBH	Subcortical heterotopia
SBL	Secondary bone lymphoma
SCM	Split cord malformation
SDAVF	Spinal dural arteriovenous fistulae
SPECT	Single-photon emission CT
SPIR	Spectral presaturation with inversion recovery
STIR	Short tau inversion recovery (MRI)
SWI	Susceptibility-weighted imaging (MRI)
T1W	T1-weighted

T2W	T2-weighted
TB	Tuberculosis
TCS	Tethered cord syndrome
TS	Tuberous sclerosis
TSCI	Traumatic spinal cord injury
WHO	World Health Organization

Part I

Brain

Congenital Brain Lesions

Sajjad G. Al-Badri, Mahmood S. Bilal, Ameer M. Aynona, Toka Elboraay, Younus M. Al-Khazaali, Osman Elamin, and Asmaa H. AL-Sharee

Case 1: Aqueductal Stenosis

Case Scenario

A 5-year-old male presented with signs and symptoms of raised intracranial pressure (ICP), including headache, vomiting, and a decreased level of consciousness (Fig. 1.1).

Imaging Description

MRI demonstrates moderate to marked dilatation of the lateral and third ventricles. There is no trans-ependymal edema, suggesting a chronic course. The fourth

S. G. Al-Badri
College of Medicine, University of Baghdad, Baghdad, Iraq

M. S. Bilal
College of Medicine, Al-Mustansiriyah University, Baghdad, Iraq

A. M. Aynona
College of Medicine, University of Babylon, Babylon, Iraq

T. Elboraay
Faculty of Medicine, Zagazig University, Elsharqia, Egypt

Y. M. Al-Khazaali
College of Medicine, Al-Nahrain University, Baghdad, Iraq

O. Elamin
Neurosurgery Department, Jordan Hospital and Medical Center, Amman, Jordan

A. H. AL-Sharee (✉)
Department of Neuroradiology, Neurosurgery Teaching Hospital, Baghdad, Iraq

S. Hoz et al. (eds.), *Neuroradiology Board's Favorites*,
https://doi.org/10.1007/978-3-031-64261-6_1

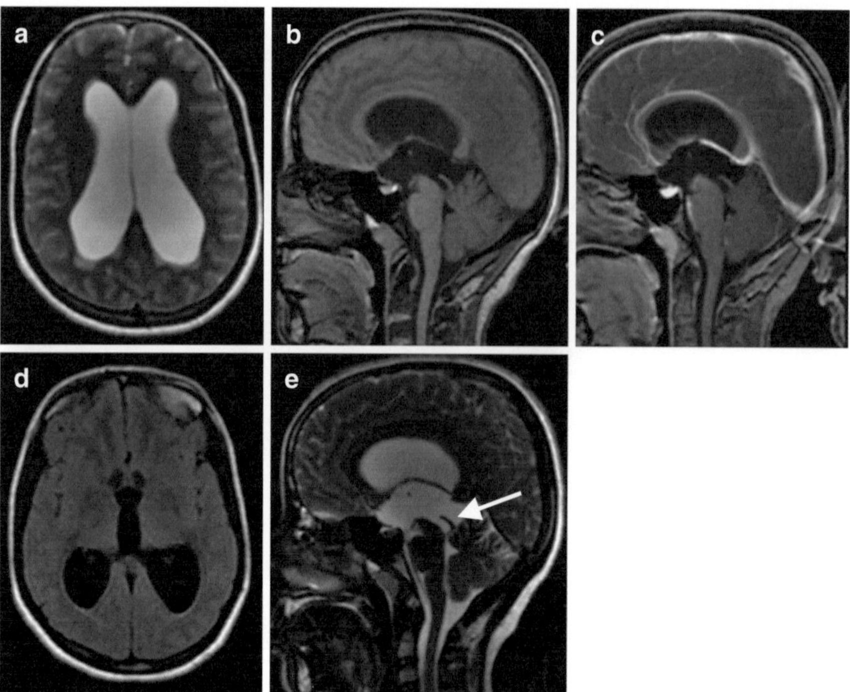

Fig. 1.1 Serial MRI images of the brain with the following: (**a**) axial T2-weighted (T2W), (**b**) sagittal T1, (**c**) sagittal T1-weighted (T1W) post-contrast, (**d**) axial fluid-attenuated inversion recovery (FLAIR), and (**e**) sagittal T2. (Figure courtesy of Dr. Samer Hoz)

ventricle, however, is within normal size. The proximal segment of the cerebral aqueduct of Sylvius is dilated. There is a small membrane below the third ventricle at the distal part of the cerebral aqueduct that is consistent with the aqueductal web (black arrow).

Aqueductal Stenosis

Aqueductal stenosis commonly presents as congenital obstructive hydrocephalus in infants, although it can also present as an acquired abnormality in adults. Causes of obstruction of the aqueduct of Sylvius can be intrinsic or extrinsic. Intrinsic obstruction may be secondary to infection or intraventricular hemorrhage, or be due to an aqueductal web. On the other hand, extrinsic compression may be due to a pineal tumor, tectal plate glioma, or other pathologies. In some cases, the etiology is idiopathic. In either case, aqueductal stenosis leads to the pathologic accumulation of cerebrospinal fluid (CSF) within the lateral and third ventricles.

Congenital aqueductal stenosis has an estimated incidence of about 1:5000 births. Fetal hydrocephalus with a near-normal posterior fossa may be observed on an antenatal exam. Secondary thinning of the cortical mantle and secondary macrocephaly can also occur. Ultrasound typically demonstrates dilatation of the third and lateral ventricles, as well as macrocephaly.

MRI scans are typically obtained from patients with suspected aqueductal stenosis as they provide more detail on the extent of obstructive hydrocephalus. The aqueduct may show superior funneling, and the distal lumen appears absent, with a normal-sized fourth ventricle. The underlying abnormality, such as a web or tumor, may also be evident in cases of secondary obstruction. While sagittal T2 is useful in detecting an absence of flow-void signal intensity at the aqueduct, sagittal constructive interference in steady state (CISS) is best at demonstrating the presence of an obstructing web. CSF flow study shows reduced aqueductal stroke volume and peak systolic velocity.

CT is of limited value in diagnosing aqueduct stenosis but may show a point of transition from a dilated to a non-dilated CSF space. Tectal tumors may appear as small, iso-dense lesions that may or may not enhance with the administration of contrast. However, CT is unreliable in ruling out the presence of brainstem tumors [1, 2].

Questions

1. **Aqueductal stenosis, the FALSE answer is:**
   A. Brain CT is of limited value in diagnosing aqueductal stenosis.
   B. Brain MRI may show proximal aqueductal funneling and a distally absent lumen.
   C. Aqueductal stenosis leads to dilation of the third and lateral ventricles on MRI.
   D. Tectal tumors are usually small and hyper-dense to the surrounding brain on CT imaging.
   E. The absence of flow-void signal intensity at the aqueductal level on sagittal T2 imaging is suggestive of aqueductal stenosis.
   The answer is D.
   Tectal tumors are usually small and iso-dense to the surrounding brain on CT imaging.

2. **Aqueductal stenosis, the FALSE answer is:**
   A. Features of fetal hydrocephalus with a near-normal posterior fossa can be observed on antenatal exams.
   B. CSF flow study can reveal decreased aqueductal stroke volume and peak systolic velocity in aqueductal stenosis.
   C. Tectal plate glioma can lead to aqueductal stenosis.
   D. Aqueductal stenosis can also occur due to extrinsic compression by a pineal tumor.
   E. Sagittal T2 imaging is the most effective sequence for demonstrating an obstructing web in aqueductal stenosis.
   The answer is E.
   Sagittal CISS imaging is the most effective sequence for demonstrating an obstructing web in aqueductal stenosis.

Case 2: Heterotopia and Cortical Lesions

Case 2.1: Subcortical Band Heterotopia (SBH) with Para-sagittal Arachnoid Cyst

Case Scenario
A 6-year-old male presented with seizures (Fig. 1.2).

Imaging Description
There are multiple large nodular heterotopic grey matter lesions composed of small gyri noted within the white matter of the left frontoparietal region. They are continuous with and extend from the ventricle to the white matter towards the cortex. They are seen compressing the frontal horn of the lateral ventricle, causing a deformation of its shape. A post-contrast study shows no abnormal enhancement. These features are consistent with SBH (arrow).

There is a well-defined, oval-shaped T1 low/T2 high signal-intensity cystic lesion noted in the left para-sagittal region. It is suppressed on FLAIR and shows no

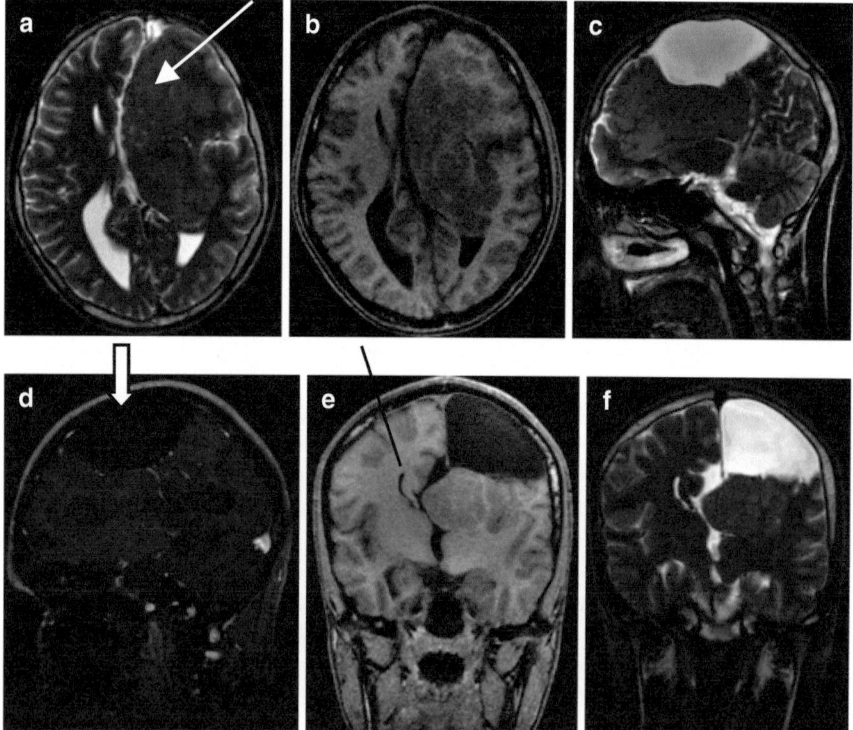

Fig. 1.2 Serial MRI images of the brain with the following: (**a**) axial T2, (**b**) axial T1, (**c**) sagittal T2, (**d**) sagittal T1 C+, (**e**) coronal T1, and (**f**) coronal T2. (Figure courtesy of Dr. Samer Hoz)

diffusion restriction on DWI/ADC. No contrast enhancement is seen on post-contrast images, suggesting an arachnoid cyst (open arrow).

On axial views, the ventricle appears similar to a racing car, consistent with corpus callosum agenesis (line).

Subcortical Band Heterotopia

Subcortical band heterotopia (SBH), also known as double cortex syndrome, is a rare neurodevelopment disorder characterized by the presence of a band of poorly organized grey matter extending from the ventricles to the cortical mantle. SBH predominantly affects females, and the pathogenesis involves mutations in the DCX or LIS1, gene resulting in the arrest of neuron migration in the subcortical or deep white matter. Three distinct subtypes exist for this condition:

1. Nodular form: extends from the ventricle into the white matter (as in this case).
2. Curvilinear form: extends from the cortex into the underlying white matter.
3. Mixed form.

The term "laminar heterotopia" is sometimes used ambiguously to refer to SBH. Furthermore, the condition is sometimes confused with lissencephaly type I-subcortical band heterotopia spectrum, which is a distinct disease entity based on imaging and genetic findings. Characteristic imaging features include an abnormally thick cortex with the decreased or absent formation of cerebral convolutions. Other differential diagnoses include Taylor dysplasia and subependymal grey matter heterotopia [3, 4].

Differential Diagnosis

Radiological features that help distinguish SBH from subependymal grey matter heterotopia and lissencephaly type I-subcortical band heterotopia spectrum include:

1. Generalized reduction in the size of the hemisphere.
2. Distorted ventricles.
3. Abnormal white matter that can be diminished or absent.
4. Thinned overlying cortex with shallow sulci.
5. Distorted basal ganglia.

Questions

1. **SBH, the FALSE answer is:**
   A. The term "laminar heterotopia" is sometimes used to refer to SBH.
   B. Generalized reduction in the size of the hemisphere can be found by imaging.
   C. Can be detected by the presence of an abnormally thick cortex with reduced or absent formation of the cerebral convolutions.
   D. Imaging may show diminished or absent white matter.
   E. It can be nodular, curvilinear, or mixed.

The answer is C.

The presence of an abnormally thick cortex with reduced or absent formation of the cerebral convolutions is not a feature of SBH. It is the characteristic feature of lissencephaly type I-subcortical band heterotopia spectrum, which has distinct imaging and genetic characteristics.

2. **SBH, the FALSE answer is:**
   A. Results from the arrested migration of neurons in the subcortical or deep white matter.
   B. Characterized by the presence of a smooth band of poorly organized neurons.
   C. Taylor dysplasia is one of the differential diagnoses for this disease.
   D. Rarely associated with corpus callosum agenesis.
   E. Associated with distorted basal ganglia on imaging.

   The answer is D.

   The heterotopic grey matter lesions are associated with corpus callosum agenesis.

Case 2.2: Subependymal Grey Matter Heterotopia

Case Scenario

A 14-year-old female presented with a history of seizure (Fig. 1.3).

Imaging Description

MRI reveals diffuse undulating subependymal nodules lining both lateral ventricles, with a greater concentration on the left side (arrow). These nodules exhibit an iso-intense signal to gray matter on all pulse sequences.

Subependymal Grey Matter Heterotopia

Subependymal grey matter heterotopia is a relatively common neuronal migration disorder characterized by the collection of grey matter neurons in an abnormal location. It results from the in-utero arrest of neuronal migration from the germinal matrix outwards to the developing cerebral cortex at 6–16 weeks of gestation. It is usually discovered on antenatal MRI or during evaluation of children with epilepsy or neurodevelopmental abnormalities. In some cases, periventricular heterotopia is discovered as an incidental finding, which is characterized by small foci of grey matter nodules located in the subependyma of the lateral ventricles. Subependymal grey matter heterotopia can be classified according to morphology into unilateral focal, bilateral focal, and bilateral diffuse subtypes. In this case, the imaging findings are consistent with the bilateral diffuse subtype, which exhibits an undulating band of grey matter around the lateral ventricles.

In males, the prognosis is generally worse with significant disability, and the decision for abortion is usually made. On the other hand, cognitive disability is relatively mild in affected females. Epilepsy typically develops later in life.

On ultrasound, subependymal heterotopic nodules may be visible if they are very large and appear slightly hyperechoic compared to normal white matter. They

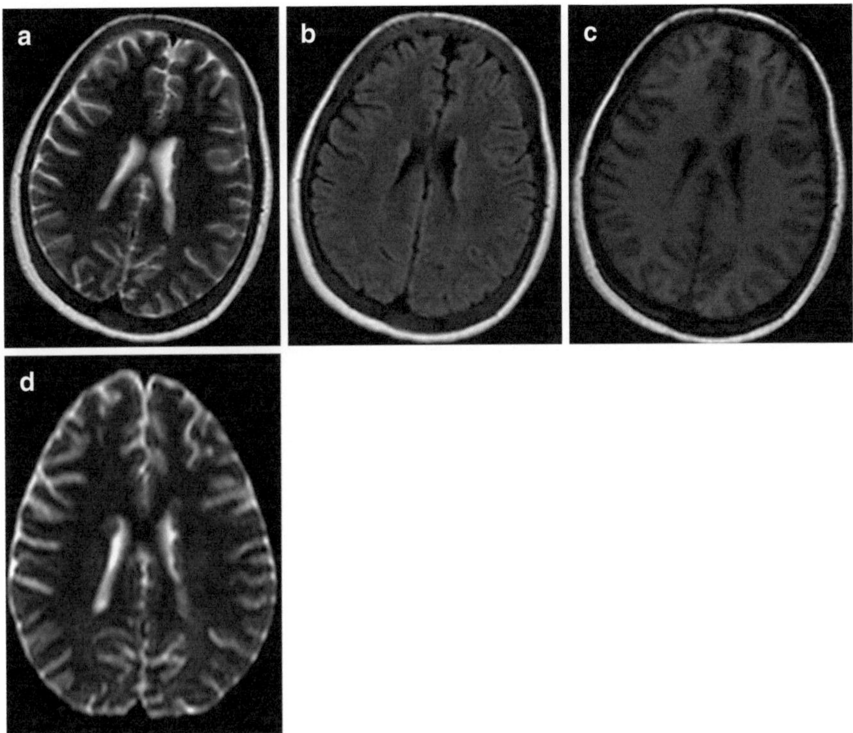

Fig. 1.3 Serial MRI images of the brain with the following: (**a**) axial T2, (**b**) axial FLAIR, (**c**) axial T1, and (**d**) axial DWI. (Figure courtesy of Dr. Samer Hoz)

protrude into the lumen of the ventricles, resulting in irregular ventricular margins. On CT, they are non-enhancing with similar attenuation to normal grey matter and do not show calcification.

MRI is the preferred imaging modality and can demonstrate features of subependymal heterotopia after 26 weeks of gestation. However, identification of these lesions can be difficult earlier during gestation due to fetal movement and the presence of telencephalon periventricular germinal matrix. Round subependymal nodules are seen underlying the ependymal lining of the lateral ventricles and bulging into the lumen, resulting in undulation of the ventricular outline. Heterotopic nodules can be variable in number and size, ranging from small nodules of grey matter to a thick layer of coalescent nodules lining the ventricular margins. They are most commonly located in the trigone and occipital horns of the lateral ventricles, but are occasionally seen in the body or frontal regions. The heterotopic nodules are non-enhancing and follow grey matter on all sequences [5, 6].

Differential Diagnosis

The differential diagnosis includes the following:

- Normal periventricular grey matter.
- Normal germinal matrix at 8–26 weeks of gestation: does not exhibit irregular ventricular margins, and heterotopic nodules may be lower in the T2 signal.
- Subependymal giant cell astrocytomas: are usually brightly enhancing and located around the foramen of Monro. They are typically present in the setting of tuberous sclerosis (TS).
- Subependymal tuber of TS: are usually calcified, with the exception of lesions that present in early childhood. They also exhibit higher T2 signal than normal grey matter.
- Subependymal hemorrhage identified on ultrasound or antenatal MRI: demonstrate evolution on follow-up imaging.

Questions

1. **Subependymal grey matter heterotopia, the FALSE answer is:**
   A. They are usually seen in the trigone and occipital horns of the lateral ventricles.
   B. CT demonstrates enhancing and calcified tissue in subependymal heterotopia.
   C. Subependymal heterotopic nodules are rarely visible on ultrasound, and only if they are very large.
   D. Subependymal nodules in subependymal heterotopia, like normal grey matter, do not enhance on post-contrast MRI sequences.
   E. Subependymal heterotopic nodules appear slightly hyperechoic compared to normal white matter on ultrasound.
   The answer is B.

   CT demonstrates non-enhancing and non-calcified tissue with similar attenuation to normal grey matter along the ventricular margins.

Case 2.3: Focal Cortical Dysplasia (FCD) Type II

Case Scenario

A 33-year-old male presented with refractory seizures (Fig. 1.4).

Imaging Description

MRI reveals two triangular foci of increased T2/FLAIR cortical signal in the right frontal and occipital lobes (open arrow), seen as blurring of the white matter-grey matter junction with the abnormal architecture of the subcortical layer. On the coronal sequences, the apex of the lesion points towards the ventricle, with a thin linear increase in T2/FLAIR extending to the ependymal surface of the right lateral ventricle (arrow). This represents the transmantle sign of Blumcke type II FCD.

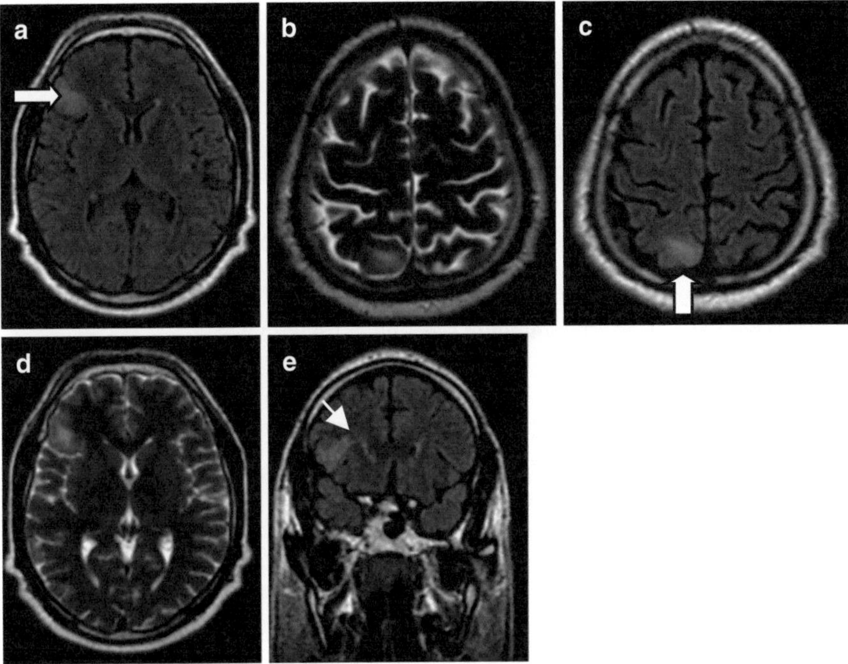

Fig. 1.4 Serial MRI images of the brain with the following: (**a**) axial FLAIR, (**b**) axial T2, (**c**) axial FLAIR, (**d**) axial T2, and (**e**) coronal FLAIR. (Figure courtesy of Dr. Samer Hoz)

Focal Cortical Dysplasia

The term focal cortical dysplasia (FCD) is used to describe abnormalities related to the formation of the cerebral cortex. It is considered one of the most frequent causes of epilepsy and can also be associated with other conditions, including hippocampal sclerosis and cortical glioneuronal neoplasms. MRI is the preferred imaging modality and typically reveals thickening of the cortex, blurring of the cortical-white matter junction, and distorted subcortical layer architecture. MRI also demonstrates white matter hyper-intensity on T2/FLAIR, with or without the transmantle sign, as well as grey matter hyper-intensity on T2/FLAIR. Additionally, an abnormal sulcal or gyral pattern may be observed, along with segmental and lobar hypoplasia or atrophy. However, features like edema, calcification, and contrast enhancement are not typically present. Blumcke's classification categorizes FCD into type 1a, 1b, 2, and 3 [7, 8].

Differential Diagnosis

The differential diagnosis of FCD includes the following:

1. Adult-type diffuse gliomas such as astrocytoma, oligodendroglioma, and glioblastoma.
2. Glioneuronal and neuronal tumors such as ganglioglioma and dysembryoplastic neuroepithelial tumors (DNETs).

Questions

1. **FCD, the FALSE answer is:**
   A. In adults, type 1 FCD is more common.
   B. MRI demonstrates white matter hyper-intensity on T2/FLAIR.
   C. Type 2 FCD is most commonly seen in the temporal lobes.
   D. For type 2 FCD, the signal of the grey matter appears slightly increased on T2.
   E. For type 1 FCD, white matter appears moderately increased on T2/FLAIR and decreased on T1.

   The answer is C.

   Type 2 FCD is most commonly seen in the frontal lobes.

Case 3: Mitochondrial Encephalomyopathy with Lactic Acidosis and Stroke-Like Episodes (MELAS)

Case Scenario

A 13-year-old female presents stroke-like episodes, seizures, and muscle weakness (Fig. 1.5).

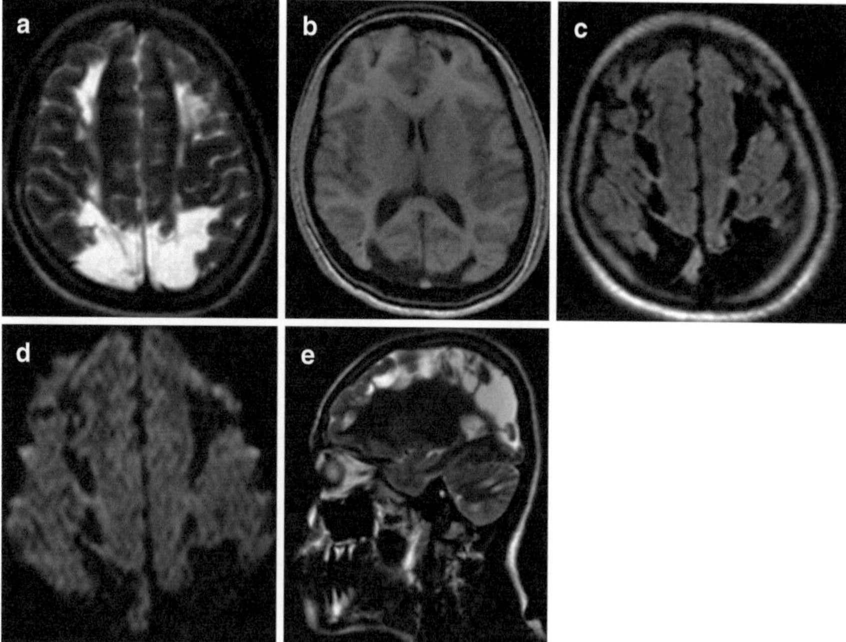

Fig. 1.5 Serial MRI images of the brain with the following: (**a**) axial T2, (**b**) axial T1, (**c**) axial FLAIR, (**d**) axial DWI, and (**e**) sagittal T2. (Figure courtesy of Dr. Samer Hoz)

Imaging Description

MRI demonstrated bilateral symmetrical cortical and subcortical lesions that extend into the deep white matter of the frontal and parieto-occipital regions. The infarction-like lesions can be seen crossing vascular boundaries with a confluent area of gliosis throughout the cerebral hemispheres, in keeping with sequelae of previous ischemic insults. The lesions exhibit high signal intensity on T2, are hypo-intense on T1 and FLAIR, and do not show restricted diffusion on DWI.

MELAS

Mitochondrial encephalomyopathy with lactic acidosis and stroke-like episodes (MELAS) is an inherited mitochondrial disease that primarily affects the muscles and nervous system. Patients with MELAS present with recurrent episodes of headache, myopathy, encephalopathy, and focal neurological deficits. The progressive nature of this condition typically results in neurological impairment by adolescence or early adulthood. Although MELAS may appear with no family history, cases resulting from sporadic mutations are rare. The pathogenesis of MELAS involves mutations in tRNA causing impairment of protein assembly in mitochondria and subsequent impairment of oxidative phosphorylation and energy production. Generally, highly metabolically active organs such as the brain and muscles are more severely affected in mitochondrial disorders.

On imaging, MELAS appears as multifocal stroke-like cortical lesions in a "shifting spread" pattern that crosses vascular boundaries. Both MRI and CT show more extensive changes in the grey matter than the white matter. Lesions may or may not be symmetrical and have a predilection for the posterior parietal and occipital lobes. MR spectroscopy may show an elevation in lactate in an otherwise normal-appearing brain. On CT, basal ganglia calcification may be seen and is more common in older patients. Digital subtraction angiography (DSA) is usually normal, but a blush correlating with enhancing gyri may be seen.

The acute and chronic phases of the condition can be differentiated on MRI. In the acute phase, the lesion begins focally in the temporal region. In 2–3 weeks, the lesion spreads to the parietal and occipital regions in a third of the patients, hence showing that the disease continues to progress following the episode of stroke.

Acute infarcts involve the subcortical white matter focally and appear as swollen gyri with increased T2 signal. They may enhance and exhibit increased signal on DWI (T2 shine through) with minimal change on ADC, suggesting vasogenic rather than cytotoxic edema. On the other hand, chronic infarcts involve multiple vascular territories in a symmetrical or asymmetrical fashion, with a predilection for the parieto-occipital and parieto-temporal regions. The black toenail sign may be seen in the subacute phase [9, 10].

Differential Diagnosis

Possible differential considerations include other mitochondrial disorders such as MERRF, Leigh syndrome, and Kearns-Sayre syndrome; status epilepticus; viral encephalitis; cerebral vasculitis; Creutzfeldt-Jakob disease; as well as infarcts due to embolism, dissection, and cerebral autosomal dominant arteriopathy.

Questions

1. **MELAS, the FALSE answer is:**
   A. The stroke-like cortical lesions in MELAS follow a "shifting spread" pattern across cerebral vascular territories.
   B. Lesion progression starts in the temporal region and may progress to the parietal and occipital regions in some cases.
   C. Basal ganglia calcification is a prominent feature in younger patients with MELAS.
   D. MRS can detect elevated lactate levels in MELAS, even in brain regions that appear normal on imaging.
   E. Both MRI and CT scans in MELAS show changes in gray matter more than white matter, particularly in the occipital, parietal, and temporal lobes.
   The answer is C.
   Basal ganglia calcification is not a prominent feature in younger patients with MELAS, but it can be more common in older patients.

Case 4: Porencephaly Cyst and Schizencephaly

Case 4.1: Porencephaly Cyst

Case Scenario
A 55-year-old female presented with long-standing seizure (Fig. 1.6).

Imaging Description
MRI demonstrated a moderate-sized cyst in the right occipital region that has a small communication with the occipital horn of the right lateral ventricle. It follows the CSF signal in all sequences, with no diffusion restriction or enhancement. It is

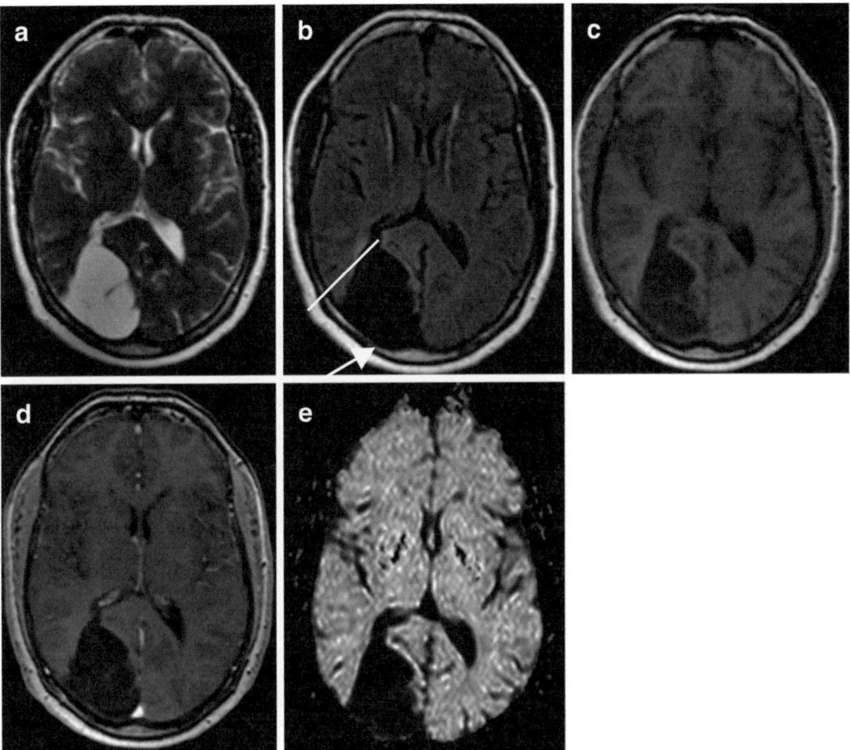

Fig. 1.6 Serial MRI images of the brain with the following: (**a**) axial T2, (**b**) axial FLAIR, (**c**) axial T1, (**d**) axial T1 C+, and (**e**) axial DWI. (Figure courtesy of Dr. Samer Hoz)

not lined by a grey matter and is associated with a small amount of adjacent FLAIR hyper-intensity (line). It exerts a mild mass effect on the adjacent parenchyma, with slight remodeling of the overlying calvarium noted (arrow).

Porencephaly

Porencephaly is a rare disease that causes the formation of a CSF-containing cavity within the cerebral hemisphere. It is usually lined by gliotic or spongiotic white matter and communicates directly with the ventricular system and/or subarachnoid space. In other cases, there may be a thin layer of brain tissue and arachnoid covering it externally. The term may be used in a broader fashion among radiologists to refer to a cleft or cystic cavity within the brain, whether or not communication exists with a CSF space, and thus overlapping with encephalomalacia. Other authors reserve the term for cysts that communicate with at least one CSF space. Still, other definitions require communication with both the ventricular system and subarachnoid space.

Porencephalic cysts can be internal (communicating with the ventricle) or external (communicating with the subarachnoid space). They are subdivided into congenital or acquired cysts. Congenital porencephalic cysts may be caused by intrauterine vascular injury resulting in cerebral ischemia or intraparenchymal hemorrhage (IPH). Intrauterine infection, with viruses such as cytomegalovirus, as well as alcohol consumption, are other potential causes. Acquired porencephalic cysts result from trauma, ischemia, infection, or surgery later in life.

Antenatal ultrasound can be used to detect porencephalic cysts. Asymmetry of the ventricles may also be seen.

On CT, they appear as hypodense intracranial cysts with distinct borders and central attenuation equivalent to that of CSF. They typically do not exert pressure on the surrounding brain parenchyma, although in some cases, local mass effect can be observed in large cysts. They do not enhance with contrast and do not contain a solid component.

Similarly, the cyst appears well-defined on MRI and commonly corresponds to a vascular territory. White matter lining the cyst wall distinguishes it from arachnoid cysts and schizencephaly, which are lined by grey matter. The lining may or may not demonstrate evidence of gliosis, depending on the timing at which the insult occurred. The signal of the cyst content is consistent with that of CSF on all sequences, exhibiting low signal intensity on T1, high signal intensity on T2, suppression on FLAIR, and no restricted diffusion on DWI [11, 12].

Differential Diagnosis

Differential considerations include neuroglia cyst, which does not communicate with the ventricles or subarachnoid space; arachnoid cyst, which is extra-axial; inter-hemispheric cyst; schizencephaly, which is lined by grey matter; as well as holoprosencephaly.

Questions

1. **Porencephaly, the FALSE answer is:**
   A. In most cases, the cavity in porencephaly attaches directly to the ventricular system.
   B. On CT, porencephalic cysts show a solid component and enhance after contrast injection.
   C. The cyst exhibits high signal intensity on T2 and demonstrates fluid signal suppression on FLAIR.
   D. Alcohol consumption during pregnancy has been linked to the development of porencephaly.
   E. Fetal MRI plays a role in differentiating intracranial cysts from extra-axial lesions.

   The answer is B.

   On CT, porencephalic cysts do not show a solid component or enhancement after contrast injection.

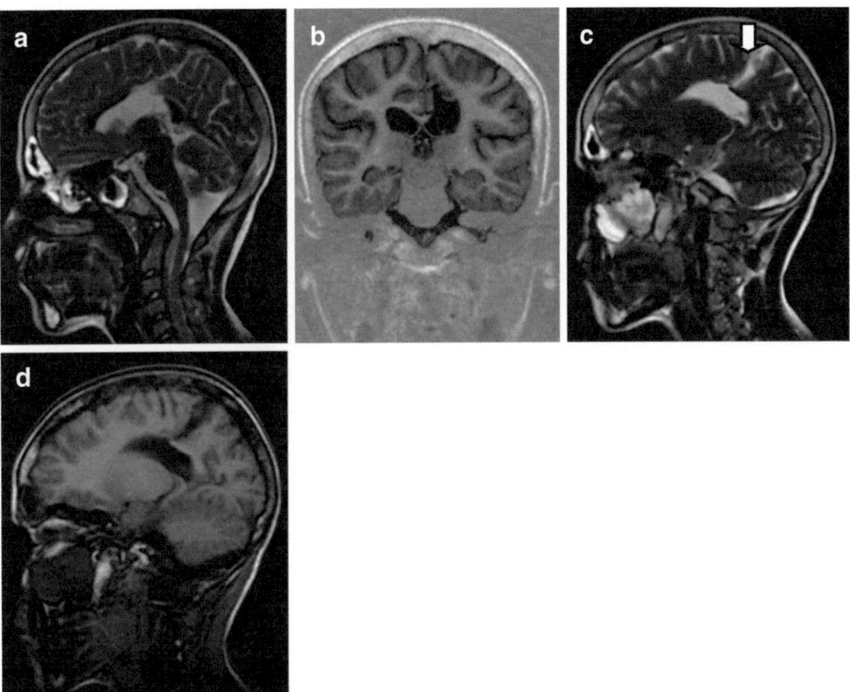

Fig. 1.7 Serial MRI images of the brain with the following: (**a**) sagittal T2, (**b**) coronal IR, (**c**) sagittal T2, and (**d**) sagittal T1. (Figure courtesy of Dr. Samer Hoz)

Case 4.2: Dysgenesis of the Corpus Callosum with Open Lip Schizencephaly

Case Scenario
A 17-year-old female presented with grand mal seizures and right spastic hemiparesis (Fig. 1.7).

Imaging Description
MRI shows a left frontoparietal parenchymal cleft, lined by grey matter and extending from the cortical surface to the lateral ventricle (arrow). The lateral ventricle is enlarged and is associated with a small ependymal defect at its lateral border as well as a large defect of the septum pellucidum (dysgenesis of the corpus callosum).

Schizencephaly
Schizencephaly is an uncommon cortical abnormality characterized by a cleft lined with grey matter that extends from the ependyma to the pia mater. Its estimated incidence is roughly 1.5 cases per 100,000 live births, with no apparent sex predilection. While the precise cause of schizencephaly remains uncertain, it is believed to be attributed to abnormal neuronal migration. Some researchers also propose that

a vascular injury during the early stages of fetal development could be a contributing factor.

Schizencephaly can occur bilaterally and is categorized into two types: In the closed-lip type, which occurs more frequently in unilateral cases, the walls of the cleft are in close proximity to each other. In contrast, the open-lip type is characterized by a cleft with separated walls and is filled with cerebrospinal fluid (CSF). Closed-lip schizencephaly is more commonly observed in unilateral cases. In 50–90% of cases, schizencephaly is accompanied by other cerebral abnormalities, including dysgenesis of the corpus callosum, as in this case, septo-optic dysplasia, grey matter heterotopia, and absence of the septum pellucidum.

Schizencephaly is increasingly being diagnosed using cranial ultrasound, either antenatally or postpartum. Ultrasound is particularly effective in identifying open-lip schizencephaly, especially when the cleft is large. However, detecting closed-lip schizencephaly with ultrasound can be more challenging. CT can be useful in detecting open-lip schizencephaly and significant grey matter heterotopia. However, they provide limited visualization compared to MRI and have limited ability to distinguish between grey and white matter.

MRI is the preferred imaging technique for diagnosing schizencephaly, as it can accurately detect the pial-ependymal cleft and visualize cortical dysplasia and heterotopic grey matter. Closed-lip schizencephaly appears as a nipple-like outpouching at the ependymal surface. Open-lip schizencephaly can be identified on MRI as a CSF-filled cleft lined with heterotopic grey matter, extending from the cortical surface to the ventricles [13, 14].

Differential Diagnoses

1. Focal cortical dysplasia (FCD): this condition may present with a cortical cleft, but it does not reach the ventricles.
2. Heterotopic grey matter: a slight cleft may be seen along the ventricular outline, but periventricular grey matter usually protrudes into the ventricle.
3. Porencephaly: an area of encephalomalacia extends from the cortical surface to the ventricular surface but is surrounded by gliotic white matter, and not grey matter. Nevertheless, some experts may use the term "true porencephaly" interchangeably with schizencephaly.

Questions

1. **Schizencephaly, the FALSE answer is:**
   A. CT imaging is the preferred modality for fully characterizing congenital abnormalities associated with schizencephaly.
   B. Closed-lip schizencephaly is more difficult to detect with this ultrasound.
   C. Closed-lip schizencephaly can be identified on MRI as a nipple-like outpouching at the ependymal surface.
   D. Schizencephaly is often associated with other cerebral anomalies.

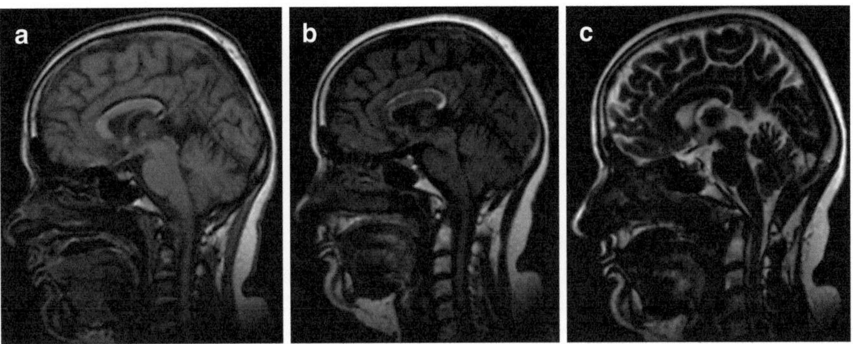

Fig. 1.8 Serial MRI images of the brain with the following: (**a**) sagittal T1 FFE, (**b**) sagittal T1, and (**c**) sagittal T2. (Figure courtesy of Dr. Samer Hoz)

 E. Dysgenesis of the corpus callosum is commonly associated with schizencephaly.

The answer is A.

MRI, and not CT, is the preferred imaging modality for fully characterizing congenital abnormalities due to its superior ability to distinguish between grey and white matter.

Case 5: Intracranial Lipomas

Case 5.1: Curvilinear Pericallosal Lipomas

Case Scenario

A 62-year-old female presented with headache and an incidental finding on MRI (Fig. 1.8).

Imaging Description

MRI demonstrates hypoplasia of the posterior body and splenium (corpus callosal dysgenesis) with a curvilinear/ribbon pericallosal lipoma extending up to the anterior commissure. The lesion is hyper-intense on T1, T2, and FLAIR, with signal attenuation observed in the fat-suppressed T1 sequence.

Pericallosal Lipomas

Pericallosal lipomas are the most common form of intracranial lipomas and are located within the interhemispheric fissure near the corpus callosum. They are rare, with an estimated incidence of 1 in 2,500 to 1 in 25,000 autopsies.

Pericallosal lipomas can be categorized into either tubulonodular or curvilinear subtypes. Tubulonodular lipomas, which are more common, manifest as rounded or lobular growths that exceed 2 cm in thickness and are located anteriorly. They are often associated with significant anomalies affecting the corpus callosum and frontal-facial structures. Furthermore, they may extend into the lateral ventricles and choroid plexus. In contrast, curvilinear pericallosal lipomas are typically elongated and appear curvilinear along the margin of the corpus callosum. They are typically situated posteriorly and measure less than 1 cm in thickness. The corpus callosum is typically hypoplastic as in these cases.

Ultrasound imaging can show typical features of fat, demonstrating a hyperechoic mass near the corpus callosum. CT is diagnostic and shows a mass with a fat density typically falling within the range of −80 to −110 Hounsfield Units (HU). Moreover, the tubulonodular type may be associated with curvilinear calcification in the periphery, appearing as a "bracket sign" on coronal sections. Cases involving vascular malformations or aneurysms warrant further investigation with CT angiography (CTA).

MRI is the most effective modality for assessment of the size and characteristics of pericallosal lipomas, as well as identifying associated corpus callosum agenesis or dysgenesis. These lesions exhibit the same signal intensity as fat across all sequences, including T1, T2, and T1 with gadolinium (Gd) with marked hyperintensity. They also show signal attenuation on fat suppression sequences but not on FLAIR, and they are non-enchancing post-contrast. Additionally, the presence of peripheral calcification causes blooming artifacts on gradient echo (GRE), echo planner imaging (EPI), or susceptibility weighted imaging (SWI) sequences. Some sequences may exhibit pronounced peripheral chemical shift artifacts as well. Since these lipomas may be concomitant with vascular abnormalities, assessment of the blood vessels is crucial, which is best done through T2 fast spin echo (FSE) sequences and magnetic resonance angiography (MRA) [15, 16].

Differential Diagnosis

The differential diagnosis for pericallosal lipomas is limited since very few intracranial masses contain significant amounts of fat. Possible differential diagnoses include intracranial dermoid cyst, intracranial teratoma, fatty falx cerebri (especially in the curvilinear type), rare lipomatous transformation of neoplasms such as neuroectodermal tumors, ependymoma, and glioma, intracranial fat, and fat-containing brain lesions.

Questions

1. **Pericallosal lipomas, the FALSE answer is:**
   A. Ultrasound imaging can reveal the typical appearance of fat in pericallosal lipomas.
   B. Pericallosal lipomas may be associated with vascular abnormalities.
   C. Tubulonodular pericallosal lipomas are usually located anteriorly.
   D. Pericallosal lipomas demonstrate the same signal intensity as fat on all MRI sequences.
   E. Curvilinear pericallosal lipomas can extend into the choroid plexus and lateral ventricles.
   The answer is E.

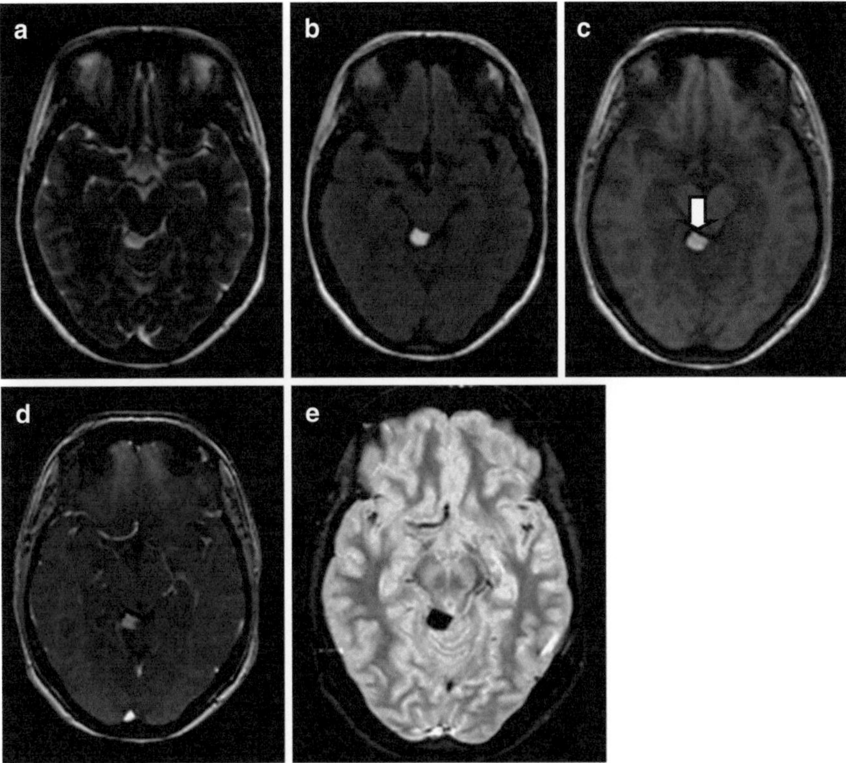

Fig. 1.9 Serial MRI images of the brain with the following: (**a**) axial T2, (**b**) axial FLAIR, (**c**) axial T1, (**d**) axial T1 C+, and (**e**) axial fat-saturation. (Figure courtesy of Dr. Samer Hoz)

This feature is typical of tubulonodular pericallosal lipomas.

Case 5.2: Intracranial Lipomas

Case Scenario
A 26-year-old female presented with headache and epilepsy (Fig. 1.9).

Imaging Description
MRI reveals a mass on the right side of the quadrigeminal plate cistern with a very high T1, T2, and FLAIR signal, which demonstrates fat saturation and chemical shift artifact, confirming that the mass is composed of fat. A peripheral signal void surrounds the lesion, indicative of calcification (arrow). After the contrast study, no enhancement is seen, consistent with a quadrigeminal cistern lipoma.

Intracranial Lipomas

Intracranial lipomas are uncommon congenital anomalies. While they can develop anywhere within the intracranial space, they tend to occur along the midline and are often associated with abnormalities in nearby structures. They arise from the abnormal persistence and maldifferentiation of the primitive meninx. Approximately 55% of intracranial lipomas are associated with other brain malformations of varying degrees. They are typically asymptomatic and are discovered incidentally, although some may be symptomatic. However, surgical removal is approached with caution due to the high vascularization of these lipomas and their strong attachment to surrounding tissues, which can lead to elevated risks of morbidity and mortality. As a result, surgical intervention is rarely recommended.

Characteristic regions for intracranial lipomas include the following:

1. Pericallosal lipoma (50%): associated with corpus callosum agenesis or dysgenesis (~50% of cases) and is divided morphologically into nodular and curvilinear types.
2. Quadrigeminal cistern lipoma (25%): associated with inferior sulcus maldevelopment.
3. Suprasellar cistern lipoma (15%)
4. Cerebellopontine angle (CPA) lipoma (10%): surgical management of lipomas in this area is challenging due to the presence of the facial nerve and vestibulocochlear nerve.
5. Sylvian fissure (5%)
6. Choroid plexus lipoma (rare)

When interpreting CT scans of intracranial lipomas, it is crucial to have a solid understanding of their common locations, as they can be easily mistaken for various other lesions with similar CT characteristics, such as dermoid cysts, teratomas, or neoplastic lipomatous transformations. On CT, intracranial lipomas typically appear as non-enhancing masses with consistent fat density (CT attenuation values around -100), exhibiting a lobulated, "soft" appearance that conforms to the neighboring anatomy. They may also exhibit peripheral calcification.

On MRI, intracranial lipomas appear as uniformly high-intensity lesions on T1-weighted images, with signal intensity ranging from iso-intense to slightly high on T2-weighted images. Administration of Gadolinium-diethylenetriamine pentaacetic acid (GdDTPA) does not result in tumor enhancement. MRI not only offers excellent visualization of the tumor's relationship with surrounding structures but also enables the detection of other lesions that may not be clearly visible on CT scans. These lipomas are often traversed by cranial nerves and adjacent vessels, which are best observed on high-resolution sequences [17, 18].

Differential Diagnosis

The differential primarily includes fat-containing masses, such as the following:

1. Intracranial dermoid: are typically milling in location and appear as multiple droplets scattered through the subarachnoid space if rupture occurs.
2. Intracranial teratoma
3. Lipomatous transformation of neoplasm: glioma, ependymoma, and neuroectodermal tumors
4. Thrombosed berry aneurysm: often exhibits a calcified rim and hemosiderin staining on GRE or SWI sequences.
5. White epidermoid: are rare and show restrictions on DWI.

Questions

1. **Intracranial lipoma, the FALSE answer is:**
   A. On MRI, an intracranial lipoma is depicted as a homogeneous low signal intensity lesion on T1WI.
   B. The presence of peripheral calcification may be detected by CT.
   C. Agenesis or dysgenesis of the corpus callosum occurs in ~50% of cases.
   D. Intracranial teratoma is one of the differential diagnoses.
   E. They are neither hematomas nor true neoplasms.
   The answer is A.

   On MRI, an intracranial lipoma is depicted as a homogeneous high-signal intensity lesion on T1WI. On T2WI, its intensity ranges from an iso- to a slightly high signal.
2. **Intracranial lipoma, the FALSE answer is:**
   A. Intracranial dermoid, if ruptured, is considered as a differential diagnosis.
   B. Homogenous enhancement on MRI.
   C. MRI can detect the presence of other lesions not well visualized on CT.
   D. Some cases tend to be located around the midline and are frequently associated with abnormal development of adjacent structures.
   E. They typically appear as a non-enhancing mass with uniform fat density.
   The answer is B.

   Administration of GdDTPA does not produce any enhancement of the tumor.

Case 6: Tuberous Sclerosis

Case Scenario

A 14-year-old female presented with seizures, intellectual disability, and multiple adenoma sebaceum (Fig. 1.10).

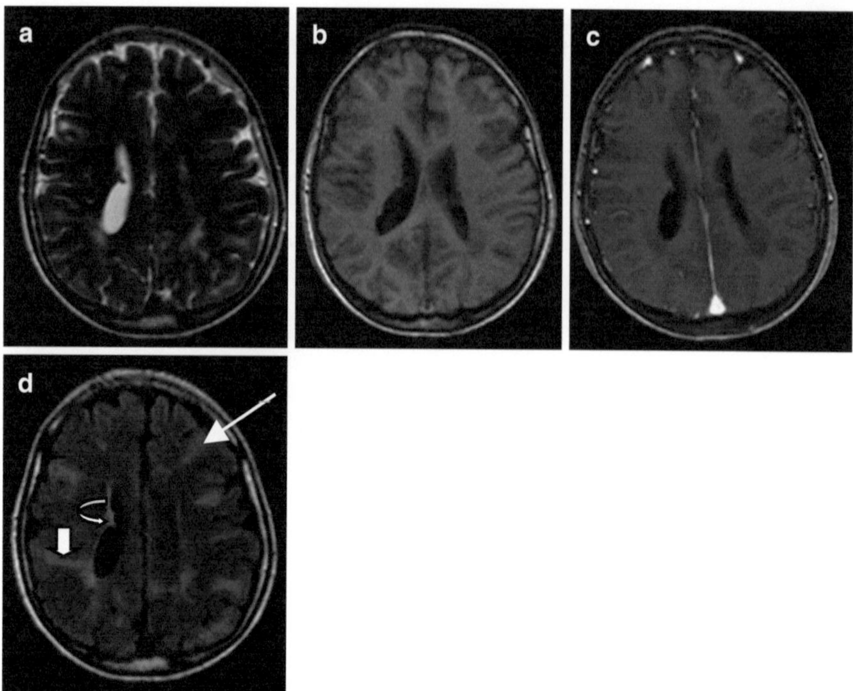

Fig. 1.10 Serial MRI images of the brain with the following: (**a**) axial T2, (**b**) axial T1, (**c**) axial T1 C+, and (**d**) axial FLAIR. (Figure courtesy of Dr. Samer Hoz)

Imaging Description

The MRI sequences reveal bilateral and extensive cortical and subcortical areas of low signal on T1 and high signal on FLAIR and T2, with no enhancement on post-contrast sequences, consistent with cortical/subcortical tubers (arrow). Additionally, there are a few linear bands of high signal on FLAIR, radiating from the periventricular white matter to the subcortical region, mainly in the right occipital lobe, indicative of the radial bands sign (open arrow). Small subependymal nodules demonstrate iso- to high signal on T1, iso-signal on T2, and high signal on FLAIR, with no significant enhancement on post-contrast sequences, characteristic of hamartomas (curve arrow).

Tuberous Sclerosis

Also referred to as tuberous sclerosis complex (TSC), tuberous sclerosis (TS) is a neurocutaneous condition characterized by the growth of benign tumors in tissues originating from the embryonic ectoderm, including the skin, eyes, and nervous system. Its incidence is roughly estimated to be between 1 in 6000 and 1 in 12,000 individuals, with most cases arising sporadically.

Imaging features of TS include the following:

1. Cortical or subcortical tubers: Approximately half of these tubers are located in the frontal lobe. They typically display high signal intensity in T2 and FLAIR sequences, and low signal intensity in T1. Only about 10% of tubers enhance with contrast. Calcification is common and often becomes evident after the age of 2.
2. Subependymal hamartomas: Around 88% are associated with calcification. Calcification usually appears within the first 6 months of life. They tend to show high T1 signal and iso- to high T2 signals, although that can vary. They can be differentiated from subependymal giant cell astrocytomas through serial growth monitoring.
3. Subependymal giant cell astrocytomas most commonly occur between the ages of 8 and 18 and tend to be large in size. They are typically strongly enhancing.
4. White matter abnormalities: their appearance can be highly variable, ranging from nodular to ill-defined, cystic, and band-like lesions, whereby radial bands are relatively specific to TS.
5. Retinal phakomas.
6. Uncommon findings include cerebral aneurysms, corpus callosum dysgenesis, cerebellar atrophy, microcephaly, Chiari malformations, arachnoid cysts, chordoma, and infarctions resulting from vascular occlusion [19, 20].

Questions

1. **TS, the FALSE answer is:**
   A. Only around 10% of tubers show enhancement on imaging.
   B. Retinal phakomas are a characteristic feature of TS.
   C. Cortical or subcortical tubers in TS typically exhibit low signal intensity on T1.
   D. Calcification is rare in subependymal hamartomas associated with TS.
   E. Cerebellar atrophy is a rare finding in TS.
   The answer is D.
   Calcification is frequently present in subependymal hamartomas associated with TS.

Case 7: Joubert Syndrome

Case Scenario

A 12-year-old female presented with a history of ataxia and cognitive impairment (Fig. 1.11).

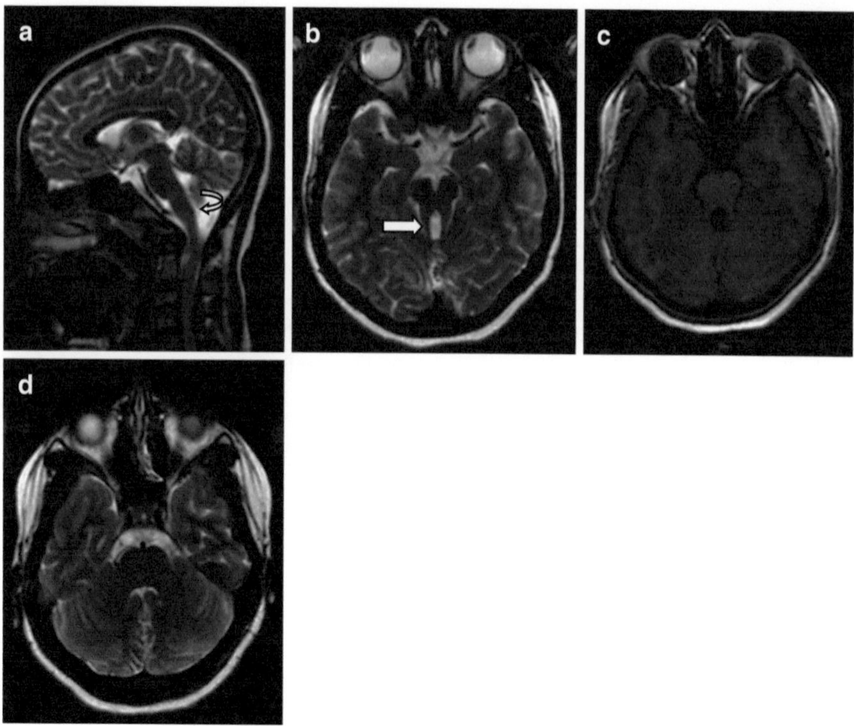

Fig. 1.11 Serial MRI images of the brain with the following: (**a**) sagittal T2, (**b**) axial T2, (**c**) axial T1, and (**d**) axial T1. (Figure courtesy of Dr. Samer Hoz)

Imaging Description

Superior cerebellar peduncles are thickened and elongated with a midline cleft at the pontomesencephalic junction, giving the appearance of a molar tooth (arrow). The inferior cerebellar vermis is hypoplastic, as seen in sagittal T2 (curve arrow), with the batwing appearance of the fourth ventricle.

Joubert Syndrome

Joubert syndrome, also known as vermian aplasia or molar tooth midbrain-hindbrain malformation, is an autosomal recessive disorder characterized by cerebellar vermian agenesis of varying degrees. When it co-occurs with eye, liver, or kidney abnormalities, it is classified as Joubert syndrome and related disorders. The estimated prevalence of Joubert syndrome is approximately 1 in 100,000 individuals.

Imaging typically demonstrates cerebellar abnormalities, including a small, dysplastic, or absent cerebellar vermis with a characteristic median cleft. Additionally, diffusion tensor imaging may identify a lack of fiber decussation in the superior cerebellar peduncles and pyramidal tracts. Other potential anomalies may involve abnormal development of the inferior olivary nucleus as well as dysplasia and heterotopia of cerebellar nuclei.

The posterior fossa often presents with a fourth ventricle that has a batwing-like shape as well as superior cerebellar peduncles that appear thick and elongated, giving rise to the molar tooth sign appearance. In 6–20% of cases, minor lateral ventriculomegaly may be seen. Corpus callosal dysgenesis may be observed in 6–10% of cases [21, 22].

Differential Diagnosis

1. Dandy-Walker malformation: in some cases, it may occur concomitantly with Joubert syndrome.
2. Rhombencephalosynapsis.
3. Mega cisterna magna: is not associated with any structural abnormalities.

Questions

1. **Joubert syndrome, the FALSE answer is:**
    A. The molar tooth sign appearance in Joubert syndrome is caused by visibly thickened, elongated inferior cerebellar peduncles.
    B. Corpus callosum dysgenesis is observed in approximately 6–10% of cases of Joubert syndrome.
    C. Minor lateral ventriculomegaly is observed in 6–20% of cases of Joubert syndrome.
    D. The batwing-shaped fourth ventricle is a common finding in Joubert syndrome.
    E. The absence of fiber decussation in the superior cerebellar peduncles and pyramidal tracts can be identified through diffusion tensor imaging.
    The answer is A.
    The molar tooth sign appearance in Joubert syndrome is caused by visibly thickened, elongated superior cerebellar peduncles.

Case 8: Capillary Hemangioma of a Lower Eyelid

Case Scenario

A 3-year-old female presented with a subcutaneous palpable lump in the left inferior periorbital region since birth (Fig. 1.12).

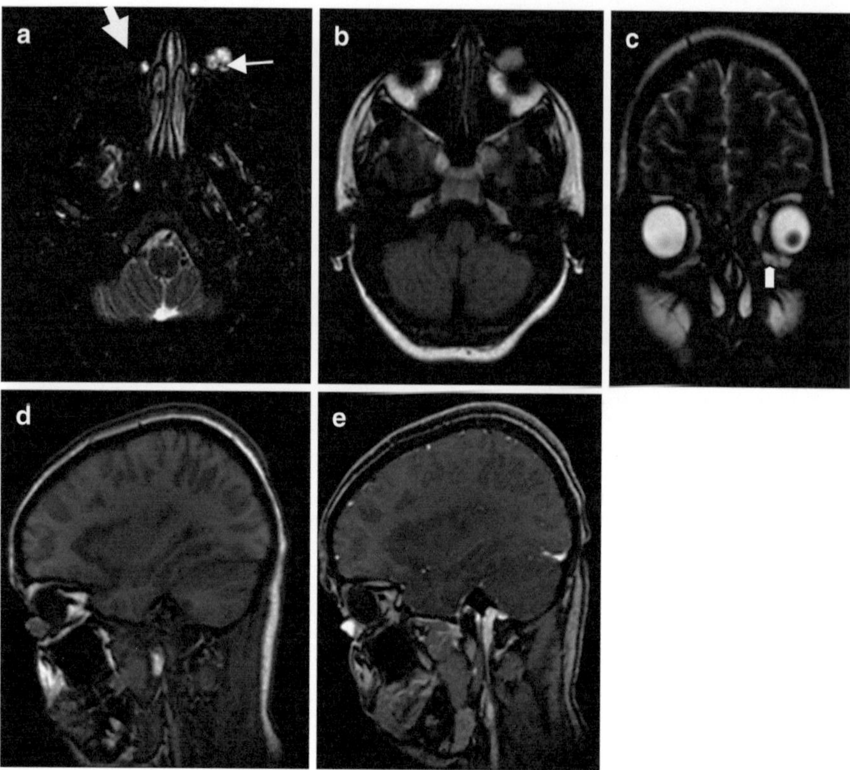

Fig. 1.12 Serial MRI images of the brain with the following: (**a**) axial short tau inversion recovery (STIR), (**b**) axial T1, (**c**) coronal T2, (**d**) sagittal T1, and (**e**) sagittal T1 C+. (Figure courtesy of Dr. Samer Hoz)

Imaging Description

MRI reveals a lobulated mass lesion in the left lower eyelid, appearing iso-intense in T1 and hyper-intense in T2, with multiple small foci exhibiting a signal void within it (arrow). The lesion demonstrates homogeneous enhancement on post-contrast study, and the left orbital structures are otherwise normal.

Capillary Hemangiomas of the Orbit

Capillary hemangiomas of the orbit are sometimes referred to as orbital infantile hemangiomas or strawberry hemangiomas due to their coloration. Unlike orbital cavernous hemangiomas, it is a neoplasm rather than a vascular malformation. It predominantly occurs in the front of the eye globe and eyelid and is often present at birth but can also develop later. In some cases, it can extend to involve extra-ocular muscles, lacrimal glands, and even pass through the optic canal or superior orbital fissure, reaching into the intracranial space. Rarely, they can be associated with systemic hemangiomas.

Ultrasound is useful for small, localized lesions and typically reveals hyperechoic and compressible lesions with high peak intra-tumoral arterial flow. On CT, they appear well-defined, homogeneous, enhancing, and lobulated. However, imaging alone may not be sufficient to distinguish these lesions from other vascular orbital abnormalities; the age of the patient and clinical presentation also play important roles in diagnosis.

MRI scans of these lesions typically show hypo-intensity on T1-weighted images and iso-intensity to hyper-intensity on T2-weighted images, often with multiple serpiginous flow voids (as seen in this case). Gadolinium-enhanced MRI reveals homogeneous enhancement with marked enhancement of intratumoral vessels. It typically appears lobulated with thin septa [23, 24].

Differential Diagnosis

The differential diagnosis of capillary hemangiomas of the orbit includes the following:

1. Vascular malformations of the orbit, such as orbital venous varix, orbital lymphangioma, and cavernous hemangioma of the orbit
2. Retinoblastoma
3. Orbital metastases
4. Rhabdomyosarcoma
5. Chloroma

Questions

1. **Capillary hemangiomas of the orbit, the FALSE answer is:**
   A. They are considered as neoplasms rather than vascular malformations.
   B. Ultrasound typically shows hyperechoic and compressible lesions with high peak intra-tumoral arterial flow.
   C. CT scan shows a strongly homogeneous, enhancing lobulated mass.
   D. MRI shows iso- to hyper-intense lesions on T2 with multiple serpiginous flow voids.
   E. MRI reveals hyper-intense lesions on T1.
   The answer is E.
   MRI reveals hypo-intense lesions on T1.

References

1. Emery SP, Narayanan S, Greene S. Fetal aqueductal stenosis: prenatal diagnosis and intervention. Prenat Diagn. 2020;40(1):58–65.
2. Kline-Fath BM, Arroyo MS, Calvo-Garcia MA, Horn PS, Thomas C. Congenital aqueduct stenosis: progressive brain findings in utero to birth in the presence of severe hydrocephalus. Prenat Diagn. 2018;38(9):706–12.
3. Oegema R, Barkovich AJ, Mancini GM, Guerrini R, Dobyns WB. Subcortical heterotopic gray matter brain malformations: classification study of 107 individuals. Neurology. 2019;93(14):e1360–73.
4. Koenig M, Dobyns WB, Di Donato N. Lissencephaly: Update on diagnostics and clinical management. Eur J Paediatr Neurol. 2021;35:147–52.
5. Donkol RH, Moghazy KM, Abolenin A. Assessment of gray matter heterotopia by magnetic resonance imaging. World J Radiol. 2012;4(3):90.
6. Dekeyzer S, Deblaere K. Subependymal gray matter heterotopia. JBR-BTR. 2015;98(3):111–2.
7. Guerrini R, Barba C. Focal cortical dysplasia: an update on diagnosis and treatment. Expert Rev Neurother. 2021;21(11):1213–24.
8. Urbach H, Kellner E, Kremers N, Blümcke I, Demerath T. MRI of focal cortical dysplasia. Neuroradiology. 2022;64:443.
9. Murakami H, Ono K. MELAS: mitochondrial encephalomyopathy, lactic acidosis and stroke-like episodes. Brain Nerve (Shinkei Kenkyu no Shinpo). 2017;69(2):111–7.
10. El-Hattab AW, Adesina AM, Jones J, Scaglia F. MELAS syndrome: clinical manifestations, pathogenesis, and treatment options. Mol Genet Metab. 2015;116(1-2):4–12.
11. Meuwissen ME, Halley DJ, Smit LS, Lequin MH, Cobben JM, De Coo R, Van Harssel J, Sallevelt S, Woldringh G, Van Der Knaap MS, De Vries LS. The expanding phenotype of COL4A1 and COL4A2 mutations: clinical data on 13 newly identified families and a review of the literature. Genet Med. 2015;17(11):843–53.
12. Tylš F, Brunovský M, Šulcová K, Kohútová B, Ryznarová Z, Kopeček M. Latent schizencephaly with psychotic phenotype or schizophrenia with schizencephaly? A case report and review of the literature. Clin EEG Neurosci. 2019;50(1):13–9.
13. Braga VL, da Costa MD, Riera R, dos Santos Rocha LP, de Oliveira Santos BF, Hondo TT, de Oliveira CM, Cavalheiro S. Schizencephaly: a review of 734 patients. Pediatr Neurol. 2018;87:23–9.
14. Nabavizadeh SA, Zarnow D, Bilaniuk LT, Schwartz ES, Zimmerman RA, Vossough A. Correlation of prenatal and postnatal MRI findings in schizencephaly. Am J Neuroradiol. 2014;35(7):1418–24.
15. Yilmaz MB, Genc A, Egemen E, Yilmaz S, Tekiner A. Pericallosal lipomas: a series of 10 cases with clinical and radiological features. Turk Neurosurg. 2016;26(3):364–8.
16. Taydas O, Ogul H, Kantarci M. The clinical and radiological features of cisternal and pericallosal lipomas. Acta Neurol Belg. 2020;120:65–70.
17. Elhend SB, Belfquih H, Hammoune N, Athmane EM, Mouhsine A. Case report author's personal copy. World Neurosurg. 2019;125:123–5.
18. Tushar K, Suhas M, Reddy LP, Prabhu AS, Purnachandra L. Neuroimaging spectrum of intracranial lipomas. Cureus. 2023;15:2.
19. Randle SC. Tuberous sclerosis complex: a review. Pediatr Ann. 2017;46(4):e166–71.
20. Portocarrero LK, Quental KN, Samorano LP, Oliveira ZN, Rivitti-Machado MC. Tuberous sclerosis complex: review based on new diagnostic criteria. An Bras Dermatol. 2018;93:323–31.
21. Devi AR, Naushad SM, Lingappa L. Clinical and molecular diagnosis of Joubert syndrome and related disorders. Pediatr Neurol. 2020;106:43–9.
22. Gana S, Serpieri V, Valente EM. Genotype–phenotype correlates in Joubert syndrome: a review. Am J Med Genet C Semin Med Genet. 2022;190(1):72–88.
23. Koka K, Patel BC. Capillary infantile hemangiomas. 2023.
24. Kreher MA, Siegel LH, Shmuylovich L. A case of large airway and orbital hemangiomas with facial capillary malformation. Pediatr Dermatol. 2023;40:1142.

Cranial Oncology

2

Maliya Delawan, Sajjad G. Al-Badri, Ameer M. Aynona,
Linah Alduraibi, Ahmed Muthana, Ali A. Dolachee,
and Asmaa H. AL-Sharee

Case 9: Intracranial Melanoma

Case 9.1: Intracranial Metastatic Melanoma: Melanotic Type

Case Scenario

A 33-year-old man presented with headaches, seizures, nausea, and vomiting (Fig. 2.1).

Imaging Description

There is a large, solitary, enhancing hemorrhagic mass lesion that is round in shape located at the gray-white matter junction of the right frontal lobe abutting the para-sagittal midline. It is surrounded by early subacute blood and a moderate amount of vasogenic edema, with a mass effect causing compression on the frontal horn of the lateral ventricle and a mild midline shift. The lesion is mainly hyperintense on T1 and hypointense on T2 and FLAIR and appears as a signal void on T2 FFE GRE

M. Delawan
College of Medicine, Gulf Medical University, Ajman, United Arab Emirates

S. G. Al-Badri · A. Muthana
College of Medicine, University of Baghdad, Baghdad, Iraq

A. M. Aynona
College of Medicine, University of Babylon, Babylon, Iraq

L. Alduraibi
Sulaiman Al Rajhi University, Qassim, Saudi Arabia

A. A. Dolachee
Department of Surgery, Al-Kindy College of Medicine, University of Baghdad, Baghdad, Iraq

A. H. AL-Sharee (✉)
Department of Neuroradiology, Neurosurgery Teaching Hospital, Baghdad, Iraq

S. Hoz et al. (eds.), *Neuroradiology Board's Favorites*,
https://doi.org/10.1007/978-3-031-64261-6_2

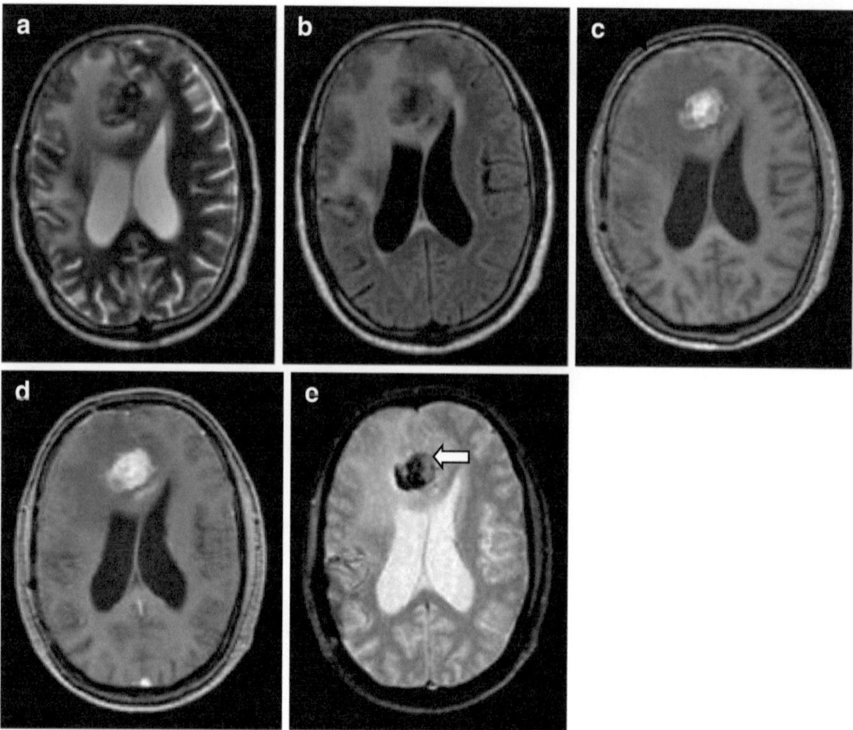

Fig. 2.1 Serial MRI images of the brain with the following: (**a**) axial T2, (**b**) axial FLAIR, (**c**) axial T1, (**d**) axial T1 C+, and (**e**) axial GRE. (Figure courtesy of Dr. Samer Hoz)

(arrow). Histopathological examination confirmed the diagnosis of melanotic metastasis.

Case 9.2: Intracranial Metastatic Melanoma: Amelanotic Type

Case Scenario

A 62-year-old woman presented with headaches, seizures, nausea, and vomiting (Fig. 2.2).

Imaging Description

There are two large hemorrhagic mass lesions at the gray-white matter junction, one located in the right frontal lobe and the other in the left occipital lobe. The lesions appear isointense on T1 and heterogeneous on T2. There is a large amount of hemosiderin staining on T2 FFE GRE, seen as a signal void. The lesions demonstrate irregular peripheral enhancement. There is minimal surrounding vasogenic edema causing a localized mass effect, but no midline shift. Histopathological examination confirmed the diagnosis of amelanotic metastasis.

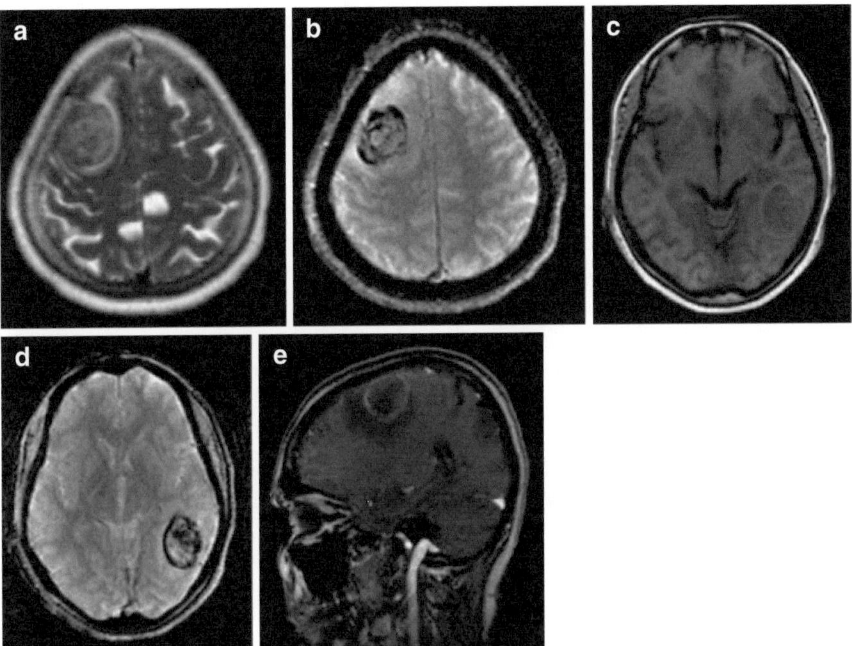

Fig. 2.2 Serial MRI images of the brain with the following: (**a**) axial T2, (**b**) axial FFE, (**c**) axial T1, (**d**) axial FFE, and (**e**) sagittal T1 C+. (Figure courtesy of Dr. Samer Hoz)

Intracranial Metastatic Melanoma

Melanoma ranks as the third most frequent cause of brain metastasis. Among brain metastases, metastatic melanomas are the most common cause for intratumoral hemorrhage apart from metastatic choriocarcinoma. They can be categorized into two subtypes:

1. Melanotic: These metastases exhibit more than 10% of the melanotic cells on histopathological examination.
2. Amelanotic: These metastases have fewer than 10% of the melanotic cells on histopathological examination.

Melanotic metastases exhibit characteristic MRI findings, appearing hyperintense on T1-weighted images and hypointense on T2-weighted images. In contrast, amelanotic metastases present with MRI findings that are non-specific. However, melanotic metastases represent only about a quarter of all intracranial melanoma metastases.

On CT scans, intracranial melanoma metastases typically appear as highly attenuating single or multiple nodules. They are usually located at the gray-white matter junction and are often associated with intratumoral hemorrhage and edema. These nodules typically enhance with contrast and are dural-based, which can make them challenging to distinguish from meningiomas. Necrosis and cystic changes are infrequent.

In MRI imaging, both melanin pigment and blood products can alter the signal in a similar manner, resulting in hyperintensity on T1-weighted images and hypointensity on T2-weighted images. However, they can be differentiated using susceptibility-weighted imaging (SWI), where melanin exhibits a weak diamagnetic effect and significant signal loss, unlike blood. On T1-weighted post-contrast images, metastatic melanomas typically demonstrate peripheral rim enhancement or a diffuse homogeneous enhancement [1, 2].

Differential Diagnosis

1. Hemorrhagic intracranial metastases: choriocarcinoma, renal cell carcinoma, thyroid carcinoma, breast carcinoma, lung carcinoma, and hepatocellular carcinoma.
2. Other melanotic tumors: melanotic meningioma and meningeal melanocytoma.
3. Other causes of hemorrhage: primary lobar hemorrhage and hemorrhage into primary brain tumor.

Questions

1. **Intracranial metastatic melanoma, the FALSE answer is:**
   A. Intracranial metastatic melanoma exhibits hyper-intensity on T1 and hypo-intensity on T2 images.
   B. Intracranial metastatic melanomas are frequently complicated by massive hemorrhage.
   C. Melanotic and amelanotic metastases exhibit the same MRI findings.
   D. Metastatic melanomas commonly show peripheral rim enhancement on MRI T1 C+.
   E. Melanotic metastases usually have non-specific MRI findings.
   The answer is C.

   Melanotic metastases exhibit characteristic MRI findings, appearing hyperintense on T1-weighted images and hypointense on T2-weighted images. In contrast, amelanotic metastases present with MRI findings that are non-specific.

Case 9.3: Choroidal Melanoma

Case Scenario

A 20-year-old man presented with decreased visual acuity. Ophthalmologic examination revealed a pigmented intraocular mass in the left eye (Fig. 2.3).

Imaging Description

MRI sequences show a left intraocular sub-retinal soft tissue mass (lateral nasal quadrant) of high signal intensity on T1 and FLAIR and low signal intensity on T2.

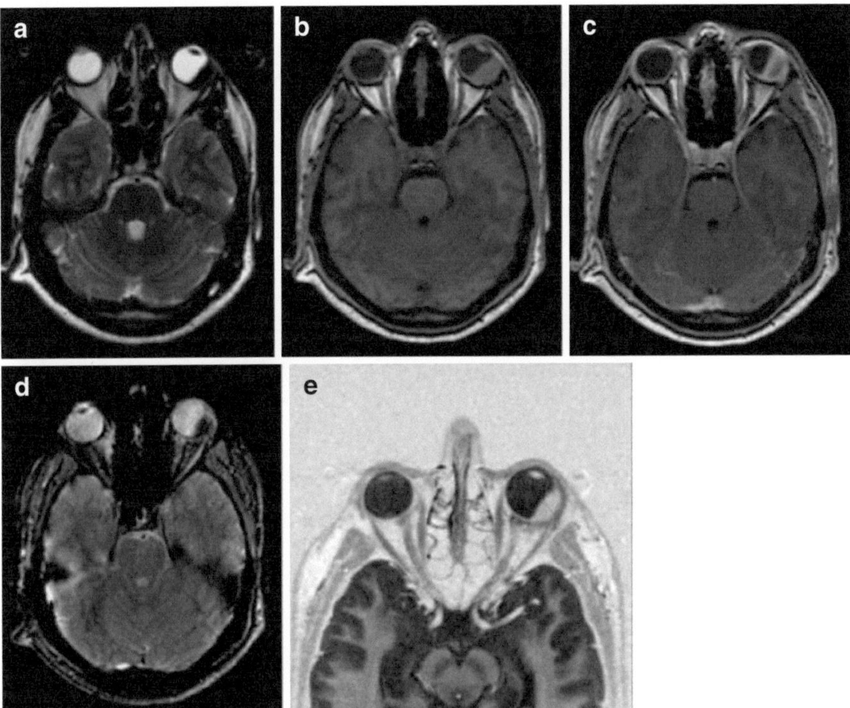

Fig. 2.3 Serial MRI images of the brain with the following: (**a**) axial T2, (**b**) axial T1, (**c**) axial T1 C+, (**d**) axial GRE, and (**e**) axial STIR. (Figure courtesy of Dr. Samer Hoz)

A signal void can be seen in the lesion on T2 FFE GRE, suggesting hemorrhage. There is an intense and relatively homogeneous enhancement following IV contrast. No extraocular extension is seen. Histopathological examination confirmed the diagnosis of uveal melanoma.

Choroidal Melanoma

Choroidal melanomas, also referred to as malignant uveal melanomas, are the most common primary eye tumors in adults, primarily affecting individuals of Caucasian descent. The incidence of these tumors increases with age, with only around 2% occurring in individuals younger than 20 years old.

Ultrasound is a valuable tool for examining small eye lesions with a thickness of less than 3 mm, which can be challenging to detect using CT or MRI. These lesions typically exhibit a solid, low-to-medium echotexture, internal vascularity, and a characteristic collar button shape. They may also present with additional features such as choroidal excavation, retinal detachment, posterior scleral bowing, and vitreous or subretinal seeding. Larger tumors often show significant sound attenuation.

CT imaging is effective for evaluating most choroidal melanomas, which appear hyperdense and have elevated, well-defined shapes resembling a lens or mushroom. These tumors enhance with contrast administration.

MRI is the preferred imaging modality for assessing choroidal melanomas. They typically appear moderately hyperintense on T1/PD scans due to the presence of melanin and hemorrhage. The exudative retinal detachment accompanying these tumors also exhibits moderate hyperintensity. On T2 scans, choroidal melanomas are moderately hypointense, while exudative retinal detachment appears moderately hyperintense. Hemorrhagic subretinal fluid exhibits variable signal patterns. Upon contrast administration with T1 C+ (Gd) MRI, the tumors enhance, and fat suppression techniques are valuable for detecting any extraocular extension [3, 4].

Differential Diagnosis
1. Retinoblastoma
2. Uveal metastases
3. Choroidal detachment
4. Choroidal hemangioma
5. Choroidal cyst
6. Uveal neurofibroma
7. Uveal schwannoma

Questions

1. **Choroidal melanoma, the FALSE answer is:**
   A. They are moderately hyperintense on T1/PD.
   B. They are hyperdense on CT.
   C. T1 C+ shows contrast enhancement.
   D. On T2, they are moderately hyperintense.
   E. They are the most common primary tumor of the eye in adults.
      The answer is D.
      On T2, they are moderately hypointense.

Case 10: Gliomas

Case 10.1: Pilomyxoid Astrocytoma

Case Scenario
A 2-year-old boy presented with complaints of headache and vomiting (Fig. 2.4).

Imaging Description
A large, lobulated, heterogeneously enhancing mass with a thick, irregular peripheral outline is centered on the hypothalamus and chiasm. The lesion extends to the third ventricle and reaches the peripontine cistern, causing obstructive hydrocephalus and a marked mass effect on adjacent tissue. There is no surrounding edema or hemorrhage. The lesion is hyperintense on T2 and hypointense on T1. MRI features are suggestive of hypothalamic-optochiasmatic glioma, particularly in this age

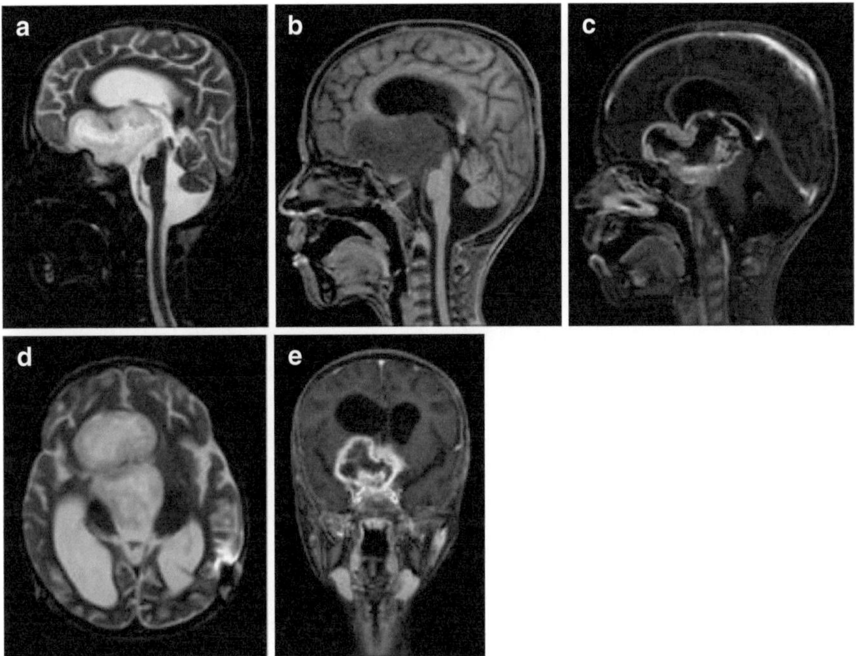

Fig. 2.4 Serial MRI images of the brain with the following: (**a**) sagittal T2, (**b**) sagittal T1, (**c**) sagittal T1 C+, (**d**) axial T2, and (**e**) coronal T1 C+. (Figure courtesy of Dr. Samer Hoz)

group. Histopathological examination confirmed the diagnosis of pilomyxoid astrocytoma.

Pilomyxoid Astrocytoma

Pilomyxoid astrocytoma is a relatively uncommon and more aggressive variant of pilocytic astrocytoma and is characterized by distinct clinical and histopathological features. It is most often encountered in infants and young children, typically presenting between the ages of 10 and 18 months. While it is estimated to account for about 2% of all childhood astrocytomas, its precise epidemiology remains uncertain since it was previously categorized as pilocytic astrocytoma. These tumors most frequently originate in the hypothalamus or optic chiasm but can also develop in other areas of the brain, including the posterior fossa and even within the spinal cord. In terms of classification, they are categorized as World Health Organization (WHO) grade 2 tumors, which distinguishes them from pilocytic astrocytomas, classified as WHO grade 1.

Pilomyxoid astrocytomas typically manifest as large, lobulated growths comprising both solid and cystic components. They tend to have well-defined borders with minimal to no surrounding edema. In some cases, they may appear H-shaped, extending from the midline into the bilateral temporal lobes. Hemorrhage and/or the dissemination through cerebrospinal fluid (CSF) can occur. On MRI scans, pilomyxoid astrocytomas typically appear isointense on T1-weighted images and typically hyperintense on T2-weighted images, reflecting the presence of a myxoid matrix. Approximately 20% of cases exhibit intratumoral hemorrhage on gradient echo (GRE) or susceptibility-weighted imaging (SWI). Enhancement on post-contrast T1-weighted images is common, typically occurring within the solid component, although peripheral enhancement may also be observed. Diffusion-weighted imaging (DWI) and apparent diffusion coefficient (ADC) maps typically show high ADC values due to the myxoid matrix within the tumor [5, 6].

Differential Diagnosis

The main differential diagnosis is pilocytic astrocytoma, which is challenging to distinguish through imaging alone. Distinctive features associated with pilomyxoid astrocytomas include the presence of hemorrhage, presentation at a young age (typically under 2 years old), involvement of the hypothalamus and optic chiasm, and the characteristic H-shaped appearance of a large suprasellar mass.

Questions

1. **Pilomyxoid astrocytoma, the FALSE answer is:**
   A. They most commonly arise in the hypothalamus or optic chiasm.
   B. On MRI, pilomyxoid astrocytomas are typically hyper-intense on T2.
   C. A relatively uncommon and more aggressive variant of pilocytic astrocytoma.
   D. Pilomyxoid astrocytoma is mostly encountered in infants and young children.
   E. Pilomyxoid astrocytoma is considered to be WHO grade 1.
   The answer is E.
   Pilomyxoid astrocytoma is considered to be WHO grade 2.

Case 10.2: Glioblastoma Multiforme (GBM)

Case Scenario

A 44-year-old female presented with complaints of headache, vomiting, and seizures (Fig. 2.5).

Imaging Description

MRI demonstrates expansion of the splenium of the corpus callosum, more on the right side. The lesion shows low signal intensity on T1W and heterogeneous but mainly high signal intensity on T2W, with areas of central necrosis. T2 FFE weighted images show areas of hemorrhagic changes, and post-contrast images show thick, irregular peripheral enhancement. The mass is surrounded by vasogenic edema and shows restricted diffusion. Histopathologic findings confirmed GBM.

Glioblastoma

Glioblastoma, previously known as glioblastoma multiforme (GBM), stands as the most common primary brain tumor among adults. These tumors are highly

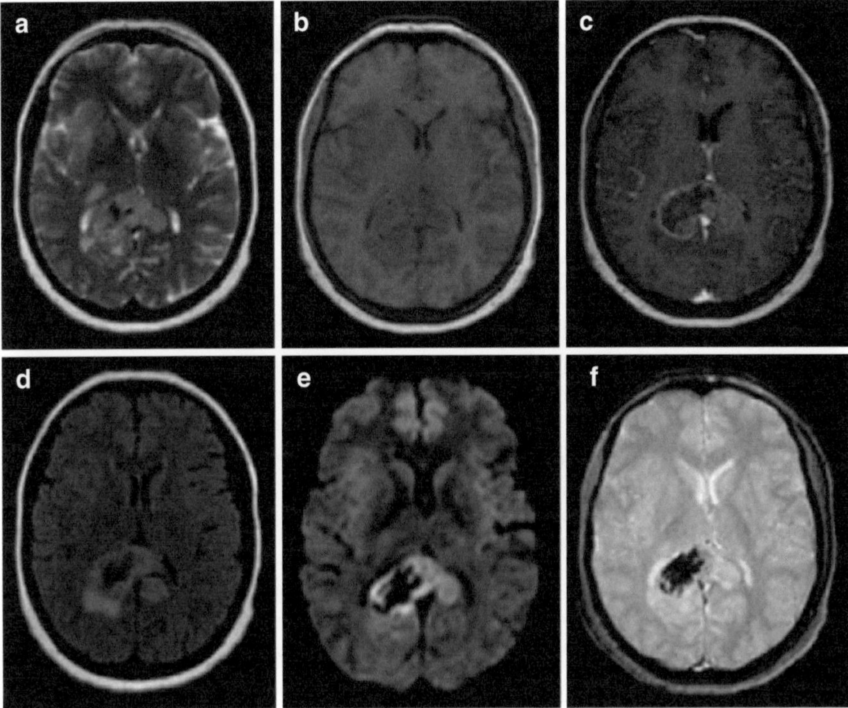

Fig. 2.5 Serial MRI images of the brain with the following: (**a**) axial T2, (**b**) axial T1, (**c**) axial T1 C+, (**d**) axial FLAIR, (**e**) axial DWI, (**f**) axial T2 FFE. (Figure courtesy of Dr. Samer Hoz)

aggressive, relatively unresponsive to treatment, and generally associated with a poor prognosis. They predominantly affect individuals over the age of 40, with the highest incidence occurring between 65 and 75 years. They exhibit a slight male predilection with a male-to-female ratio of 3:2.

GBMs can manifest in two forms: In 90% of cases, it is primary, arising de novo. In the remaining 10% of cases, it is secondary, arising from a pre-existing lower-grade tumor. In rare instances, they may be linked to prior radiation exposure or specific syndromes such as neurofibromatosis type 1 (NF1), Maffucci syndrome, Turcot syndrome, and Ollier disease.

Typically, glioblastomas are diagnosed when they have already reached a considerable size. These tumors are typically situated within white matter, often displaying thick, irregular peripheral enhancement, central necrosis, and occasional hemorrhagic components. Surrounding edema usually contains infiltrating neoplastic cells. Approximately 20% of cases present with multifocal patterns, where multiple areas of enhancement are connected by abnormal white matter signal. This is distinct from multi-centric disease, where no such connections exist.

MRI imaging reveals glioblastomas as lesions that appear hypo- to isointense on T1-weighted images, featuring central necrosis and irregular peripheral enhancement upon contrast administration (T1 C+). T2-weighted and FLAIR images depict a hyperintense lesion with associated vasogenic edema in the surrounding area. On gradient echo/susceptibility-weighted imaging (GE/SWI), these tumors display susceptibility artifacts on T2* images due to the presence of blood products. The solid or enhancing component often exhibits an elevated signal on diffusion-weighted imaging (DWI). In comparison to normal brain tissue and lower-grade tumors, glioblastomas exhibit an increased relative cerebral blood volume (rCBV) on MR perfusion imaging. Magnetic resonance spectroscopy (MRS) findings typically include elevated choline (Cho), lactate, and lipids, as well as decreased N-acetylaspartate (NAA) and myoinositol levels [7, 8].

Differential Diagnosis

1. Grade 4 IDH-mutant astrocytoma: typically occurs in younger patients.
2. Cerebral metastases: often found at the gray-white matter junction, sparing the adjacent cortex. Unlike GBM, relative cerebral blood volume (rCBV) in the surrounding edema is reduced.
3. Primary CNS lymphoma: while typically showing homogeneous enhancement, central necrosis may develop, particularly in patients with acquired immunodeficiency syndrome (AIDS).
4. Cerebral abscess: the presence of central restricted diffusion can aid in differentiation, although GBM with hemorrhage can present similarly. Additional distinctive features include the dual rim sign and a smooth, complete low-intensity rim on susceptibility-weighted imaging (SWI).
5. Tumefactive demyelination: usually encountered in younger patients, it often displays an open ring pattern of enhancement.
6. Subacute cerebral infarction: a patient's medical history is valuable for diagnosis, and elevations in Cho and rCBV are typically absent.
7. Cerebral toxoplasmosis: should be considered, especially in patients with AIDS.

Questions

1. **GBM, the FALSE answer is:**
   A. GBM shows irregular peripheral enhancement on T1 C+.
   B. MRS demonstrates decreased choline, lactate, and lipids.
   C. GBM is T1 hypo- to iso-intense.
   D. GBM is the most common primary brain tumor in adults.
   E. T2/FLAIR shows a hyperintense lesion.
   The answer is B.
   MRS demonstrates increased choline, lactate, and lipids.

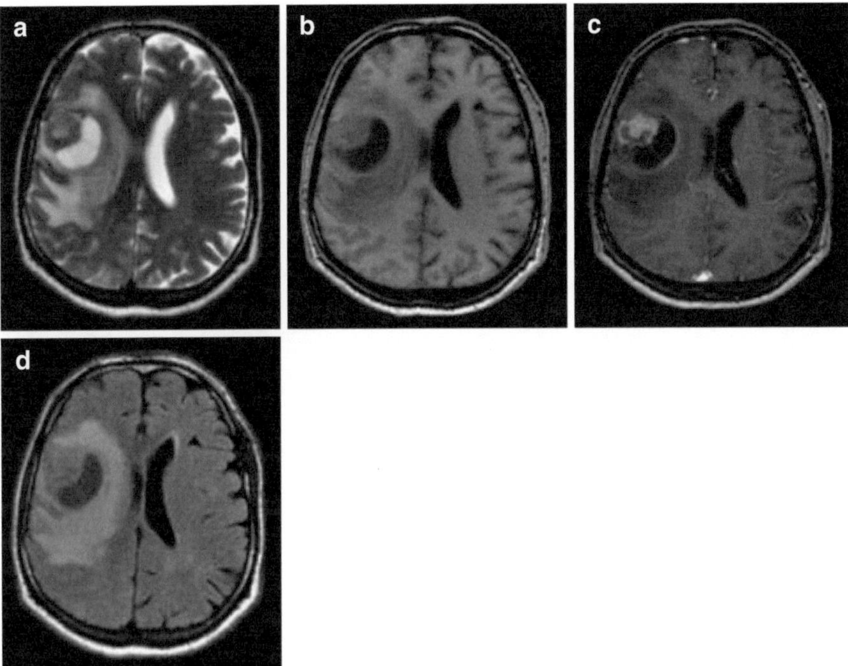

Fig. 2.6 Serial MRI images of the brain with the following: (**a**) axial T2, (**b**) axial T1, (**c**) axial T1 C+, and (**d**) axial FLAIR. (Figure courtesy of Dr. Samer Hoz)

Case 10.3: Pleomorphic Xanthoastrocytoma (PXA)

Case 10.3.1

Case Scenario
A 24-year-old female presented with complaints of headache and seizures (Fig. 2.6).

Imaging Description
A large solitary intra-axial multi-lobulated mass lesion can be seen in the cortical/subcortical region centered within the lateral aspect of the right fronto-parietal lobe. The lesion is a predominantly solid mass with cystic components. The solid component is hypo- to iso-intense on T1 and has a central part that is slightly hyper-intense on T2 imaging. The cystic component is hypo-intense on T1 and hyper-intense on T2. There is a T2/FLAIR signal abnormality consistent with vasogenic edema grade II. On post-contrast imaging, the solid component of the lesion enhances vividly with non-enhancement of the cystic component. There is no evidence of blood products within the lesion. There is a local mass effect and compression of the frontal horn of the lateral ventricle with minimal midline shift. Histopathological examination confirmed the diagnosis of PXA.

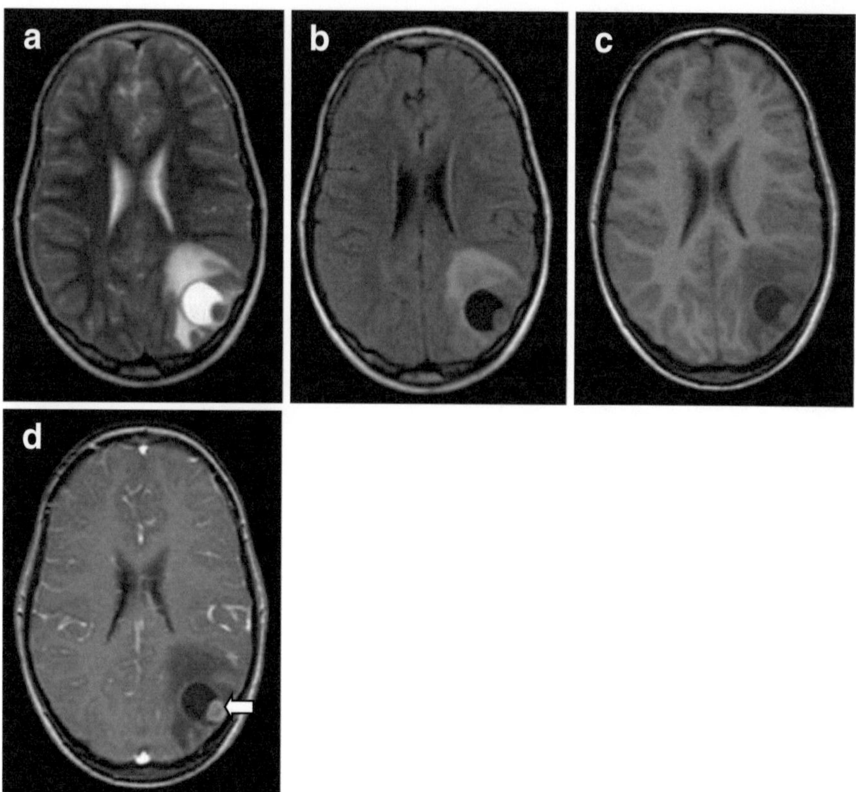

Fig. 2.7 Serial MRI images of the brain with the following: (**a**) axial T2, (**b**) axial FLAIR, (**c**) axial T1, and (**d**) axial T1 C+. (Figure courtesy of Dr. Samer Hoz)

Case 10.3.2

Case Scenario
A 22-year-old man presented with seizures (Fig. 2.7).

Imaging Description
A brain MRI demonstrated a well-defined left occipital lobe cystic lesion with a peripherally located enhancing nodule. The nodule appears slightly hypointense on T2 compared to the gray matter and isointense on T1. It demonstrates bright contrast enhancement (arrow). Histopathological examination confirmed the diagnosis of PXA.

Pleomorphic Xanthoastrocytomas
Pleomorphic xanthoastrocytomas (PXAs) are uncommon astrocytic tumors categorized as WHO grade 2 or 3, comprising just 1% of all primary brain tumors. They tend to primarily affect children and young adults, with the highest incidence

occurring in individuals aged 10–30 years. Approximately 98% of PXAs are supratentorial, typically manifesting as brightly enhancing superficial cortical tumors that often contain a peripheral eccentric cystic portion (50–60%). The temporal lobe is the site of about half of these tumors, with the remaining cases more frequently occurring in the frontal lobe, followed by the parietal lobe.

On CT scans, PXAs generally appear as hypo- or iso-dense lesions and may exhibit either well-defined or ill-defined borders. Due to their slow growth, some characteristic signs of slow-growing tumors may be present, such as scalloping of the adjacent skull bone, and they may lack the typical surrounding edema seen in more aggressive lesions. Calcification is a rare occurrence. On T1-weighted MRI, the solid component of PXAs appears iso- to hypo-intense compared to normal gray matter, while the cystic part appears hypo-intense. Leptomeningeal involvement is observed in over 70% of cases, although it rarely results in direct invasion of the dura. Nonetheless, a reactive dural tail sign may occasionally be visible due to the tumor's peripheral location and its association with the leptomeninges. The solid component typically demonstrates intense enhancement on contrast-enhanced T1-weighted images (T1 C+ Gd). On T2-weighted images, the solid component appears iso- to hyper-intense relative to gray matter, while the cystic regions appear hyper-intense. This hyper-intensity in the cystic areas on T2/FLAIR sequences is due to a higher protein content and, in part, the absence of significant surrounding vasogenic edema. PXAs are typically avascular when assessed using digital subtraction angiography (DSA) [9, 10].

Differential Diagnosis
1. Ganglioglioma: while they may appear quite similar on imaging, they typically exhibit less intense contrast enhancement, and calcification is present in around 50% of cases. A dural tail sign is typically absent.
2. Dysembryoplastic neuroepithelial tumor (DNET): these tumors usually do not show contrast enhancement and often display a characteristic "bubbly appearance."
3. Oligodendroglioma: calcification is frequently observed, and this type of tumor generally affects older individuals in their fourth and fifth decades of life.
4. Desmoplastic infantile ganglioglioma: primarily impacting younger children, this tumor type often presents with multiple lesions and notable dural involvement.
5. Cystic meningioma.

Questions

1. **PXA, the FALSE answer is:**
   A. On MRI T1, the solid component is iso- to hypo-intense to gray matter, while the cystic component appears hypointense.
   B. The solid component typically enhances vividly on T1 C+ (Gd).
   C. PXAs are usually WHO grade 2 or 3 tumors.

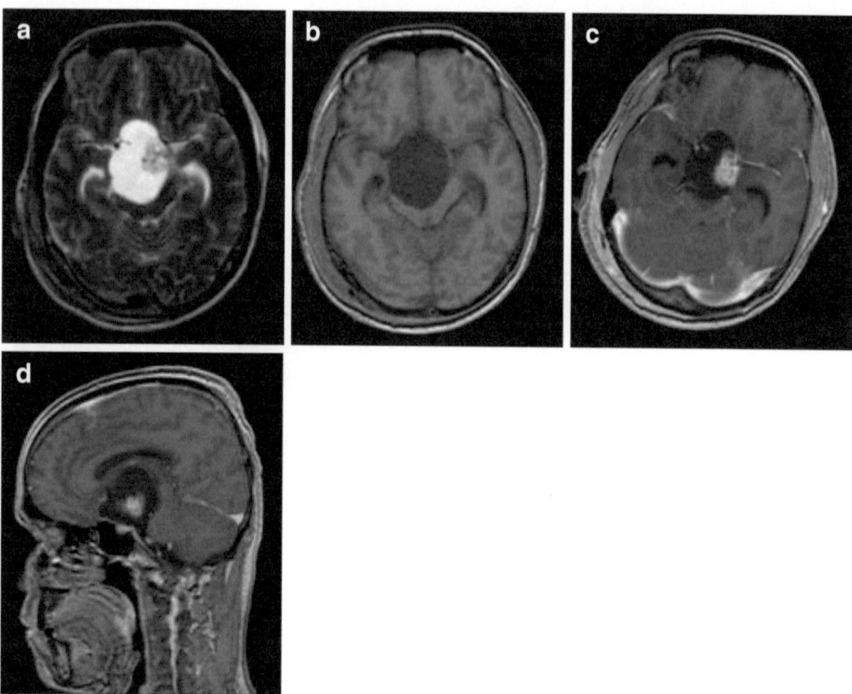

Fig. 2.8 Serial MRI images of the brain with the following: (**a**) axial T2, (**b**) axial T1, (**c**) axial T1 C+, and (**d**) sagittal T1 C+. (Figure courtesy of Dr. Samer Hoz)

    D. PXAs are mostly found in the frontal lobe.
    E. Calcification is rare in PXA.
       The answer is D.
       PXAs are mostly found in the temporal lobe.

Case 10.4: Suprasellar Pilocytic Astrocytoma

Case Scenario
A 32-year-old male presented with visual disturbance and dim vision (Fig. 2.8).

Imaging Description
A large, well-defined cystic lesion with a bright, intensely enhancing mural nodule is seen in the suprasellar region, involving the optic chiasma. Superiorly, the lesion extends into the third ventricle. The lesion is causing a mass effect on adjacent brain tissue, but without edema or midline shift. Histopathological examination confirmed the diagnosis of pilocytic astrocytoma.

Pilocytic Astrocytoma

Sometimes referred to as juvenile pilocytic astrocytomas, these are well-defined astrocytic gliomas that typically impact young individuals. According to the current WHO classification of central nervous system (CNS) tumors, they are categorized as grade 1 tumors, which generally implies a favorable prognosis. These tumors can vary in appearance on imaging, but most are recognized as large cystic masses with an enhancing mural nodule. Calcification may be evident in approximately 20% of cases.

Pilocytic astrocytomas most commonly occur in the cerebellum, optic pathways, and spinal cords. They are typically found in young individuals, with 75% occurring during the first two decades of life, usually around the age of 9–10, without a significant gender predilection. These tumors are strongly associated with neurofibromatosis type 1 (NF1), and NF1-related tumors are more likely to affect the optic nerves and chiasm. In fact, up to 20% of NF1 patients will develop these tumors, often in early childhood. Conversely, around one-third of pilocytic astrocytomas involving the optic nerves are linked to NF1.

Pilocytic astrocytomas typically arise from midline structures, with the cerebellum being the most common location (approximately 60% of cases), followed by the optic pathway (optic nerve, optic chiasm, hypothalamus, and optic radiation) in about 25–30% of cases. With regards to radiological characteristics, these tumors can have various appearances on imaging. The majority (67%) present as large cystic lesions with a prominently enhancing nodule on the wall. Almost all of these tumors exhibit enhancement (about 95%), and roughly 20% may contain calcifications. Hemorrhage is a rare complication.

On T1-weighted MRI images, the solid portion of the tumor is typically iso- to hypointense compared to the surrounding brain, while the cystic part displays a fluid signal, unless hemorrhage has occurred. After the administration of gadolinium, there is vivid contrast enhancement of the solid component and the cyst wall in approximately 50% of cases. In T2-weighted or susceptibility-weighted images, the solid component appears hyperintense compared to the adjacent brain, and the cystic component shows a high signal. Finally, in T2-weighted or susceptibility-weighted images, the signal is lost if calcification or hemorrhage is present [11, 12].

Differential Diagnosis

1. High-grade astrocytoma with piloid features is more commonly found in adults and individuals with NF1, and it is associated with an unfavorable prognosis.
2. Hemangioblastoma is typically encountered in adults, though it can be associated with von Hippel-Lindau disease in children. This type of tumor usually does not show enhancement in the cyst wall or contain calcifications. Instead, it often presents with a smaller mural nodule displaying angiographic contrast blush.
3. Medulloblastoma typically occurs at midline structures, particularly the vermis and the roof of the fourth ventricle, rather than the cerebellar hemisphere. It is more frequently seen in younger patients aged 2–6 years.
4. Atypical teratoid/rhabdoid tumor is characterized by a larger and heterogeneously enhancing mass.

5. Ependymoma tends to occupy the fourth ventricle and extend out of the foramen of Luschka and the foramina of Magendie. It less commonly presents as a large cystic component.
6. Ganglioglioma.
7. PXA.
8. Cerebellar hemorrhage.

Questions

1. **Supra-sellar optic pathway pilocytic astrocytomas, the FALSE answer is:**
    A. Almost all pilocytic astrocytomas show enhancement on imaging.
    B. Pilocytic astrocytomas usually present as a large cystic lesion.
    C. The solid component of a pilocytic astrocytoma typically appears hyperintense compared to the surrounding brain on T1W imaging.
    D. In T2W imaging, the solid component appears hyperintense relative to the adjacent brain.
    E. The most common location of pilocytic astrocytomas is the cerebellum.
    The answer is C.

    In T1W imaging, the solid component of the tumor is typically iso- to hypointense compared to the surrounding brain.

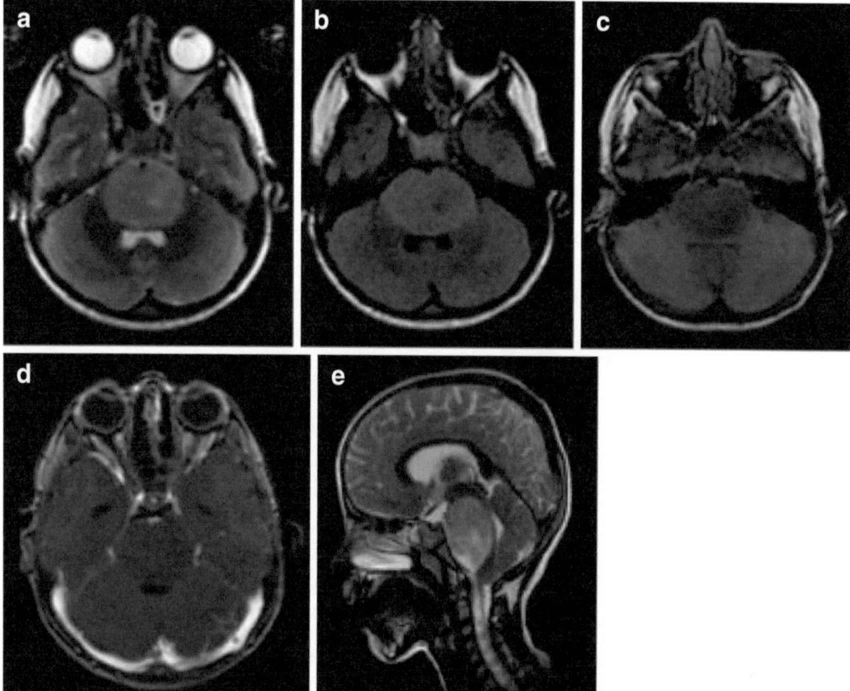

Fig. 2.9 Serial MRI images of the brain with the following: (**a**) axial T2, (**b**) axial FLAIR, (**c**) axial T1, (**d**) axial T1 C+, and (**e**) sagittal T2. (Figure courtesy of Dr. Samer Hoz)

Case 10.5: Brainstem Gliomas

Case Scenario
An 8-year-old girl presented with a history of bulbar palsy (Fig. 2.9).

Imaging Description
MRI reveals marked diffuse expansion of the pons with extension into the medulla. This expansion is centered in the pons, causing elevation of the midbrain and effacement of the fourth ventricle. It is relatively well-circumscribed, with low T1 and high T2 and FLAIR signals but no apparent enhancement.

Brainstem Gliomas
The terms "diffuse brainstem gliomas" and "diffuse intrinsic pontine gliomas" were previously used to describe distinct brainstem astrocytomas. However, they are no longer considered a separate entity. According to the WHO classification of CNS tumors in its 2016 update, these terms have been replaced by a range of distinct categories based on molecular characteristics. The majority of these tumors are now classified as "diffuse midline glioma H3 K27M–mutant." These tumors are commonly seen in children, with the peak incidence occurring between the ages of 3 and 10 years, accounting for 10–15% of all pediatric brain tumors. Interestingly, when associated with NF1, this type of tumor often has a more favorable prognosis and tends to follow a less aggressive course.

Under the WHO classification, "diffuse midline glioma H3 K27M-mutants" is now considered a distinct disease entity. The remaining tumors are categorized differently, such as diffuse brainstem gliomas without K27M mutations, which fall into the groups of diffuse pediatric-type high-grade gliomas, H3-wildtype, and IDH-wildtype.

Regarding their radiological characteristics, only 25–40% of all cases are found outside the pons. Less commonly, they can occur in the medullary or mesencephalic regions. Radiographic features of diffuse intrinsic pontine gliomas include pons enlargement, displacement of the basilar artery anteriorly against the clivus, a flattened floor of the fourth ventricle, and obstructive hydrocephalus. The tumor can extend outward into the basal cisterns or centrally into the fourth ventricle. Typically, the tumor appears homogeneous before treatment, but areas of necrosis may be present.

CT usually shows a hypodense lesion with minimal enhancement. MRI is the preferred diagnostic method and reveals decreased intensity on T1-weighted images and heterogeneous increased intensity on T2-weighted images. These tumors typically enhance on post-contrast T1-weighted images, but enhancement may occur or increase after radiotherapy. In some cases, MRI may appear normal, with a slight restriction seen on diffusion-weighted imaging (DWI) [13, 14].

Differential Diagnosis
Other potential diagnoses for brainstem gliomas that should be considered include rhombencephalitis, acute disseminated encephalomyelitis (ADEM), neurofibromatosis type 1 (NF1), tuberous sclerosis (TS), osmotic demyelination, Langerhans cell histiocytosis, hamartoma, as well as other tumors like medulloblastoma and ependymoma.

Questions

1. **Brainstem gliomas, the FALSE answer is:**
   A. Most of these tumors are now classified as diffuse midline glioma H3 K27M-mutant.
   B. Only 25–40% of all cases occur outside the pontine.
   C. CT scan usually shows a hypodense lesion with low enhancement.
   D. Flattening of the floor of the fourth ventricle (the flat floor of the fourth ventricle sign) and obstructive hydrocephalus can be seen.
   E. MRI reveals an increase in intensity on T1 and a heterogeneous decrease on T2.
   The answer is E.
   MRI reveals a decrease in intensity on T1 and a heterogeneous increase on T2.

Case 10.6: Pilocytic Astrocytoma

Case 10.6.1

Case Scenario

A 17-year-old girl presented with headache and cranial nerve dysfunction (Fig. 2.10).

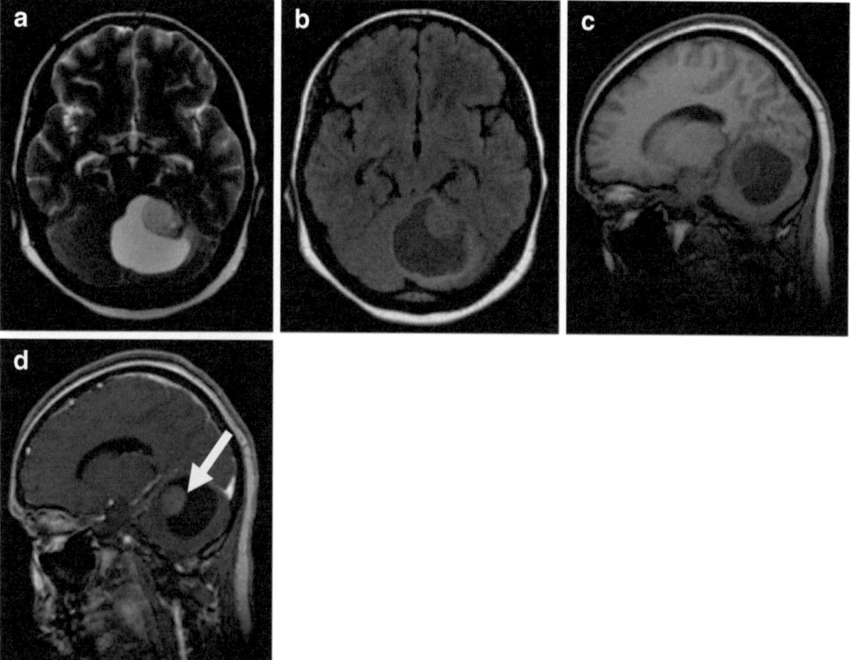

Fig. 2.10 Serial MRI images of the brain with the following: (**a**) axial T2, (**b**) axial T1, (**c**) sagittal T1, and (**d**) sagittal T1 C+. (Figure courtesy of Dr. Samer Hoz)

Imaging Description

MRI demonstrates a large, partially cystic mass with its epicenter in the left cerebellar hemisphere, displacing and effacing the fourth ventricle and resulting in hydrocephalus. The mass has a solid component that appears nodular and slightly hypointense on T1 compared to the cerebellum and hyperintense on T2. It demonstrates bright contrast enhancement (arrow). Histopathological examination confirmed the diagnosis of pilocytic astrocytoma.

Case 10.6.2

Case Scenario

A 5-year-old girl presented with headache and signs of increased ICP (Fig. 2.11).

Imaging Description

MRI shows a large midline posterior fossa lesion centered on the cerebellar vermis. The lesion is cystic and has a predominantly high T2 signal and a low T1 signal, with a heterogeneous signal within the lower part of it seen as a solid component that appears as nodular and slightly hypointense on T1 compared to the cerebellum and hyperintense on T2. It demonstrates bright contrast enhancement (arrow). The lesion is exerting a mass effect on the brainstem and causing the effacement of the

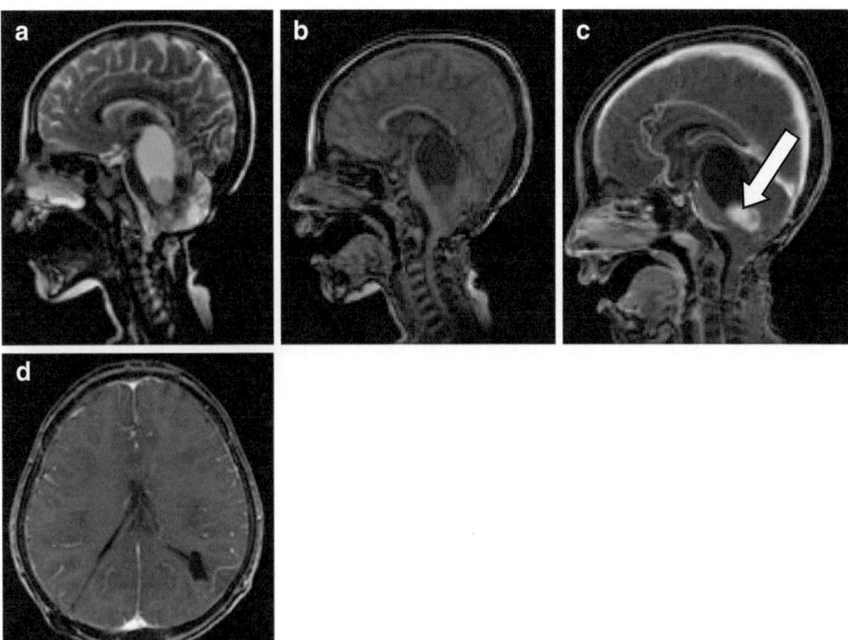

Fig. 2.11 Serial MRI images of the brain with the following: (**a**) sagittal T2, (**b**) sagittal T1, (**c**) sagittal T1 C+, and (**d**) axial T1 C+. (Figure courtesy of Dr. Samer Hoz)

fourth ventricle. There is no hydrocephalus due to the placement of bilateral ventriculoperitoneal shunts. There is no transependymal edema. There is a slight inferior displacement of the cerebellar tonsils around the brainstem. Histopathological examination confirmed the diagnosis of pilocytic astrocytoma.

Pilocytic Astrocytoma

Also referred to as juvenile pilocytic astrocytomas, these are astrocytic gliomas classified as WHO grade 1. They are more prevalent in younger individuals, with 75% of cases occurring during the first two decades of life. Although they account for only a small percentage of all intracranial neoplasms (0.6–5.1%), they are the most common primary brain tumor among children (15%). Most of these tumors originate in the cerebellum (60%), followed by the optic pathway (25–30%), which is particularly common in individuals with neurofibromatosis type 1 (NF1). Less common sites include the brainstem, cerebral hemispheres (more common in adults), cerebral ventricles, velum interpositum, and spinal cord. There is no notable sex predilection, and the prognosis is generally favorable.

Pilocytic astrocytomas are strongly linked to NF1, with NF1-associated tumors showing a specific tendency to affect the optic nerves and chiasm. Up to 20% of all NF1 patients develop pilocytic astrocytomas, and approximately one-third of pilocytic astrocytomas involving the optic nerves are associated with NF1.

On imaging, pilocytic astrocytomas can exhibit various appearances. Most commonly, they present as a large cystic component with a prominently enhancing nodule (67%). This nodule may have an enhancing (46%) or non-enhancing (21%) cyst wall. Other presentations include entirely solid lesions in 17% of cases or heterogeneous masses with mixed solid portions and multiple cysts, often accompanied by central necrosis in the remaining 16% of cases. Calcification is observed in up to one-fifth of cases, but hemorrhage is infrequent.

The solid component typically appears iso- to hypointense relative to the surrounding brain tissue on T1-weighted images, while the cystic component shows fluid signal unless complicated by hemorrhage. Contrast-enhanced T1-weighted images reveal marked enhancement, with approximately 50% of cases showing enhancement of the cyst wall. On T2-weighted images, the solid component appears hyperintense, while the cystic portion displays a high signal. Calcification or hemorrhage, when present, results in signal loss on T2*-weighted, GRE, or SWI sequences [15, 16].

Differential Diagnosis

1. High-grade astrocytoma with piloid features: should be considered as a differential, especially in adults and those with NF1. The prognosis is very poor.
2. Hemangioblastoma: usually seen in adults. In pediatric cases, hemangioblastoma commonly occurs in association with Von Hippel-Lindau disease. In hemangioblastomas, there is usually no enhancement of the cyst wall. Calcification is not a typical feature, and angiography often reveals the presence of a small, enhancing nodule with contrast blush.

3. Medulloblastoma: typically found along the midline, especially in the vermis and the roof of the fourth ventricle, rather than in the lateral parts of the cerebellum. This type of tumor tends to manifest in younger patients aged between 2 and 6 years old.
4. Atypical teratoid/rhabdoid tumor presents as a larger heterogeneously enhancing mass.
5. Ependymoma: typically occupies the fourth ventricle and extends out of the foramina of Luschka and Magendie. A large cystic component is less common in ependymomas.
6. Ganglioglioma.
7. PXA.
8. Cerebellar hemorrhage.

Questions

1. **Pilocytic astrocytoma, the FALSE answer is:**
   A. The majority present as solid lesions.
   B. They are the most common primary brain tumor among children.
   C. The solid component is iso- to hypointense relative to the adjacent brain on T1.
   D. Hemorrhage is uncommon.
   E. Most arise from the cerebellum.
      The answer is A.
      The majority present as a large cystic lesion with a brightly enhancing mural nodule.

Case 10.7: Multifocal Glioblastoma

Case Scenario
A 60-year-old man presented with severe headache and vomiting (Fig. 2.12).

Imaging Description
MRI shows a large region of high T2 signal with mass effect, and a midline shift can be seen involving the majority of the right temporal lobe, with expansion to the frontal and parietal lobes. Within this region, two lesions are seen with thick, vivid peripheral enhancement and non-enhancing central areas suggestive of necrosis. The lesions exhibit heterogeneous signal intensity; they are mainly hypointense in T2 and T1 with central necrosis and signal dropout in T2 FFE GRE, suggestive of the presence of hemosiderin (hemorrhage). The patient refused to undergo operation. The last image, which was taken after 1 year, shows that the lesion became more aggressive with an increase in midline shift, with enhancing and abnormal white matter between lesions. Histopathological examination confirmed the diagnosis of multifocal glioblastoma.

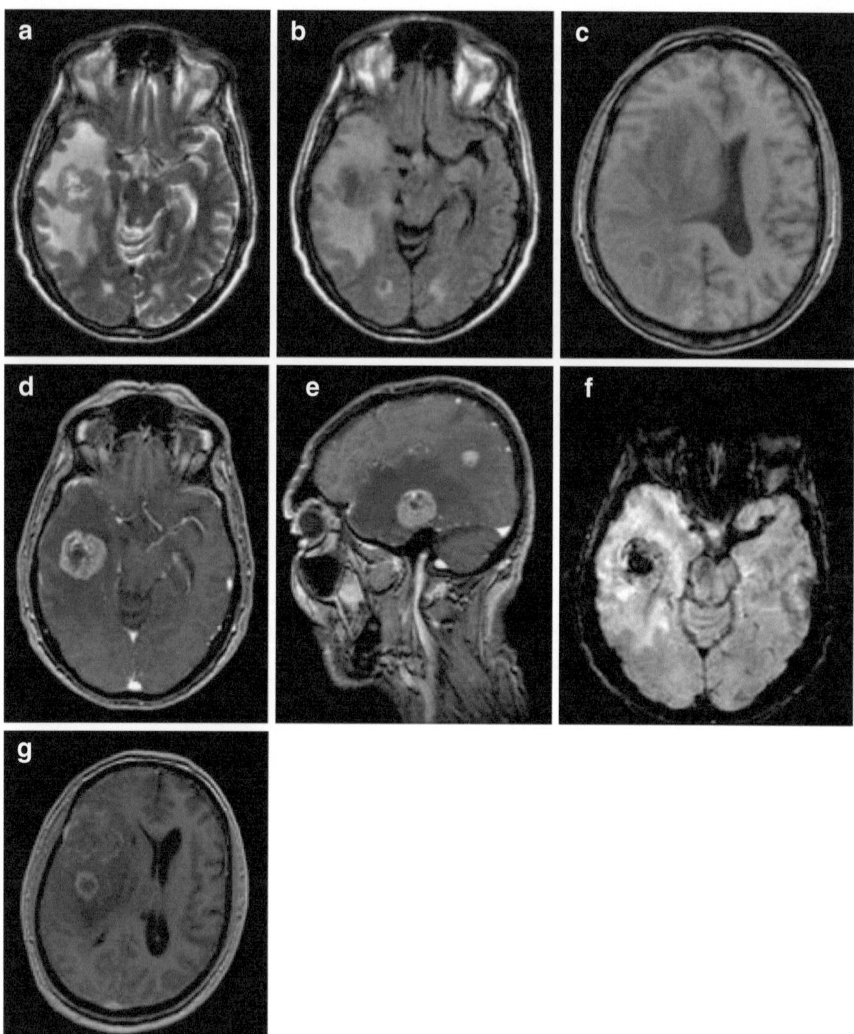

Fig. 2.12 Serial MRI images of the brain with the following: (**a**) axial T2, (**b**) axial FLAIR, (**c**) axial T1, (**d**) axial T1 C+, (**e**) sagittal T1 C+, (**f**) axial GRE, and (**g**) axial T1 C+ after 1 year. (Figure courtesy of Dr. Samer Hoz)

Multifocal Glioblastomas

These are tumors characterized by multiple distinct regions of contrast enhancement, which are either embedded within or connected by areas of abnormal T2/FLAIR signal. These tumors are considered to be part of one tumor. Multifocal glioblastomas account for 2–20% of all glioblastomas and have a poorer prognosis compared to solitary tumors. In contrast, multicentric glioblastomas feature enhancing regions within normal brain tissue in between, and they are believed to represent separate tumors that have developed simultaneously [17, 18].

Questions

1. **Multifocal glioblastomas, the FALSE answer is:**
   A. Do not show contrast enhancement.
   B. Are encountered in approximately 2–20% of all glioblastomas.
   C. Have a worse prognosis than solitary tumors.
   D. Are connected by areas of abnormal T2/FLAIR signal.
   E. Multicentric glioblastomas differ from them in that they are distinct tumors that have developed simultaneously.

   The answer is A.

   They are tumors that exhibit several distinct areas of contrast-enhancing tumors that are either embedded within or connected by T2/FLAIR signal abnormalities.

Case 11: Diffuse Leptomeningeal Glioneuronal Tumor

Case Scenario

A 5-year-old child presented with ataxia and seizures (Fig. 2.13).

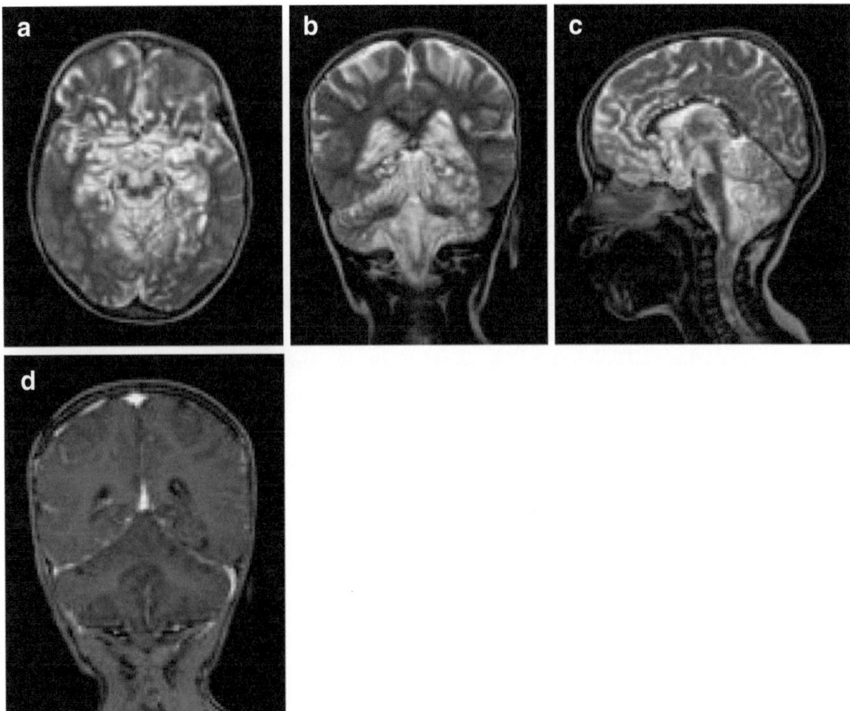

Fig. 2.13 Serial MRI images of the brain with the following: (**a**) axial T2, (**b**) coronal T2, (**c**) sagittal T2, (**d**) coronal T1 C+. (Figure courtesy of Dr. Samer Hoz)

Imaging Description

There are diffuse "cyst-like" foci involving the inferior part of the cerebral hemispheres (temporal lobes and inferior frontal lobes). The coronal T2W image of the brain demonstrates numerous foci located predominantly along the leptomeningeal surfaces of the posterior fossa (cerebellum and brainstem) and spine (upper part appearing in the image). These did not enhance after contrast administration or suppress on FLAIR. This patient has a history of hydrocephalus and underwent a ventriculoperitoneal shunt. Transependymal edema is evident, as indicated by abnormal periventricular T2 and FLAIR. A sagittal T2W image of the cervicothoracic spine shows numerous small cervicothoracic intramedullary cyst-like lesions expanding the spinal cord.

Diffuse Leptomeningeal Glioneuronal Tumor

Previously referred to as disseminated oligodendroglial-like leptomeningeal tumor of childhood, this is a recently recognized CNS tumor that is uncommon. Typically, it manifests as prominent leptomeningeal enhancement without a readily identifiable parenchymal component. This type of tumor primarily affects children and teenagers, with the majority of reported cases involving individuals under the age of 18. However, it can also occur in young to middle-aged adults, with a slightly higher incidence in males.

The primary clinical symptom observed is hydrocephalus, resulting from the significant accumulation of the tumor in the subarachnoid space. Occasionally, patients may display focal neurological symptoms like cranial nerve dysfunction, ataxia, spinal cord compression, or seizures.

Most of these tumors exhibit low-grade histology, and in the latest WHO classification of CNS tumors (2021), they have not been officially assigned a grade. However, the classification notes that the majority of low-grade tumors share similarities with other WHO grade 2 tumors, while those displaying high-grade characteristics resemble WHO grade 3 tumors.

Regarding the radiographic features of these tumors, MRI is the preferred imaging method, although CT may reveal several similar features with some limitations. On MRI, the primary distinguishing feature of this tumor is the thick nodular leptomeningeal enhancement, most prominently in the basal cisterns and extending over the brain and spinal cord's surface. Another specific finding of this tumor is the presence of numerous small subpial cysts exhibiting high T2, low T1, and FLAIR attenuating signals. These cysts are situated on the surface of the inferior regions of the cerebral hemispheres, posterior fossa, and spinal cord. Often, no primary parenchymal mass is identified. However, discrete intraparenchymal lesions can be detected in some patients, most frequently in the spinal cord [19, 20].

Differential Diagnosis

The primary considerations for the differential diagnosis include other conditions that lead to leptomeningeal enhancement, such as leptomeningeal carcinomatosis, tuberculous leptomeningitis, and leptomeningeal seeding from a primary CNS tumor.

Questions

1. **Diffuse leptomeningeal glioneuronal tumor, the FALSE answer is:**
   A. Intraparenchymal lesions are commonly found in the cerebral hemispheres.
   B. Hydrocephalus is a common symptom of this tumor.
   C. The presence of numerous small subpial cysts is considered somewhat specific to diffuse leptomeningeal glioneuronal tumor.
   D. The primary characteristic on MRI is thick nodular leptomeningeal enhancement.
   E. The subpial cysts are usually located on the surface of the inferior portions of the cerebral hemispheres, posterior fossa, and spinal cord.

   The answer is A.

   Intraparenchymal lesions are not commonly found in the cerebral hemispheres in diffuse leptomeningeal glioneuronal tumor. However, discrete intraparenchymal lesions can be detected in some patients, most commonly in the spinal cord.

Case 12: Colloid Cyst of the Third Ventricle

Case Scenario

A 50-year-old female presented with positional headaches, nausea, and vomiting (Fig. 2.14).

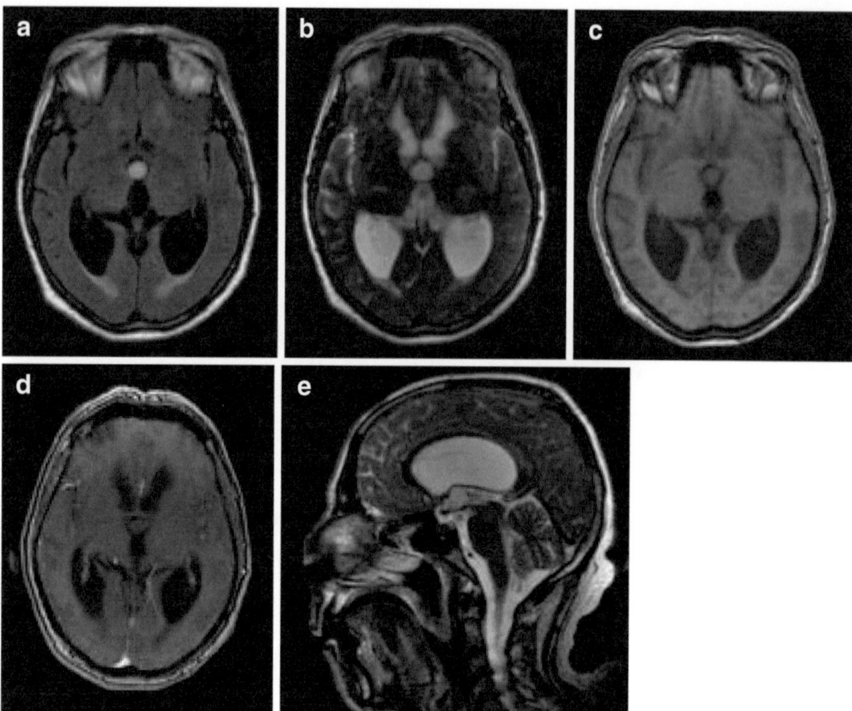

Fig. 2.14 Serial MRI images of the brain with the following: (**a**) axial FLAIR, (**b**) axial T2, (**c**) axial T1, (**d**) axial T1 C+, and (**e**) sagittal T2. (Figure courtesy of Dr. Samer Hoz)

Imaging Description

MRI demonstrates a rounded, unilocular, sharply demarcated cyst lesion at the roof of the third ventricle with high signal intensity on T2 and FLAIR sequences and isointensity on T1. Mild dilatation of both lateral ventricles with mild periventricular CSF permeation can be seen. Histopathological examination confirmed the diagnosis of a colloid cyst.

Colloid Cysts of the Third Ventricle

Colloid cysts located in the third ventricle are cysts lined with epithelial cells and typically display benign characteristics on imaging. Although they generally do not produce symptoms, in rare cases, they can lead to abrupt and severe hydrocephalus. These cysts are typically visualized as a well-defined hyperattenuating mass on unenhanced CT scans, situated at the anterosuperior part of the third ventricle. On MRI, they typically appear hyperintense on T1-weighted images and isointense to brain tissue on T2-weighted images. In some instances, peripheral rim enhancement may be present. Colloid cysts account for approximately 2% of primary brain tumors and make up 15–20% of intraventricular masses. They are most commonly found at the foramen of Monro, accounting for 99% of cases. While most diagnoses occur in early middle age, typically between 30 and 40 years old, colloid cysts can also be identified in pediatric cases, making up 8% of all cases.

On all imaging modalities, colloid cysts manifest as rounded, well-defined lesions located at the foramen of Monro, with sizes ranging from a few millimeters to 3–4 cm. CT scans generally reveal a well-defined, round lesion at the roof of the third ventricle that is usually hyperdense. Isodense and hypodense cysts are uncommon, and calcification is infrequent. In terms of MR signal characteristics, approximately 50% of colloid cysts exhibit a high T1 signal, while the remainder appear either isointense or hypointense compared to nearby brain tissue. T1 C+ (Gd) imaging occasionally reveals subtle rim enhancement, possibly indicating enhancement of stretched septal veins located adjacent to the cyst. T2 signal intensity varies, with most cysts displaying a low T2/T2* signal due to their thick "motor oil" fluid composition. However, some cysts may exhibit a peripheral high T2 signal with a low central T2 signal or a uniformly high signal. On FLAIR imaging, cysts with low T2 signals appear similar to attenuated cerebrospinal fluid (CSF) and can be challenging to differentiate [21, 22].

Differential Diagnosis

Other masses that can arise in the region of the foramen of Monro include calcified or hyperdense meningioma, giant cell astrocytoma, pilocytic astrocytoma, or the presence of blood in that region.

Questions

1. **Colloid cysts of the third ventricle, the FALSE answer is:**
   A. MRI is more effective than CT at fully characterizing the lesion.
   B. Calcification is uncommon on CT scans.
   C. Colloid cysts typically exhibit high T1 signals on MRI in approximately 50% of cases.
   D. Colloid cysts are mostly located at the foramen of Monro (in 99% of cases).
   E. Colloid cysts appear hypointense on T1W images.
   The answer is E.
   Colloid cysts are typically hyperintense on T1W images and isointense to brain tissue on T2W images.

Case 13: Hypothalamic Hamartoma

Case Scenario

A 4-year-old boy presented with mild cognitive impairment (Fig. 2.15).

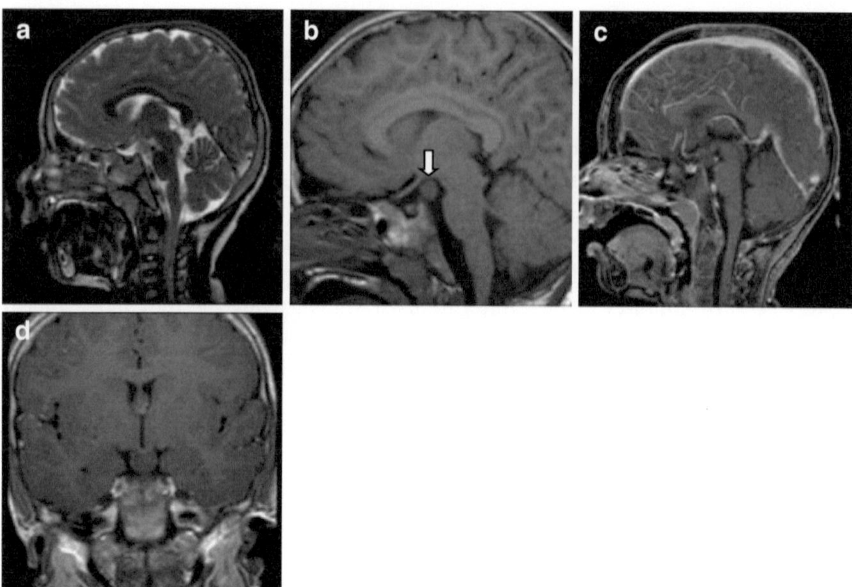

Fig. 2.15 Serial MRI images of the brain with the following: (**a**) sagittal T2, (**b**) sagittal T1, (**c**) sagittal T1 C+, and (**d**) coronal T1 C+. (Figure courtesy of Dr. Samer Hoz)

Imaging Description

There is a well-defined, small, rounded nodule located anteriorly in contact with the floor of the third ventricle and posterior to the pituitary stalk. Compared to white matter, the nodule is T1 isointense and T2/FLAIR hyperintense with no contrast enhancement (see arrow).

Hypothalamic Hamartomas

Also referred to as tuber cinereum hamartomas, these are benign heterotopias located in the hypothalamus and originating from the tuber cinereum, which is located between the optic chiasm and mammillary bodies. The presence of these anomalies can lead to various symptoms, including gelastic seizures, behavioral problems, visual disturbances, and precocious puberty. Diagnosis is typically confirmed through histopathological examination.

Hypothalamic hamartomas can be sessile or pedunculated. Sessile hamartomas can cause distortion or displacement of the mammillary bodies and displace the columns of the fornix anterolaterally. The degree of extension below the third ventricle is variable. In contrast, pedunculated hamartomas are attached to the tuber cinereum and project into the suprasellar cistern.

On CT, the hamartoma appears as a non-enhancing soft tissue nodule with similar density to the surrounding brain. Calcification is rare but may be seen in the suprasellar region. An eroded dorsum sellae, or an enlarged pituitary fossa may also be observed in some cases.

MRI is the preferred imaging modality. Hypothalamic hamartomas usually appear hypointense on T1-weighted images compared to the surrounding brain tissue and do not show contrast enhancement. On T2-weighted images, they appear hyperintense compared to the cerebral cortex in the majority of cases (approximately 93%). This hyperintensity is more pronounced on FLAIR imaging. Magnetic resonance spectroscopy (MRS) reveals an increase in myoinositol, a decrease in the NAA/Creatine (Cr) ratio, and an increase in the Cho/Cr ratio relative to the amygdala [23, 24].

Differential Diagnosis

Hypothalamic hamartomas can be confused with other lesions in the suprasellar/hypothalamic region. The main differential is hypothalamic-chiasmatic glioma, as other lesions in this area usually exhibit different signal intensities or may demonstrate contrast enhancement.

Questions

1. **Hypothalamic hamartomas, the FALSE answer is:**
   A. Hypothalamic hamartomas typically appear hypointense on T1W images compared to the surrounding brain tissue.
   B. Hypothalamic hamartomas are always benign and non-neoplastic.
   C. MRI is the preferred imaging method for assessing the hypothalamic region.
   D. Hypothalamic hamartomas can cause gelastic seizures, visual problems, early onset of puberty, and behavioral problems.
   E. On T2W MRI images, hypothalamic hamartomas typically appear hypointense.
   The answer is E.
   On T2W images, hypothalamic hamartomas appear hyperintense compared to the cerebral cortex in the majority of cases.

Case 14: Adamantinomatous Craniopharyngioma

Case Scenario

A 12-year-old boy presented with a history of visual disturbances (Fig. 2.16).

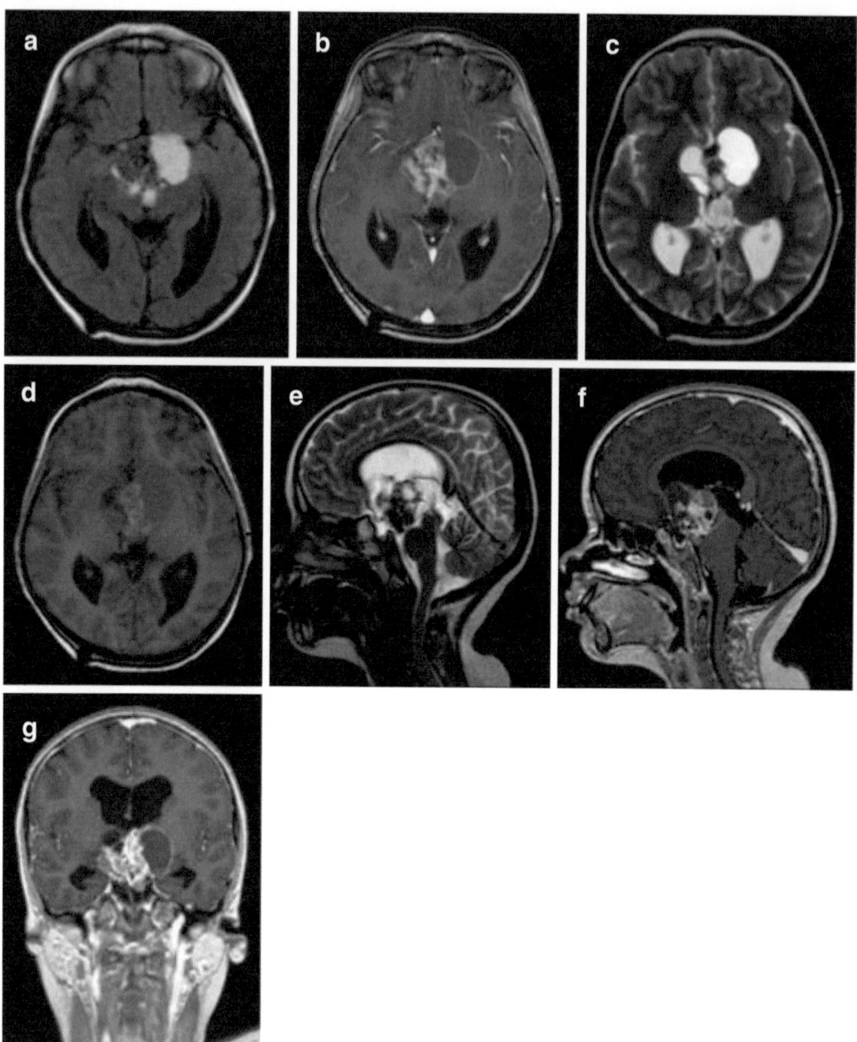

Fig. 2.16 Serial MRI images of the brain with the following: (**a**) axial FLAIR, (**b**) axial T1 C+, (**c**) axial T2, (**d**) axial T1, (**e**) sagittal T2, (**f**) sagittal T1 C+, and (**g**) coronal T1 C+. (Figure courtesy of Dr. Samer Hoz)

Imaging Description

MRI reveals a well-defined large suprasellar solid-cystic mass lesion. The solid component elicits a low signal on T1, a slightly high signal on FLAIR, and T2 with heterogeneous enhancement on post-contrast sequences. The cystic component shows an iso signal to the cortical gray matter on T1 and a very high signal on FLAIR and T2. No restricted diffusion is seen, but a mass effect is noted on the optic chiasma, which is displaced superiorly. The floor of the third ventricle is slightly elevated with ventricular dilatation, and a ventriculoperitoneal shunt is seen on the right side. The histopathological examination confirmed the diagnosis of adamantinomatous craniopharyngiomas.

Adamantinomatous Craniopharyngioma

Adamantinomatous craniopharyngioma typically manifests as a cystic mass arising in the pituitary area and is often accompanied by peripheral calcification. These tumors are categorized as WHO classification grade 1 tumors. They predominantly appear in children, unlike papillary craniopharyngiomas, which are more commonly found in adults. However, adamantinomatous craniopharyngioma can be observed across all age groups, affecting both males and females equally and showing no racial predilection.

Regarding their radiographic features, only about 20–25% of cases include a small intrasellar component, while the majority are primarily suprasellar tumors. Purely intrasellar tumors are rare, accounting for less than 5% of cases, and may lead to pituitary fossa expansion. Large tumors, although rare, can result in obstructive hydrocephalus due to optic chiasm distortion or midbrain compression. Ectopic lesions can also be found in the nasopharynx, posterior fossa, and cervical spine.

On CT scans, adamantinomatous craniopharyngioma cysts are typically large, with a density similar to CSF in about 90% of cases. Solid components exhibit soft tissue density and enhance in 90% of cases. Calcifications, often stippled and located peripherally, are present in 90% of cases. On MRI, cysts appear iso- to hyperintense to gray matter (referred to as "motor oil cysts" due to their high protein content) on T1-weighted images. On T2-weighted images, they are variably hyperintense, appearing completely hyper-intense in 80% of cases and partially hyperintense in the remaining cases. The solid component shows vivid enhancement on post-contrast T1-weighted images and has a variable or mixed appearance on T2-weighted images. Conventional imaging is not optimal for visualizing calcifications, and susceptibility sequences may be more useful. MRA may reveal displacement of the A1 segment of the anterior cerebral artery (ACA). In MRS, a broad lipid spectrum is observed, with a flat baseline due to the content of the cyst [25, 26].

Differential Diagnosis

1. Rathke cleft cyst: is typically unilocular, intrasellar, and does not contain a solid portion or exhibit enhancement on imaging.
2. Pituitary macroadenoma (with cystic degeneration or necrosis): epicenter is located in the intrasellar region rather than in the suprasellar region, leading to pituitary fossa enlargement. While bright cystic areas may occasionally appear on T1-weighted images, they lack calcification, unlike adamantinomatous craniopharyngiomas.

3. Pituitary apoplexy: typically presents acutely.
4. Intracranial teratoma: the presence of fat can serve as a helpful distinguishing feature.

Questions

1. **Adamantinomatous craniopharyngioma, the FALSE answer is:**
   A. Typically presents in children, in contrast to papillary craniopharyngioma, which occurs in adults.
   B. Displacement of the A1 segment of the ACA may be demonstrated on MRA.
   C. Calcification can be seen on CT in 90% of cases.
   D. Typically demonstrates hypointensity on T2.
   E. The solid component shows vivid enhancement on T1 C+ (Gd) and has a variable or mixed appearance on T2.
   The answer is D.

   Adamantinomatous craniopharyngioma typically demonstrates variable hyperintensity on T2, either completely in 80% of cases or partially. The solid component exhibits vivid enhancement on T1 with contrast (Gd) and has a variable or mixed appearance on T2.

Case 15: Pineal Germinoma

Case Scenario

A 21-year-old male presented with headaches and up-gaze palsy (Fig. 2.17).

Imaging Description

MRI reveals a large, well-defined enhancing solid mass lesion centered in the pineal region. It is heterogeneous with small areas of cystic change and demonstrates restricted diffusion. There is marked compression of the tectum, resulting in obstructive hydrocephalus with transependymal edema. The histopathological examination confirmed the diagnosis of germinoma.

Pineal Germinoma

Intracranial germinomas, also referred to as dysgerminomas or extra-gonadal seminomas, belong to the category of germ cell tumors. They are most frequently observed in children, with the peak incidence occurring between the ages of 3 and 10 years, accounting for 3–5% of pediatric intracranial tumors. In adults, these tumors are exceptionally rare, constituting less than 1% of intracranial tumors in

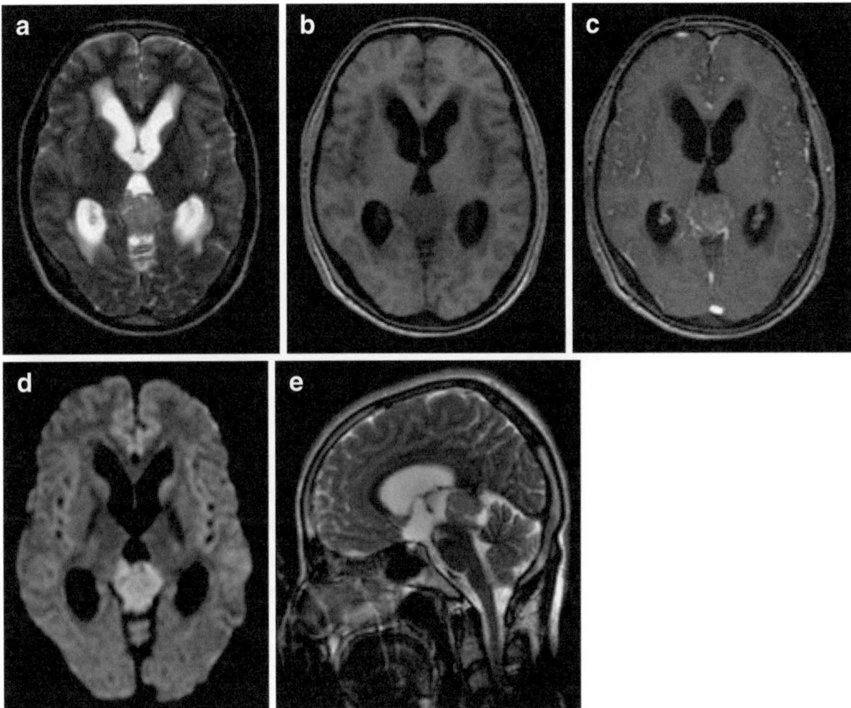

Fig. 2.17 Serial MRI images of the brain with the following: (**a**) axial T2, (**b**) axial T1, (**c**) axial T1 C+, (**d**) axial DWI, (**e**) sagittal T2. (Figure courtesy of Dr. Samer Hoz)

this age group. Germinomas are particularly common in the pineal region, representing roughly 50% of all tumors in this area and approximately 80% of all intracranial germ cell tumors. While germinomas typically occur along the midline in the pineal region, they can also manifest along the floor of the third ventricle or in the suprasellar region. There appears to show a male predominance (M:F = 5–22:1) for germinomas in the pineal region, whereas germinomas in the suprasellar region are more evenly distributed between males and females (M:F = 1:1.3).

With regards to radiographic characteristics, germinomas exhibit soft tissue density and typically enhance on imaging. In the pineal region, they engulf normal pineal tissue and may be associated with central calcification. This central calcification differs from the scattered foci of calcification seen in pineocytoma and pineoblastoma. Cystic components are present in approximately 45% of cases. On CT scans, germinomas generally appear hyperdense compared to the surrounding brain tissue and exhibit prominent enhancement. When these lesions occur in the floor of the third ventricle or suprasellar region, they can fill and expand the infundibular and supraoptic recesses. Initial imaging may appear normal, but if there is clinical suspicion, such as in cases of idiopathic hypothalamic diabetes insipidus, close monitoring is crucial to detect potential abnormal enhancement and thickening of the pituitary stalk. The presence of calcification in the pineal region is an important indicator of underlying tumors in the pediatric age group. However, it may be absent in children under 6.5 years old and in approximately 10% of children between 11 and 14 years old.

On MRI, germinomas typically present as a soft tissue mass with an ovoid or lobulated shape, often engulfing the calcified pineal gland. They can appear isointense or slightly hyperintense compared to adjacent brain tissue on T1-weighted images. Some germinomas may contain cystic or hemorrhagic areas, which appear as low signal regions, and have a tendency to infiltrate adjacent brain tissue, causing edema. Central calcification within the tumor exhibits low signal intensity on T2-weighted images. On post-contrast T1-weighted images, germinomas demonstrate vivid and uniform enhancement [27, 28].

Differential Diagnosis

1. Pineal cyst: most common benign lesion in this area.
2. Germ cell tumors: embryonal carcinoma, choriocarcinoma, yolk sac carcinoma (endodermal sinus tumor), and teratoma.
3. Pineal parenchymal tumors (contributing to approximately 30% of primary pineal area tumors): pineocytoma and pineoblastoma.
4. Glioma: typically arises from the tectal region and exhibits astrocytic histology.
5. Pineal metastasis.
6. Primary pineal malignant melanoma.
7. Inclusion cysts (dermoid/epidermoid).
8. Meningioma near the pineal region.
9. Rare vascular lesions: cavernoma in pineal region, and vein of Galen aneurysmal malformation.

Questions

1. **Pineal germinoma, the FALSE answer is:**
   A. It is more common in females than males.
   B. CT scan usually shows hyperdense lesions compared to the adjacent brain area.
   C. The infundibular and supraoptic recesses usually fill and expand when the lesion occurs in the floor of the third ventricle/suprasellar region.
   D. Typically appear isointense or in some cases hyperintense to the adjacent brain on T1.
   E. MRI reveals isointense or some hyperintense lesion to the adjacent brain on T2.

   The answer is A.

   There appears to show a male predominance (M:F = 5–22:1).

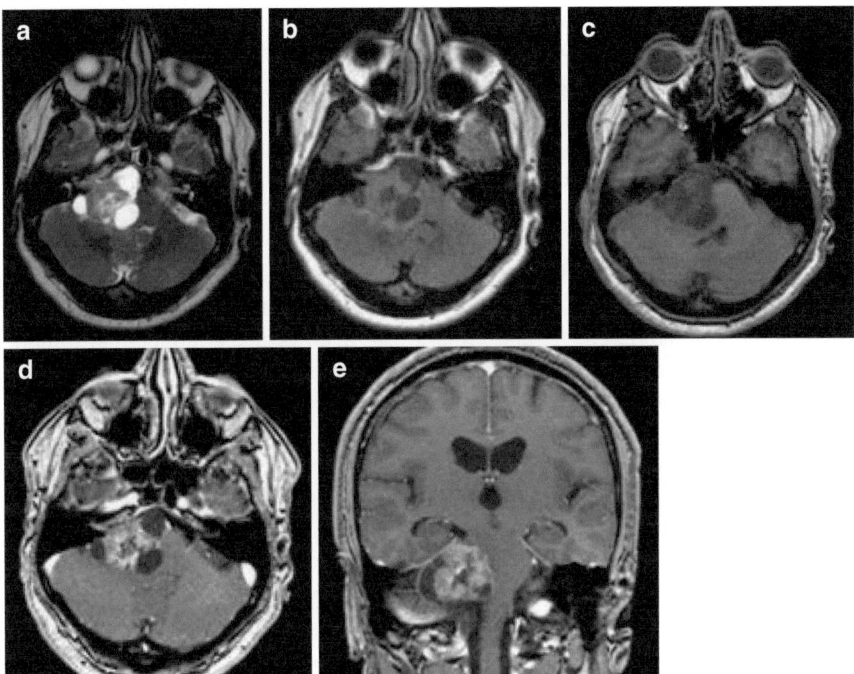

Fig. 2.18 Serial MRI images of the brain with the following: (**a**) axial T2, (**b**) axial FLAIR, (**c**) axial T1, (**d**) axial T1 C+, and (**e**) coronal T1 C+. (Figure courtesy of Dr. Samer Hoz)

Case 16: Vestibular Schwannoma

Case Scenario

A 48-year-old man presented with tinnitus in the right ear, vertigo, and ataxia (Fig. 2.18).

Imaging Description

There is a well-defined large extra-axial complex mass in the right CPA cistern and centered in the internal auditory canal (IAC), which has low signal intensity to gray matter on T1W MR image and exhibits heterogeneous hyperintensity on T2W. Axial contrast-enhanced T1W reveals irregular enhancement of the mass. It shows multiple internal, non-enhancing cystic changes. DWI revealed no diffusion restriction of the tumor. There is a disproportionately small amount of edema, given the size of the mass. The tumor extends from the medial aspect of the IAC, which is enlarged and continues through the porus acusticus into the CPA cistern. A sizable mixed tissue component is present more medially in the CPA. This mass exerts a mass effect on the right side of the pons, the ipsilateral middle cerebellar peduncle, and the anteromedial part of the right cerebellar hemisphere. The fourth ventricle is compressed and displaced to the left side. Histopathological examination confirmed the diagnosis of schwannoma.

Vestibular Schwannoma

Also referred to as acoustic neuromas, vestibular schwannomas are benign tumors classified as WHO grade 1. These tumors originate from Schwann cells ensheathing the vestibulocochlear nerve. They account for approximately 80% of masses in the cerebellopontine angle (CPA) and around 8% of all primary intracranial tumors. Most solitary vestibular schwannomas (95%) are sporadic, but bilateral occurrences can happen and are strongly linked to neurofibromatosis type 2 (NF2). However, bilateral vestibular schwannomas can also occur in the familial form of vestibular schwannoma without other NF2-related features. The highest incidence is observed in individuals between their fourth and sixth decades of life, with a median age of 50 years. However, in cases associated with NF2, patients tend to present earlier, typically in their third decade of life.

On CT scans, signs of bone erosion and widening of the internal auditory canal (IAC) and porus acusticus may be visible. Vestibular schwannomas can be challenging to detect and often display varying densities on non-contrast CT. Contrast enhancement is usually present but can be minimal, especially in the presence of cystic components, which can develop in larger tumors. These cysts may represent CSF-filled pockets adjacent to the tumor or indicate cystic degeneration within the tumor itself.

On T1-weighted MRI images, vestibular schwannomas typically appear slightly hypointense (63%) or isointense (37%) compared to the surrounding brain. Any cystic components, if present, appear hypointense. On T2-weighted images, they exhibit heterogeneous hyperintensity, with fluid-filled regions appearing hyperintense. Occasionally, peritumoral arachnoid cysts may be observed. Contrast-enhanced T1-weighted images usually show intense enhancement, except in larger tumors, where enhancement may be more variable. Hemorrhagic areas can also be occasionally detected, although calcifications are not typically present in vestibular schwannomas [29, 30].

Differential Diagnosis

1. Meningiomas: typically appear more homogenous on imaging, while vestibular schwannomas often display signal variations due to the presence of hemorrhagic or cystic components. Meningiomas are characterized by a broad dural base and do not exhibit the characteristic "trumpeted" appearance in the IAC. Large meningiomas are often asymmetrical to the IAC and are more likely to contain calcifications.

2. Epidermoid tumors: usually lack an enhancing component and appear with very high signal intensity on DWI. They do not lead to the enlargement of the IAC.
3. Metastatic lesions: metastasis involving the IAC is uncommon and typically does not result in the widening of the IAC.
4. Ependymomas: tend to occur in younger patients, are primarily centered on the fourth ventricle, and do not extend into the IAC.

Questions

1. **Vestibular schwannoma, the FALSE answer is:**
   A. On T1, they appear slightly hypointense or isointense, and heterogeneously hyperintense on T2.
   B. Ependymomas are commonly centered on the fourth ventricle and can extend into the IAC.
   C. It is not usually associated with calcification.
   D. The peak incidence is in the fourth to sixth decades of life for solitary lesions.
   E. Bilateral vestibular schwannomas are highly suggestive of NF2.
   The answer is B.

   Ependymomas, usually presenting in younger patients, are centered on the fourth ventricle and do not extend into the IAC.

Case 17: Epidermoid Cyst

Case 17.1: Suprasellar Epidermoid Cyst

Case Scenario
A 23-year-old woman presented with galactorrhea and decreased vision (Fig. 2.19).

Imaging Description
MRI demonstrates a well-defined cystic mass lesion in the suprasellar region that is hyperintense on T2, hypointense on T1, and heterogeneous hypointense with a dirty appearance on FLAIR. There is internal, restricted diffusion within the lesion. The lesion markedly compresses the optic chiasm, which is bowed superiorly around the lesion. There is a moderate mass effect on the floor of the third ventricle, hypothalamus, and the cavernous sinuses, causing partial encasement of bilateral ICA. Histopathological examination confirmed the diagnosis of epidermoid cyst.

Case 17.2: White Epidermoid Cyst

Case Scenario
A 45-year-old man presented with an incidental finding of a brain mass (Fig. 2.20).

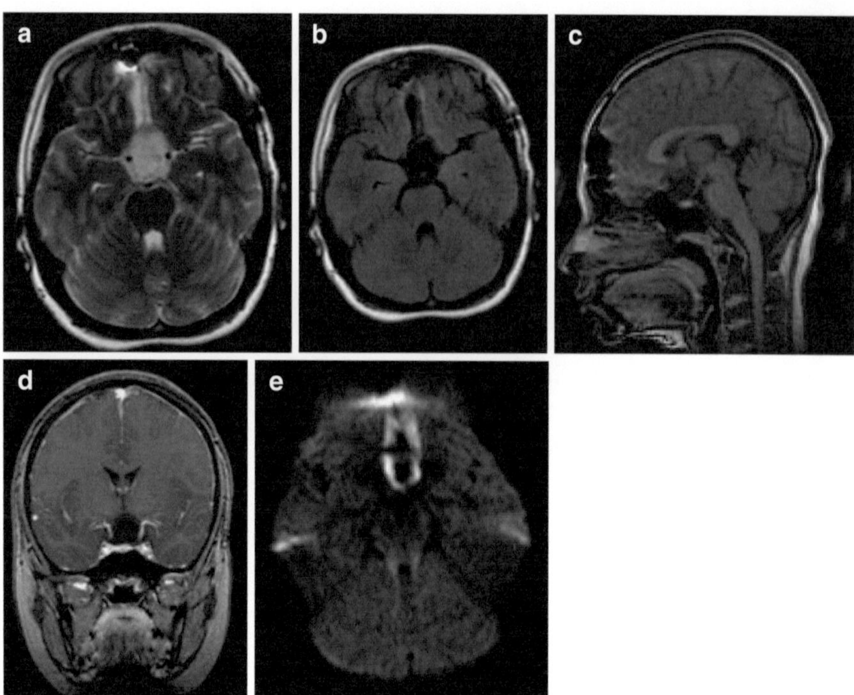

Fig. 2.19 Serial MRI images of the brain with the following: (**a**) axial T2, (**b**) axial FLAIR, (**c**) sagittal T1, (**d**) coronal T1 C+, and (**e**) axial DWI. (Figure courtesy of Dr. Samer Hoz)

Imaging Description

MRI reveals a large right-sided extra-axial posterior fossa tumor that appears to be predominantly cystic, showing heterogeneous signal intensity. The more solid portions have an intense intrinsic high T1 signal and patchy low T2 signal with no convincing internal enhancement. Heterogeneous restriction diffusion is noted throughout the lesion. The lesion causes a significant mass effect in the posterior fossa. Histopathological examination confirmed the diagnosis of an epidermoid cyst.

Epidermoid Cyst

Epidermoid cysts located within the brain are uncommon congenital anomalies that develop as a result of the inclusion of ectodermal elements during neural tube closure. They account for only 1% of all brain tumors. These tumors typically grow slowly and are diagnosed between the ages of 20 and 40 due to mass effect. With regards to sex predilection, some studies have suggested a higher occurrence among males. The standard treatment is surgical removal, and patients generally have a favorable prognosis.

These cysts are irregularly shaped growths that occupy CSF spaces, surrounding and gradually exerting pressure on nearby structures. In particular, epidermoid cysts in the posterior fossa are known to displace the basilar artery away from the pons.

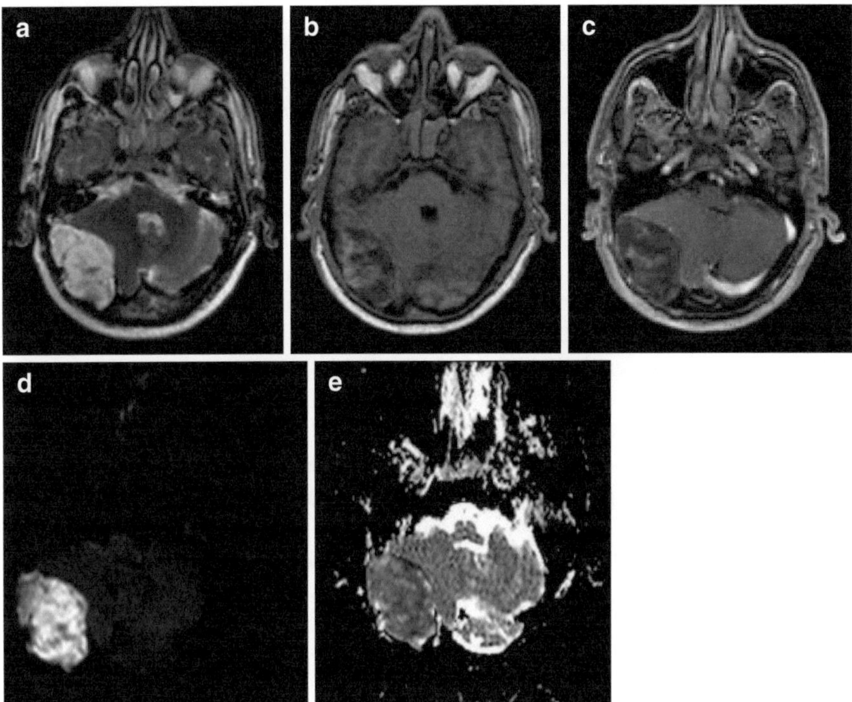

Fig. 2.20 Serial MRI images of the brain with the following: (**a**) axial T2, (**b**) axial T1, (**c**) axial T1 C+, (**d**) axial DWI, and (**e**) axial ADC. (Figure courtesy of Dr. Samer Hoz)

Since epidermoids contain skin cells, they appear similar to CSF on CT and MRI scans, except on DWI, which shows restriction. White epidermoid cysts are a rare subtype, distinguished by their "white" appearance in T1W imaging. Unlike typical epidermoid cysts, they do not exhibit a density on CT scans or an intensity on MRI scans that closely resembles CSF, except for DWI, where they display restricted diffusion. The majority (90%) of epidermoids are situated inside the dura mater: approximately 40–50% are found in the cerebellopontine angle (CPA), 10–15% in the suprasellar cistern (as in this case), about 17% in the fourth ventricle, and rarely in the middle cranial fossa, interhemispheric region, or spine. The remaining 10% are found outside the dura mater, with most of them within the skull.

On CT scans, the presence of cellular debris and a high cholesterol content in epidermoid cysts makes them appear to have the same density as CSF, making them difficult to distinguish from arachnoid cysts. Calcification is seen in a minority of cases (10–25%), with interhemispheric epidermoid cysts sometimes showing peripheral calcification. Occasionally, epidermoid cysts can appear hyperdense due to saponification, hemorrhage, or a high protein content, as is the case with white epidermoid cysts. They do not enhance with contrast, and enhancement of their walls is extremely rare.

Similar to CT, epidermoid cysts appear very similar to dilated CSF spaces and arachnoid cysts on most MRI sequences, except on DWI/ADC, where they exhibit restricted diffusion. On T1-weighted images, they have the same intensity as CSF, but the periphery often shows a higher signal intensity compared to CSF. On rare occasions, white epidermoid cysts or internal bleeding can produce high signal intensity. Contrast-enhanced T1-weighted MRI scans may occasionally reveal a thin enhancing rim around the periphery and, in rare cases of malignant transformation, increased intensity. Their appearance on T2-weighted images can vary; in most cases, they are isointense to CSF (65%), but they can also be slightly more intense than gray matter (35%). Rarely, they can be hypointense to gray matter, particularly in the context of white epidermoid cysts. On FLAIR imaging, they often appear heterogeneous, with generally higher signal intensity than CSF [31, 32].

Differential Diagnosis

1. CSF collections, such as an arachnoid cyst or mega cisterna magna, are less lobulated. They displace vessels and follow CSF on all sequences, including FLAIR and DWI.
2. Dermoid cysts often exhibit fat density due to sebum and are usually located along the midline.
3. Inflammatory cysts such as neurocysticercosis are smaller and can be multiple. They may enhance peripherally and have associated edema with typically no restricted diffusion.
4. Cystic tumors like acoustic schwannoma and craniopharyngioma usually have a solid-enhancing component.
5. Neurenteric cysts.

Questions

1. **Intracranial epidermoid cyst, the FALSE answer is:**
   A. Most are located intradurally, rather than extradurally.
   B. DWI/ADC can be helpful in differentiating between epidermoid cysts and arachnoid cysts or dilated CSF spaces.
   C. Calcification seen on CT is present in a minority of cases only.
   D. On FLAIR, they generally exhibit a lower signal than CSF.
   E. White epidermoid cysts have a high signal intensity on T1.
   The answer is D.
   On FLAIR, they generally exhibit a higher signal than CSF.

Case 18: Dysembryoplastic Neuroepithelial Tumor (DNET)

Case Scenario

A 35-year-old woman presented with seizure (Fig. 2.21).

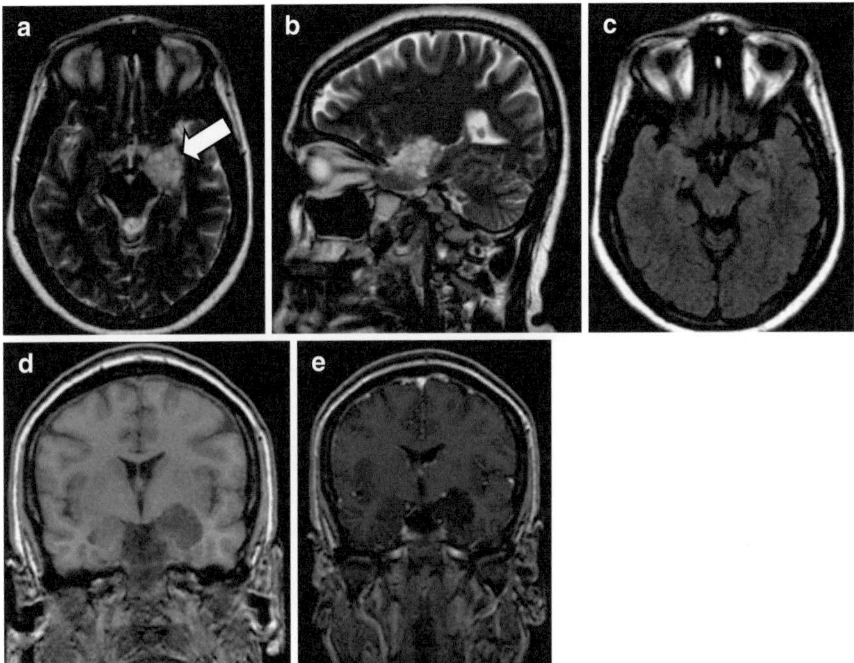

Fig. 2.21 Serial MRI images of the brain with the following: (**a**) axial T2, (**b**) sagittal T2, (**c**) axial FLAIR, (**d**) coronal T1, and (**e**) coronal T1 C+. (Figure courtesy of Dr. Samer Hoz)

Imaging Description

MRI shows a well-demarcated intracortical cystic lesion can be seen in the left temporal lobe, following CSF in all sequences. The lesion has a soap bubble appearance with partial suppression in FLAIR, with no surrounding edema (arrow). No enhancement is noted on T1W images after the administration of gadolinium. A few delicate septa-like structures are visible within the lesion. A mild mass effect is present on the left side of the midbrain. Histopathological examination confirmed the diagnosis of DNET.

Dysembryoplastic Neuroepithelial Tumor (DNET)

DNETs are WHO grade 1 glioneuronal tumors arising from the cortical or deep gray matter. They originate from secondary germinal layers, and up to 80% of cases are associated with cortical dysplasia. Intractable focal seizures are a classic presenting feature, and most instances are diagnosed in children or young adults. There is a slight male predilection, and some studies have suggested an association with Noonan syndrome. The most common locations for DNETs are the temporal lobe (65%) and the frontal lobe (20%). Other less common sites include the caudate nucleus, pons, and cerebellum. Frequently, DNETs present concomitantly with focal cortical dysplasia (FCD). If FCD is distinct from the tumor cells, it is categorized as Blumcke classification IIIb FCD.

DNETs typically manifest as well-defined tumors situated within the cortical gray matter. On CT scans, they appear as areas of reduced density, often with minimal or no enhancement. When these tumors are cortical, as is frequently the case, they may cause an indentation or reshaping of the inner surface of the skull vault without causing erosion. Occasionally, they can lead to a slight enlargement of the cranial fossa. Approximately 30% of cases exhibit calcification on CT scans, typically located in the deepest regions of the tumor, especially near areas showing enhancement or hemorrhage.

On MRI, DNETs appear hypointense in comparison to the surrounding brain tissue on T1-weighted images, and enhancement is observed in about 20–30% of cases on T1-weighted images with contrast. This enhancement may be heterogeneous or involve a mural nodule. These tumors usually do not produce significant surrounding edema. On T2-weighted images, they generally exhibit high signal intensity with a "bubbly" appearance. FLAIR sequences reveal mixed signal intensity and the presence of the bright rim sign. Partial suppression of some of the "bubbles" may also be observed on FLAIR. T2*-weighted imaging frequently shows calcification, but there is typically no evidence of hemosiderin staining since hemorrhage occurs infrequently. DWI demonstrates no restricted diffusion. Magnetic resonance spectroscopy (MRS) findings are nonspecific, although lactate may be detected in some cases [33, 34].

Differential Diagnosis

1. Ganglioglioma: tends to display contrast enhancement more frequently and lacks the characteristic "bubbly" appearance seen in DNETs. Approximately 50% of cases show visible calcification.
2. PXA: typically exhibits pronounced contrast enhancement, and the presence of the dural tail sign is often observable.
3. Diffuse low-grade astrocytoma: have an IDH mutation and do not possess the specific glioneuronal elements found in some other tumors.
4. Oligodendroglioma: are characterized by an IDH mutation and 1p19q codeletion, with absent specific glioneuronal elements.
5. Desmoplastic infantile astrocytomas and gangliogliomas primarily affect young children and frequently involve the dura extensively. They tend to be large in size and often present as multiple lesions.
6. Multinodular and vacuolating neuronal tumors: can also have a "bubbly" appearance but are located in the juxtacortical white matter rather than in the cortex.

Questions

1. **DNETs, the FALSE answer is:**
   A. They are hypointense compared with the adjacent brain on T1.
   B. They are WHO grade 1 glioneuronal tumors.
   C. The majority arise in deep gray matter.
   D. They have a "bubbly appearance" on T2.
   E. On CT, they are hypodense with minimal or no enhancement.
      The answer is C.
      The majority arise in cortical gray matter.

Case 19: Medulloblastoma

Case 19.1: Pediatric Medulloblastoma

Case Scenario
A 7-year-old boy presented with an incidental finding by a barber of a swelling in the right occipital region that increases in size every time he goes to shave his hair (Fig. 2.22).

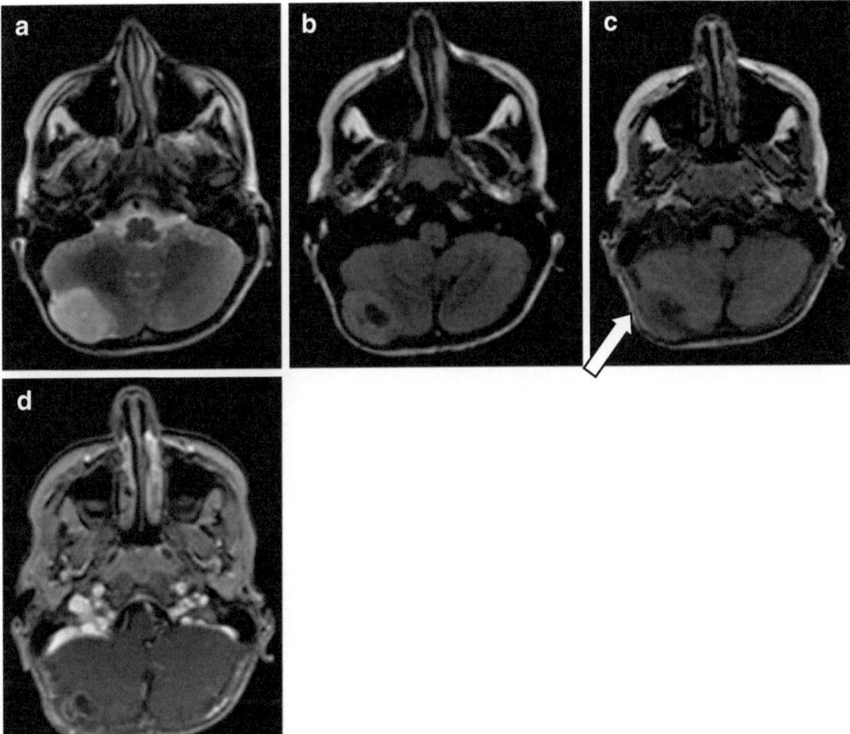

Fig. 2.22 Serial MRI images of the brain with the following: (**a**) axial T2, (**b**) axial FLAIR, (**c**) axial T1, and (**d**) axial T1 C+. (Figure courtesy of Dr. Samer Hoz)

Imaging Description

MRI shows a well-defined mass lesion that is hypointense on T1 and hyperintense on T2/FLAIR. There is no associated vasogenic edema, but the lesion exerts a mild mass effect on the adjacent parenchyma with slight remodeling of the overlying calvarium of occipital bone (arrow), suggesting a long-standing pathology. The lesion contains small cystic components. There is restricted diffusion, and heterogeneous peripheral enhancement can be seen following Gadolinium. Histopathological examination confirmed the diagnosis of medulloblastoma.

Case 19.2: Recurrent medulloblastoma

Case Scenario

A 14-year-old boy had a history of medulloblastoma that was operated on (Fig. 2.23).

Imaging Description

Contrast-enhanced MRI shows recurrence of multiple nodular mass lesions at the site of the previous operation (fourth ventricle) and intraventricular extension to the suprasellar (line) and pineal regions (open arrow). In addition, there are multiple diffuse and relatively thick nodular mass lesions at the

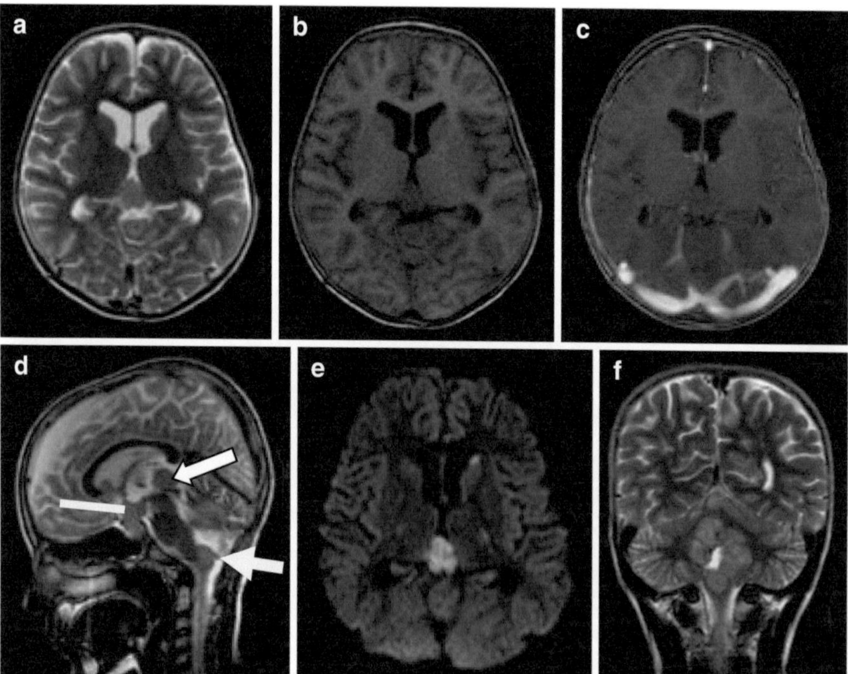

Fig. 2.23 Serial MRI images of the brain with the following: (**a**) axial T2, (**b**) axial T1, (**c**) axial T1 C+, (**d**) sagittal T2, (**e**) axial DWI, and (**f**) coronal T2. (Figure courtesy of Dr. Samer Hoz)

subependymal region of the lateral, third, and fourth ventricles (arrow). They appear isointense on T1 and T2, with no apparent enhancement and restricted diffusion on DWI. Histopathologic findings of the tumor that was operated on 11 months ago confirmed the diagnosis to be the desmoplastic nodular type of medulloblastoma.

Medulloblastomas

Medulloblastomas are the second most common malignant brain tumors in children, following high-grade gliomas. They account for 12–25% of all central nervous system tumors and 30–40% of tumors located in the posterior fossa among children. In adults, medulloblastomas make up only 0.4–1.0% of all brain tumors. This condition is generally more prevalent in males than in females and typically peaks between the ages of 3 and 7, with 77% of cases diagnosed before the age of 19. In adults, medulloblastomas tend to present in the third and fourth decades of life, are more likely to develop in unusual locations, and have a better prognosis. The most common presentation involves a midline mass in the roof of the fourth ventricle, which often leads to obstructive hydrocephalus and mass effect. All medulloblastomas are classified as WHO grade 4 tumors and are further categorized into five subtypes based on tumor genomics: WNT, SHH-activated TP53-wildtype, SHH-activated TP53-mutant, group 3, and group 4. The standard treatment typically involves surgical resection, radiation therapy, and chemotherapy.

Approximately 94% of medulloblastomas originate in the cerebellum, with the vermis being the most common site (75%). They often extend into the fourth ventricle from its roof and can even infiltrate the brainstem. This pattern is particularly common in group 3 and 4 medulloblastomas, as well as certain SHH-activated tumors. Conversely, only 28% of cases occur in the vermis in adults, typically in the case of SHH-activated tumors, and they are usually located laterally in the cerebellar hemispheres. Tumors arising outside the vermis tend to have less well-defined margins and may exhibit larger cystic formations.

On CT scans, medulloblastomas typically appear hyperdense (90%), and the presence of cysts or necrotic areas is frequent (40–50%), especially in older patients. Calcifications are observed in about 10–20% of cases. More than 90% of cases show contrast enhancement. On MRI, these tumors are hypointense compared to gray matter on T1-weighted images. T1-weighted images with contrast reveal enhancement in 90% of cases, which is often heterogeneous. WNT-activated tumors tend to exhibit vivid enhancement, while group 4 tumors show less enhancement. On T2-weighted and FLAIR images, they are isointense to hyperintense compared to gray matter and appear heterogeneous due to the presence of necrosis, cysts, and calcifications. Surrounding edema is commonly observed. DWI and ADC mapping show restricted diffusion due to their high cellularity, resulting in low ADC values.

MRS shows elevated choline (Cho) and decreased N-acetylaspartate (NAA) levels [35, 36].

Differential Diagnosis

In the pediatric population, the differential diagnoses include the following:

1. Ependymoma: typically arises from the floor of the fourth ventricle and extends through the foramen of Luschka, reaching into the basal cisterns. It usually does not lead to significant diffusion restriction.
2. Atypical teratoid/rhabdoid tumor: commonly found in very young children and is known for its aggressive nature.
3. Pilocytic astrocytoma: characterized by a large cystic component and a prominently enhancing mural nodule.
4. Brainstem glioma (exophytic).
5. Choroid plexus papilloma is more frequently located in the lateral ventricles in children.

In the adult population, the following alternative diagnoses should be considered:

1. Cerebellar metastasis
2. Hemangioblastoma
3. Choroid plexus papilloma
4. Ependymoma

Questions

1. **Medulloblastoma, the FALSE answer is:**
   A. Is usually hyperdense on CT.
   B. Is more common in males than females and has a peak incidence between 3 and 7 years of age.
   C. Appears hyperintense to gray matter on T1 MRI with no contrast enhancement.
   D. Is considered to be a WHO grade 4 tumor.
   E. Usually originates from the roof of the fourth ventricle.
   The answer is C.

   On MRI, medulloblastoma typically appears hypointense to gray matter on T1. T1 C+ (Gd) shows enhancement in 90% of cases, which is often heterogeneous.

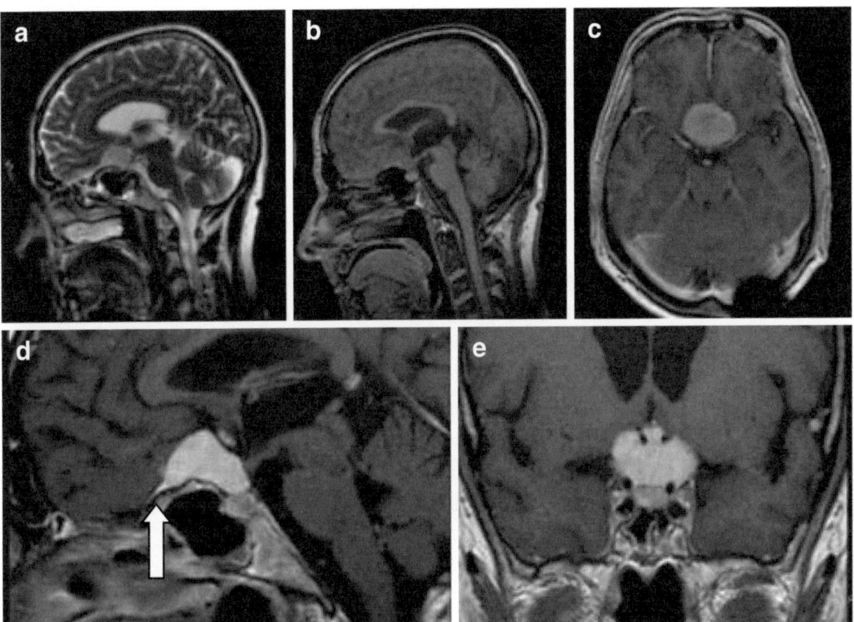

Fig. 2.24 Serial MRI images of the brain with the following: (**a**) sagittal T2, (**b**) sagittal T1, (**c**) axial T1 C+, (**d**) sagittal T1 C+, and (**e**) coronal T1 C+. (Figure courtesy of Dr. Samer Hoz)

Case 20: Suprasellar Meningioma

Case Scenario

A 60-year-old man presented with headache and progressive diminution of vision (Fig. 2.24).

Imaging Description

MRI revealed a mass lesion in the suprasellar region, appearing isointense in T1 and hyperintense in T2, with marked homogenous enhancement and a noticeable dural tail (arrow). The lesion exerted marked compression on the optic chiasm, causing it to bow superiorly around the mass. There is a moderate mass effect on the floor of the third ventricle, hypothalamus, and cavernous sinuses, as well as partial encasement of both ICAs. Histopathological examination confirmed the diagnosis of meningioma.

Meningiomas

Meningiomas are the most common tumors originating from the meninges and have a non-glial origin, arising from arachnoid cap cells. Typically, meningiomas manifest as masses anchored to the dura mater and appear similar in signal intensity to gray matter on both T1-weighted and T2-weighted imaging. They exhibit strong enhancement on both CT and MRI scans, although certain subtypes can show considerable variations in their appearance on imaging.

These tumors are more prevalent among females, with a ratio of 2:1 within the cranial cavity and 4:1 within the spinal cord. Meningiomas can exert mass effects depending on their location, with supratentorial sites being the most common (85–90%), followed by juxtasellar (5–10%), infratentorial (5–10%), and other intradural locations (<5%). In some cases, meningiomas can extend through bone, leading to pronounced hyperostosis and local mass effects such as proptosis. While venous invasion and occlusion can occur, they typically progress slowly, and most instances of venous invasion are asymptomatic due to the gradual enlargement of collateral veins.

Plain radiographs are no longer considered useful for identifying meningiomas. In the past, certain characteristics, such as enlarged grooves in meningeal arteries, areas of hyperostosis or lysis, calcification, and movement of the calcified pineal gland or choroid plexus due to pressure from the tumor, were noted.

CT is often the initial imaging tool used to evaluate neurological symptoms and can also detect incidental lesions. Approximately 60% of non-contrast CT scans show meningiomas as slightly hyperdense compared to normal brain tissue, while the rest appear isodense. About 20–30% of cases may exhibit some calcification, and over 50% may display varying degrees of adjacent edema. In post-contrast CT scans, around 72% of meningiomas show intense and uniform enhancement, while malignant or cystic variants exhibit less intense enhancement and greater heterogeneity. Hyperostosis, occurring in 5% of cases, may be seen adjacent to the base of the skull, and it is essential to distinguish reactive hyperostosis from direct invasion of the skull vault by the neighboring meningioma or primary intraosseous meningioma.

Typical meningiomas exhibit specific signal characteristics on MRI. On T1-weighted images, they typically appear isointense to gray matter (60–90%), although they can also be hypointense (10–40%), particularly in fibrous and psammomatous variants. On T1-weighted contrast-enhanced images, meningiomas usually demonstrate uniform and intense enhancement. On T2-weighted images, meningiomas are isointense to gray matter in about 50% of cases, but they can also be hyperintense (35–40%), in which case they are associated with a softer texture and hypervascular tumors. Harder and more fibrous meningiomas typically appear hypointense relative to gray matter (10–15%). While grade 2 and 3 meningiomas may exhibit restricted diffusion on DWI, this is not a consistent finding [37, 38].

Differential Diagnosis

Solitary fibrous tumors of the dura are the main differential that should be considered for dural masses. These tumors tend to be more aggressive, exhibit extensive peripheral vascularity, and display increased microlobulation. Dural metastases, particularly from breast cancer, should also be on the differential diagnosis list.

Location-specific differentials are relevant as well. For instance, in the cerebellopontine angle (CPA), acoustic schwannoma is a potential diagnosis to consider. In the pituitary region, pituitary macroadenoma and craniopharyngioma may be among the differential diagnoses. At the base of the skull, hypertrophic pachymeningitis, extramedullary hematopoiesis, chondrosarcoma, and chordoma are additional possibilities.

Questions

1. **Meningioma, the FALSE answer is:**
   A. Is typically isointense to gray matter on T1W images.
   B. Is rarely calcified.
   C. Hyperostosis occurs in about 5% of cases.
   D. Approximately 72% of meningiomas enhance brightly and homogeneously on post-contrast CT.
   E. The most common type is supratentorial (85–90%), followed by juxtasellar (5–10%), infratentorial (5–10%), and miscellaneous intradural meningiomas (<5%).

   The answer is B.
   Calcification is a common finding.

Case 21: Arachnoid Cyst

Case Scenario

A 52-year-old woman presented with complaints of limb weakness with no other specific symptoms (Fig. 2.25).

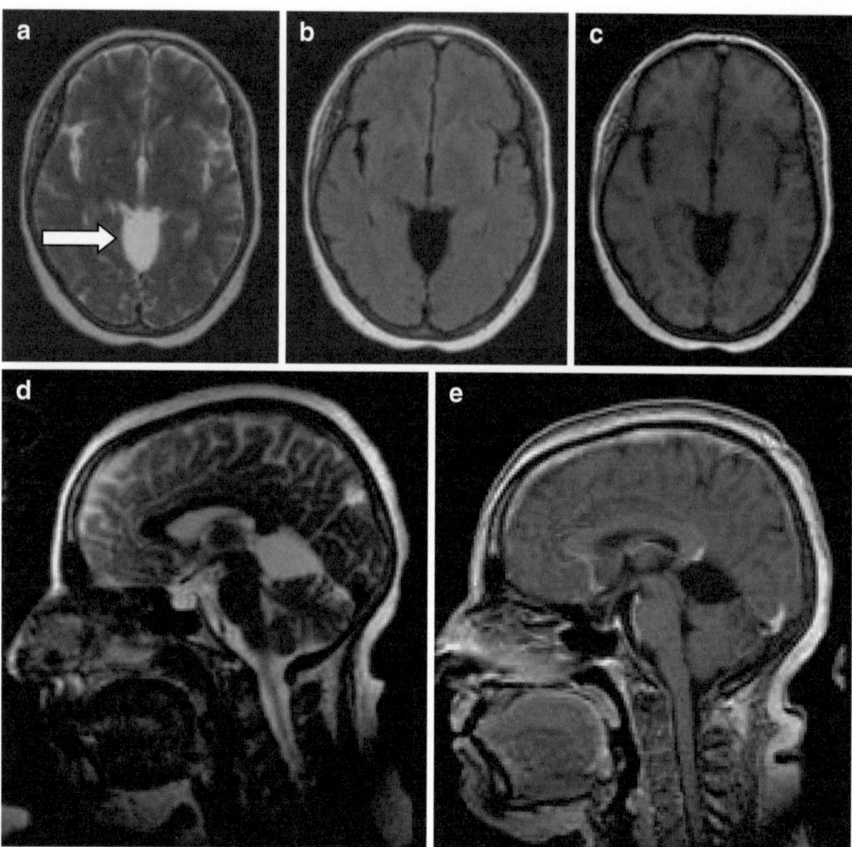

Fig. 2.25 Serial MRI images of the brain with the following: (**a**) axial T2, (**b**) axial FLAIR, (**c**) axial T1, (**d**) sagittal T2, and (**e**) sagittal T1 C+. (Figure courtesy of Dr. Samer Hoz)

Imaging Description

MRI with gadolinium shows a small cyst located at the quadrigeminal cistern (arrow) that follows the CSF signal in all pulse sequences. It is homogeneously hypointense on T1W MR and FLAIR images and hyperintense on T2W MR, with no contrast enhancement and no restriction on diffusion-weighted images. The cyst is exerting slight compression over the cerebellum. There is no compression on the aqueduct and no hydrocephalus. Histopathological examination confirmed the diagnosis of an arachnoid cyst.

Arachnoid Cysts

Arachnoid cysts are relatively common, benign, and typically asymptomatic lesions that can occur within the central nervous system. They are most frequently found intracranially but can also occur within the spinal canal. These cysts are typically located in the subarachnoid space and contain CSF. Arachnoid cysts make up approximately 1% of all intracranial masses and tend to occur more frequently in males. While most cases are sporadic, they are more often seen in individuals with mucopolysaccharidoses.

With regards to their radiographic characteristics, arachnoid cysts can manifest anywhere within the CNS. However, they are most commonly found in the middle cranial fossa, where they can cause the Sylvian fissure to bulge and widen, which is observed in approximately 50–60% of cases. Additionally, around 30–40% of arachnoid cysts are located in the retrocerebellar region.

On CT scans, arachnoid cysts have well-defined margins and can displace adjacent structures, sometimes even causing reshaping of the surrounding bone if they are large and left untreated. CT cisternography, involving the introduction of contrast into the subarachnoid space, is a valuable diagnostic technique for demonstrating the connection between the cyst and the subarachnoid space. Since this communication is often slow, the cyst may fill later, and contrast might accumulate within it, outlining its dependent portion.

Because arachnoid cysts are filled with CSF, they exhibit similar signal characteristics to CSF on all MRI sequences, including FLAIR and DWI. This helps differentiate them from other types of cysts, such as epidermoid cysts, as arachnoid cyst walls are typically thin and not easily visible on imaging, and there is no enhancement of any solid component. Phase-contrast imaging can be used to identify the cyst's location and its communication with the subarachnoid space. High-resolution sequences like constructive interference in steady state (CISS) and fast imaging employing steady-state acquisition (FIESTA) in magnetic resonance cisternography can help delineate the cyst wall and adjacent anatomical structures.

Furthermore, it is worth noting that a rare complication of arachnoid cysts is their spontaneous rupture into the subdural space [39, 40].

Differential Diagnosis

1. Enlarged CSF space, such as mega cisterna magna.
2. Epidermoid cysts: typically exhibit a heterogeneous or "dirty" signal on FLAIR, limited diffusion, a more lobulated shape, and may encase nearby arteries and cranial nerves.
3. Leptomeningeal cysts, which can occur post-trauma.
4. Subdural hygroma/chronic subdural hemorrhage: do not usually display the CSF signal intensity on MRI and may have an enhancing membrane.
5. Cystic tumors: often have a solid-enhancing component and are intra-axial.
6. Pilocytic astrocytoma.
7. Hemangioblastoma.
8. Non-neoplastic cysts, such as neurenteric cysts, neuroglial cysts, and porencephalic cysts. These often follow a history of trauma or stroke and are surrounded by gliotic brain.

9. Tumefactive perivascular spaces, particularly anterior temporal lobe perivascular spaces.
10. Neurocysticercosis: is usually present in multiples when located in the subarachnoid space.

Questions

1. **Arachnoid cysts, the FALSE answer is:**
   A. Are usually asymptomatic and benign.
   B. Can be found in both the intracranial compartment and within the spinal canal.
   C. Have a solid component.
   D. Follow CSF on all MRI sequences, which distinguishes them from other types of cysts.
   E. Phase contrast imaging can be used to identify the location and communication of arachnoid cysts with the subarachnoid space.
   The answer is C.
   Unlike epidermoid cysts, arachnoid cysts do not have a solid component, and the wall is usually thin and not easily visible on imaging.

Case 22: Dysplastic Cerebellar Gangliocytoma (Lhermitte-Duclos Disease)

Case Scenario

A 50-year-old male presented with a history of headache and ataxia (Fig. 2.26).

Imaging Description

MRI demonstrated an expansive lesion in the right cerebellar hemisphere that is hyperintense in T2 and hypointense in T1, with faint enhancement but no restricted diffusion in DWI. There is a mass effect on the brain stem and fourth ventricle, causing a shift and effacement. The lesion has a characteristic striate/tigroid appearance. Histopathological examination confirmed the diagnosis of dysplastic cerebellar gangliocytoma.

Lhermitte-Duclos Disease

This tumor, also known as dysplastic cerebellar gangliocytoma, is a rare type of cerebellar tumor characterized by a distinctive striped appearance resulting from

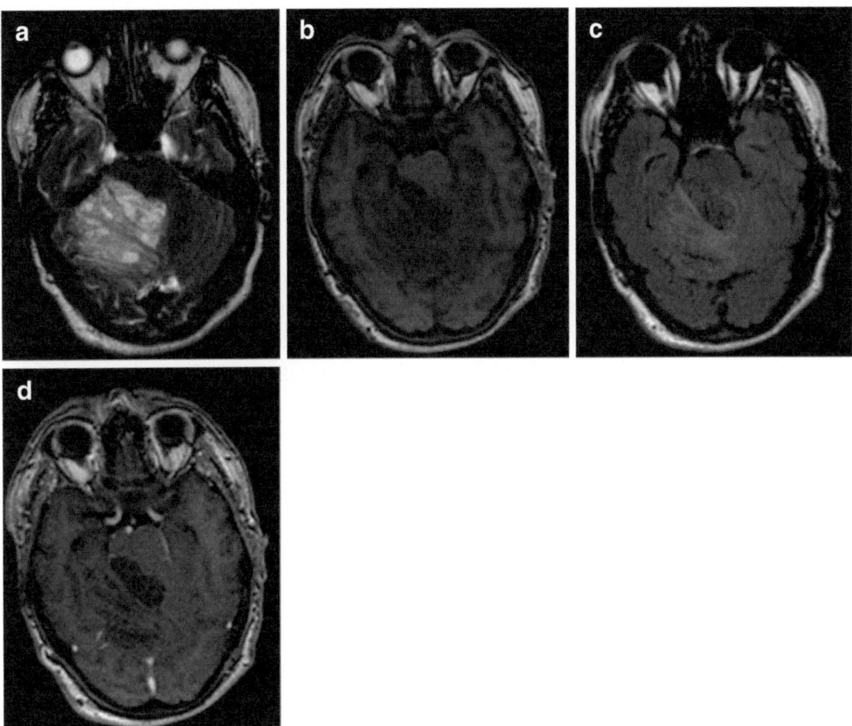

Fig. 2.26 Serial MRI images of the brain with the following: (**a**) axial T2, (**b**) axial T1, (**c**) axial FLAIR, and (**d**) axial T1 C+. (Figure courtesy of Dr. Samer Hoz)

thickening and increased T2 signal intensity of the cerebellar folia. While these tumors are typically present in young adults, they can occur at any age. Dysplastic cerebellar gangliocytomas are classified as grade 1 tumors by the World Health Organization (WHO) and are categorized as one of several glioneuronal and neuronal tumors in the current WHO classification of CNS tumors.

The lesion primarily affects the cerebellar cortex and is usually limited to one hemisphere. It rarely extends to the vermis or the opposite hemisphere. CT scans may reveal a cerebellar mass with non-specific hypoattenuation, and calcification may be visible in some cases. On MRI, the cerebellar folia appear widened and display a striped or tiger shaped, often described as "corduroy" or "laminated." T1-weighted images appear hypointense, while T2-weighted images appear hyperintense with preserved cortical striations. Diffusion-weighted imaging (DWI) may show hyperintensity due to the T2 shine-through effect. The presence of contrast enhancement on T1-weighted contrast-enhanced MRI is rare, and if it occurs, it is usually superficial, possibly due to vascular proliferation. Magnetic resonance spectroscopy (MRS) reveals elevated lactate, slightly reduced NAA (by approximately 10%), reduced myo-inositol (by 30–80%), reduced Cho (by 20–50%), and a reduced

Cho/Cr ratio. Positron emission tomography (PET) and single-photon emission (SPECT) imaging demonstrate increased uptake on F-fluorodeoxyglucose (FDG)-PET and Tl-201 SPECT [41, 42].

Differential Diagnosis

With the distinctive appearance of dysplastic cerebellar gangliocytoma, diagnosis is typically straightforward, especially when the radiological features are typical. However, in situations involving infection or sudden deterioration, one should consider the possibilities of cerebellitis or subacute cerebellar infarction. Additionally, the appearance may resemble that of extensively nodular medulloblastoma.

Questions

1. **Lhermitte-Duclos disease, the FALSE answer is:**
   A. The cerebellar cortex tends to be limited to one hemisphere.
   B. Appears hyperintense on T1W images.
   C. On MRI, the cerebellar folia appear widened with a striated or tiger shaped.
   D. PET/SPECT imaging demonstrates increased uptake on FDG-PET.
   E. Extensively nodular medulloblastoma may mimic the appearance of Lhermitte-Duclos disease.
   The answer is B.
   Appears hypointense on T1W images.

Case 23: Angiomatous Meningioma

Case Scenario

An 11-year-old boy presented with complaints of headache and vertigo (Fig. 2.27).

Imaging Descriptions

MRI reveals a well-defined, round, vividly enhancing extra-axial mass with a dural tail situated at the midline of the occipital region within the infra-tentorial region and attached to the tentorium cerebri. It is causing a mass effect and compressing the adjacent brain tissue. The lesion is isointense on T1 and hyperintense on T2 and FLAIR. There is a bright enhancement after the IV contrast study with invasion of the adjacent cerebral venous sinuses, including the end of the straight sinus, the lower part of the superior sagittal sinus, as well as both sides of the transverse sinus. Histopathological examination confirmed the diagnosis of angiomatous meningioma.

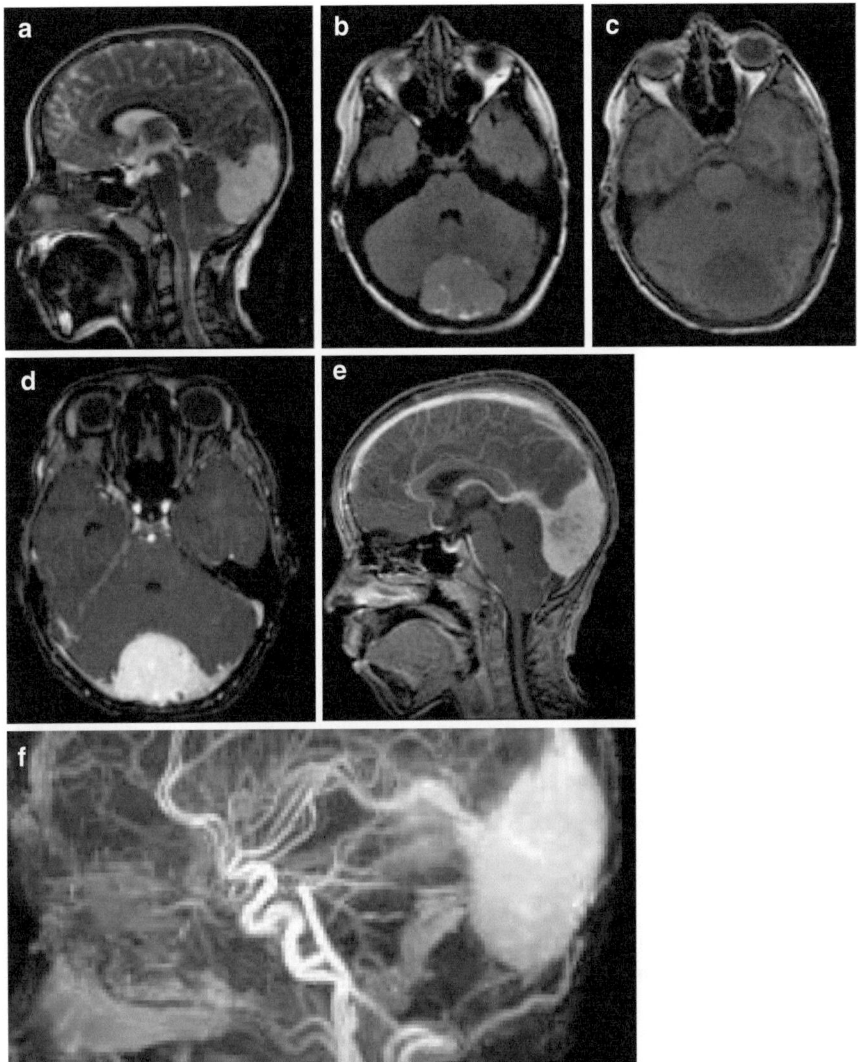

Fig. 2.27 Serial MRI images of the brain with the following: (**a**) sagittal T2, (**b**) axial FLAIR, (**c**) axial T1, (**d**) axial T1 C+, (**e**) sagittal T1 C+, and (**f**) MRA. (Figure courtesy of Dr. Samer Hoz)

Angiomatous Meningiomas

Angiomatous meningiomas are a rare subtype, constituting only 2.1% of all meningiomas. In terms of their epidemiology and clinical presentation, they do not significantly differ from more common histological subtypes. However, distinguishing them from hemangiopericytomas, which were previously considered a form of meningioma and were once known as "angioblastic meningioma," can be a challenging task. Angiomatous meningiomas are characterized by a substantial presence of blood vessels, accounting for over half of the tumor composition, along with regions showing typical meningothelial differentiation. They fall under the classification of WHO grade 1 tumors and do not exhibit cellular atypia or anaplasia.

On CT scans, angiomatous meningiomas share a similar appearance with more common subtypes of typical meningiomas. They appear slightly denser than the adjacent brain parenchyma and exhibit strong enhancement with contrast administration. On MRI, they can be challenging to distinguish from microcystic or chordoid meningiomas, but they often display prominent flow voids and may exhibit a dural tail. Additionally, these tumors sometimes show signs of cystic changes. They appear hypointense on T1-weighted images and hyperintense on T2-weighted images compared to gray matter. Peritumoral edema is frequently present, and they demonstrate vivid enhancement on contrast-enhanced T1-weighted images. DWI/ADC indicates facilitated diffusion, and in some cases, feeding vessels may be visible on MRA [43, 44].

Differential Diagnosis

1. Solitary fibrous tumors of the dura: can closely resemble angiomatous meningiomas, although they are less likely to exhibit a dural tail, more likely to display heterogeneous enhancement, and tend to have an irregular contour or microlobulation. Additionally, their ADC values may be higher than those observed in angiomatous meningiomas, and they are more prone to bone and brain invasion. These tumors are somewhat more prevalent in younger males compared to older females.
2. Other meningioma variants, such as microcystic meningioma, chordoid meningioma, secretory meningioma, angiomatous meningioma, and cartilaginous metaplastic meningioma: typically do not exhibit flow voids as prominently.
3. Hemangioblastoma: usually found in the posterior fossa and typically presents as a cyst with an enhancing mural nodule.

Questions

1. **Angiomatous meningioma, the FALSE answer is:**
   A. Is a rare type of meningioma, comprising only 2.1% of all meningiomas.
   B. On CT, it is slightly hypodense to the adjacent brain parenchyma.
   C. Shows vivid enhancement on T1 C+.
   D. On MRI, it can be difficult to differentiate from microcystic or chordoid meningiomas.
   E. Shows facilitated diffusion on DWI/ADC.
   The answer is B.
   On CT, angiomatous meningiomas appear slightly hyperdense to the adjacent brain parenchyma.

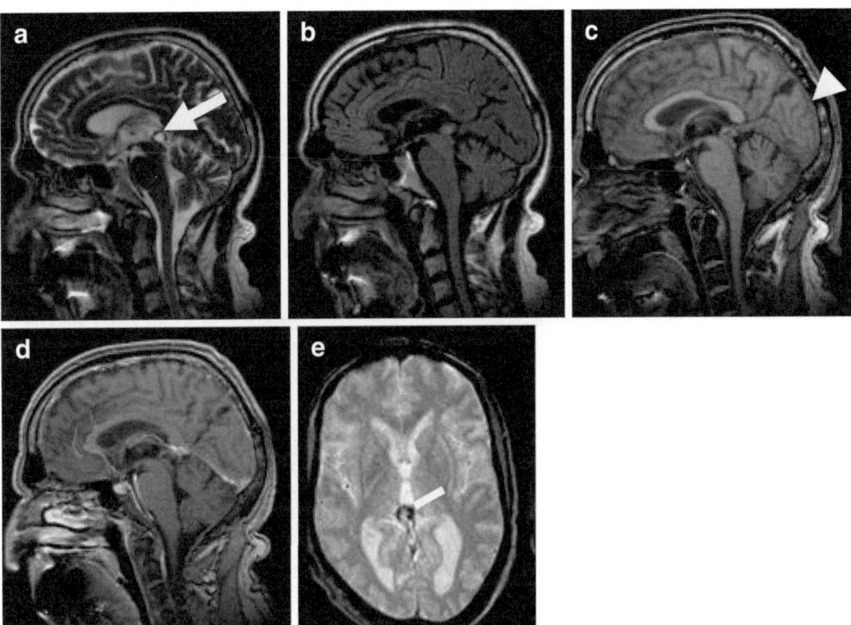

Fig. 2.28 Serial MRI images of the brain with the following: (**a**) sagittal T2, (**b**) sagittal FLAIR, (**c**) sagittal T1, (**d**) sagittal T1 C+, and (**e**) axial FFE GRE. (Figure courtesy of Dr. Samer Hoz)

Case 24: Pineal Cyst

Case Scenario

A 35-year-old man presented with headaches (Fig. 2.28).

Imaging Description

MRI demonstrated a well-defined, small, rounded pineal cyst with rim enhancement but no nodule or solid component (arrow). It is hyperintense in T2 and FLAIR, and hypointense in T1. It minimally distorts the tectal plate, and there is no hydrocephalus. Phase imaging (GRE) confirms the presence of a peripherally located calcification (line) (on phase imaging, the peripherally located calcium is black). The remainder of the brain is unremarkable in appearance, with no intra extra-axial masses or regions of abnormal contrast enhancement.

Pineal Cyst

Pineal cysts are common and typically remain asymptomatic. They are often discovered incidentally and can be difficult to differentiate from cystic tumors, especially when they reach significant sizes or exhibit unusual characteristics. Consequently, many patients undergo extended follow-up for these lesions, which can lead to increased anxiety among patients.

These cysts within the pineal gland usually present as single-chamber cysts, varying in density or fluid signal, ranging from being similar to CSF to approximately 60% of cases displaying slight hyperintensity on T1-weighted images. Most cases exhibit a thin and smooth ring of contrast enhancement, and about 25% of cases contain calcifications. Pineal cysts are typically identified in young adults, with a higher prevalence among females (a female-to-male ratio of 3:1). They are detected in around 5% of brain MRIs and are found in 20–40% of autopsy studies. A study using high-resolution MRIs revealed asymptomatic cysts in 23% of healthy individuals.

On CT scans, pineal cysts appear as well-defined lesions with fluid density, often accompanied by a thin rim of calcification in approximately 25% of cases. Peripheral enhancement is typically observed, and the cysts can compress and displace the internal cerebral veins.

MRI scans indicate that pineal cysts are generally iso- to hypointense on T1-weighted images, with 55–60% of cases showing slight hyperintensity compared to CSF. They usually exhibit homogeneous signal characteristics. On T2-weighted scans, they appear as high-signal lesions, slightly hypointense in comparison to CSF. Thin septations or small internal cysts may be present. On FLAIR scans, the cysts typically display a high signal and are not completely suppressed. DWI and ADC maps demonstrate no restricted diffusion.

Approximately 60% of pineal cysts show enhancement on post-contrast T1-weighted scans. The enhancement is typically limited to the rim of the cyst, appearing thin and measuring less than 2 mm. It is worth noting that if post-contrast imaging is delayed by 60–90 minutes, gadolinium may diffuse into the cyst fluid, causing the mass to appear solid. In atypical cases, enhancement may present as nodular, or evidence of previous hemorrhage within the cyst may be visible [45, 46].

Differential Diagnosis

1. Pineal parenchymal tumors, including pineocytoma, pineal parenchymal tumor with intermediate differentiation, and papillary tumor of the pineal region.
2. Epidermoid cyst.
3. Arachnoid cyst: typically found behind the pineal gland.
4. Germ cell tumors, including germinoma, embryonal carcinoma, choriocarcinoma, and teratoma.
5. Cerebral metastasis.
6. Vein of Galen aneurysm.

Questions

1. **Pineal cyst, the FALSE answer is:**
   A. Is typically iso- to hypointense on T1 MRI scans.
   B. May contain thin septations or small internal cysts.
   C. May cause splaying of the internal cerebral veins.
   D. Can cause restricted diffusion on DWI/ADC scans.
   E. Is usually asymptomatic and found incidentally.
   The answer is D.
   Pineal cysts do not typically demonstrate restricted diffusion on DWI/ADC scans.

Case 25: Trigeminal Schwannoma

Case Scenario

A 50-year-old woman presented with right trigeminal neuralgia (Fig. 2.29).

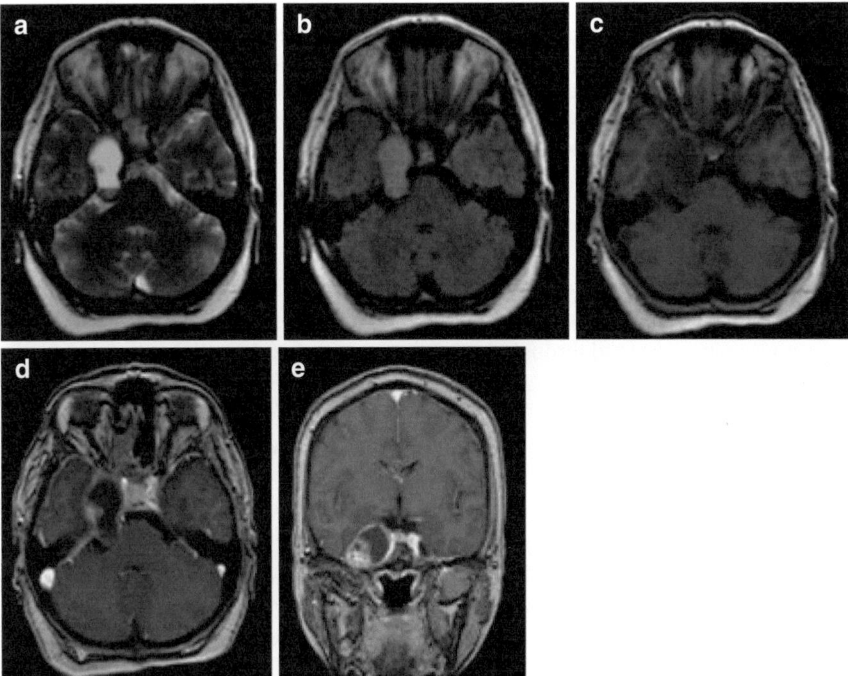

Fig. 2.29 Serial MRI images of the brain with the following: (**a**) axial T2, (**b**) axial FLAIR, (**c**) axial T1, (**d**) axial T1 C+, and (**e**) coronal T1 C+. (Figure courtesy of Dr. Samer Hoz)

Imaging Description

A right-sided extra-axial lesion can be seen in the medial aspect of the temporal region with posterior fossa extension. It is extending to the Meckel's cave and is causing a mass effect on the brain stem due to its extension into the CPA cistern. The lesion has solid and cystic components, as shown in the post-gadolinium axial image, and is mainly hyperintense on T2 and FLAIR and hypointense on T1, with heterogeneous thick enhancement. Histopathological examination confirmed the diagnosis of trigeminal schwannoma.

Trigeminal Schwannoma

Trigeminal schwannomas are slow-growing tumors that develop from schwann cells. While they are less common than vestibular schwannomas, accounting for less than 0.2% of all intracranial tumors, they are still the second most common intracranial schwannoma and make up a third of tumors in Meckel's cave. These tumors are typically seen in individuals in their third to fourth decades of life, with a slight preference for females. They can also be associated with NF2.

Trigeminal schwannomas share imaging features similar to other schwannomas in different anatomical locations. Depending on their extent, they may have a dumbbell appearance if they extend into both the cisternal and Meckel's cave regions, with a narrowing at the Gasserian ganglion. In smaller cases, they can be limited to specific compartments of the trigeminal nerve as follows:

1. Preganglionic (cisternal) trigeminal schwannomas: limited to the prepontine and CPA cisterns.
2. Ganglionic (trigeminal ganglion) trigeminal schwannomas: found exclusively within Meckel's cave and are the most prevalent subtype.
3. Postganglionic trigeminal schwannomas: typically either restricted to the cavernous sinus or extend through the cranial foramina, although they rarely remain solely within the extracranial compartment. This subtype is the least common, accounting for 10% of all cases, and may affect any of the three divisions of the trigeminal nerve, with the ophthalmic division being the most frequently involved.

On CT scans, schwannomas can be challenging to identify due to their location, typically appearing isodense to brain tissue. Larger tumors may exhibit hypoattenuating cystic areas. Contrast-enhanced CT scans often show moderate and heterogeneous enhancement, and bone algorithms can help assess tumors extending through foramina. These tumors usually cause smooth bone remodeling.

MRI is the preferred imaging modality for detailed anatomical visualization and better contrast resolution. Trigeminal schwannomas typically appear isointense to the brain on T1-weighted images, with cystic areas appearing hypointense if present. On T2-weighted images, they are somewhat hyperintense compared to brain tissue, including any cystic regions. In Meckel's cave, they sharply contrast with the contralateral side filled with CSF. T1-weighted post-contrast images show prominent enhancement, which is often heterogeneous in around 70% of cases. These tumors exhibit high signal intensity on both DWI and ADC maps, indicating T2 shine-through rather than restricted diffusion [47, 48].

Differential Diagnosis

The potential differential diagnoses vary depending on the size and site of the tumor. For larger tumors that extend into the CPA, the following possibilities should be considered:

1. Vestibular schwannoma: ~80% of CPA masses.
2. Meningioma: ~10% of CPA masses.
3. Ependymoma.
4. Metastasis.
5. Chondrosarcoma.

In cases where the tumor is small and localized within Meckel's cave, the list of differentials also encompasses ICA aneurysms, vascular malformations, and pituitary macroadenoma.

Questions

1. **Trigeminal schwannoma, the FALSE answer is:**
   A. Ganglionic trigeminal schwannomas are confined to the prepontine and CPA cisterns.
   B. Is isointense to brain on T1.
   C. T1 C+ shows prominent enhancement.
   D. Is isodense to brain on CT.
   E. Cystic areas, which develop in larger tumors, appear hypodense on CT scans.
   The answer is A.
   Ganglionic trigeminal schwannomas are confined to the Meckel's cave.

References

1. Ma J, Zhang Z, Li S, Chen X, Wang S. Intracranial amelanotic melanoma: a case report with literature review. World J Surg Oncol. 2015;13(1):1–7.
2. Splavski B, Muzevic D, Kovacevic M, Splavski B Jr, Bajek G. Management and prognosis of primary cerebral melanocytic tumors. A case report and systematic review. J Neurol Surg A Cent Eur Neurosurg. 2015;76(2):144–8.
3. Hamza HS, Elhusseiny AM. Choroidal melanoma resection. Middle East Afr J Ophthalmol. 2018;25(2):65.

4. Desjardins L. Le mélanome de la choroïde.

5. Mbekeani JN, Abdel Fattah M, Ul Haq A, Al Shail E, Ahmed M. Pediatric pilomyxoid astrocytoma–ophthalmic and neuroradiologic manifestations. Eur J Ophthalmol. 2022;32(5):2604–14.

6. Kulac I, Tihan T. Pilomyxoid astrocytomas: a short review. Brain Tumor Pathol. 2019;36:52–5.

7. Batash R, Asna N, Schaffer P, Francis N, Schaffer M. Glioblastoma multiforme, diagnosis and treatment; recent literature review. Curr Med Chem. 2017;24(27):3002–9.

8. Davis ME. Glioblastoma: overview of disease and treatment. Clin J Oncol Nurs. 2016;20(5):S2.

9. Shaikh N, Brahmbhatt N, Kruser TJ, Kam KL, Appin CL, Wadhwani N, Chandler J, Kumthekar P, Lukas RV. Pleomorphic xanthoastrocytoma: a brief review. CNS Oncol. 2019;8(3):CNS39.

10. Detti B, Scoccianti S, Maragna V, Lucidi S, Ganovelli M, Teriaca MA, Caini S, Desideri I, Agresti B, Greto D, Buccoliero AM. Pleomorphic Xanthoastrocytoma: a single institution retrospective analysis and a review of the literature. Radiol Med. 2022;127(10):1134–41.

11. Salles D, Laviola G, Malinverni AC, Stávale JN. Pilocytic astrocytoma: a review of general, clinical, and molecular characteristics. J Child Neurol. 2020;35(12):852–8.

12. Bond KM, Hughes JD, Porter AL, Orina J, Fang S, Parney IF. Adult pilocytic astrocytoma: an institutional series and systematic literature review for extent of resection and recurrence. World Neurosurg. 2018;110:276–83.

13. Eisele SC, Reardon DA. Adult brainstem gliomas. Cancer. 2016;122(18):2799–809.

14. Purohit B, Kamli AA, Kollias SS. Imaging of adult brainstem gliomas. Eur J Radiol. 2015;84(4):709–20.

15. Chourmouzi D, Papadopoulou E, Konstantinidis M, Syrris V, Kouskouras K, Haritanti A, Karkavelas G, Drevelegas A. Manifestations of pilocytic astrocytoma: a pictorial review. Insights Imaging. 2014;5(3):387–402.

16. Xia JG, Yin B, Liu L, Lu YP, Geng DY, Tian WZ. Imaging features of pilocytic astrocytoma in cerebral ventricles. Clin Neuroradiol. 2016;26:341–6.

17. Li Y, Zhang ZX, Huang GH, Xiang Y, Yang L, Pei YC, Yang W, Lv SQ. A systematic review of multifocal and multicentric glioblastoma. J Clin Neurosci. 2021;83:71–6.

18. Baro V, Cerretti G, Todoverto M, Della Puppa A, Chioffi F, Volpin F, Causin F, Busato F, Fiduccia P, Landi A, d'Avella D. Newly diagnosed multifocal GBM: a monocentric experience and literature review. Curr Oncol. 2022;29(5):3472–88.

19. Cho HJ, Myung JK, Kim H, Park CK, Kim SK, Chung CK, Choi SH, Park SH. Primary diffuse leptomeningeal glioneuronal tumors. Brain Tumor Pathol. 2015;32:49–55.

20. Abongwa C, Cotter J, Tamrazi B, Dhall G, Davidson T, Margol A. Primary diffuse leptomeningeal glioneuronal tumors of the central nervous system: report of three cases and review of literature. Pediatr Hematol Oncol. 2020;37(3):248–58.

21. Yadav YR, Yadav N, Parihar V, Yatin KH, Ratre S. Management of colloid cyst of third ventricle. Turk Neurosurg. 2015;25(3):362.

22. O'Neill AH, Gragnaniello C, Lai LT. Natural history of incidental colloid cysts of the third ventricle: a systematic review. J Clin Neurosci. 2018;53:122–6.

23. Cohen NT, Cross JH, Arzimanoglou A, Berkovic SF, Kerrigan JF, Miller IP, Webster E, Soeby L, Cukiert A, Hesdorffer DK, Kroner BL. Hypothalamic hamartomas: evolving understanding and management. Neurology. 2021;97(18):864–73.

24. Alomari SO, El Houshiemy MN, Bsat S, Moussalem CK, Allouh M, Omeis IA. Hypothalamic hamartomas: a comprehensive review of the literature—part 1: neurobiological features, clinical presentations and advancements in diagnostic tools. Clin Neurol Neurosurg. 2020;197:106076.

25. Otte A, Müller HL. Childhood-onset craniopharyngioma. J Clin Endocrinol Metabol. 2021;106(10):e3820–36.

26. Müller HL. The diagnosis and treatment of craniopharyngioma. Neuroendocrinology. 2020;110(9–10):753–66.

27. Cohen D, Litofsky NS. Diagnosis and management of pineal germinoma: from eye to brain. Eye Brain. 2023;15:45–61.

28. Stephens S, Kuchel A, Cheuk R, Alexander H, Robertson T, Rajah T, Tran Q, Inglis PL. Management trends and outcomes of pineal germinoma in a multi-institutional Australian cohort. J Clin Neurosci. 2021;90:1–7.
29. Connor SE. Imaging of the vestibular schwannoma: diagnosis, monitoring, and treatment planning. Neuroimaging Clin. 2021;31(4):451–71.
30. Gupta VK, Thakker A, Gupta KK. Vestibular schwannoma: what we know and where we are heading. Head Neck Pathol. 2020;14:1058–66.
31. Law EK, Lee RK, Ng AW, Siu DY, Ng HK. Atypical intracranial epidermoid cysts: rare anomalies with unique radiological features. Case Rep Radiol. 2015;2015:528632.
32. Jamjoom DZ, Alamer A, Tampieri D. Correlation of radiological features of white epidermoid cysts with histopathological findings. Sci Rep. 2022;12(1):2314.
33. Suh YL. Dysembryoplastic neuroepithelial tumors. J Pathol Transl Med. 2015;49(6):438–49.
34. Phi JH, Kim SH. Dysembryoplastic neuroepithelial tumor: a benign but complex tumor of the cerebral cortex. Brain Tumor Res Treat. 2022;10(3):144–50.
35. Millard NE, De Braganca KC. Medulloblastoma. J Child Neurol. 2016;31(12):1341–53.
36. Keil VC, Warmuth-Metz M, Reh C, Enkirch SJ, Reinert C, Beier D, Jones DT, Pietsch T, Schild HH, Hattingen E, Hau P. Imaging biomarkers for adult medulloblastomas: genetic entities may be identified by their MR imaging radiophenotype. Am J Neuroradiol. 2017;38(10):1892–8.
37. Chaudhry SK, Raza R, Naveed MA, Rehman I. Suprasellar meningiomas: an experience of four cases with brief review of literature. Cureus. 2021;13(1):e12470.
38. Kwancharoen R, Blitz AM, Tavares F, Caturegli P, Gallia GL, Salvatori R. Clinical features of sellar and suprasellar meningiomas. Pituitary. 2014;17:342.
39. Mustansir F, Bashir S, Darbar A. Management of arachnoid cysts: a comprehensive review. Cureus. 2018;10(4):e2458.
40. Sarwar S, Rocker J. Arachnoid cysts in paediatrics. Curr Opin Pediatr. 2023;35(2):288–95.
41. Zhang HW, Zhang YQ, Liu XL, Mo YQ, Lei Y, Lin F, Feng YN. MR imaging features of Lhermitte–Duclos disease: case reports and literature review. Medicine. 2022;101(4):e28667.
42. Liu Z, He Y, Fu J, Wu J, Song T, Wang Y, Huang T. Lhermitte-Duclos disease: a case report and literature review. J Cent South Univ Med Sci (Zhong nan da xue xue bao Yi xue ban). 2021;46(2):195–9.
43. Buerki RA, Horbinski CM, Kruser T, Horowitz PM, James CD, Lukas RV. An overview of meningiomas. Future Oncol. 2018;14(21):2161–77.
44. Ahmed Z, Prayson RA. Angiomatous meningioma in Sturge–Weber syndrome. J Clin Neurosci. 2015;22(6):1066–8.
45. Berhouma M, Ni H, Delabar V, Tahhan N, Salem SM, Mottolese C, Vallee B. Update on the management of pineal cysts: case series and a review of the literature. Neurochirurgie. 2015;61(2–3):201–7.
46. Favero G, Bonomini F, Rezzani R. Pineal gland tumors: a review. Cancers. 2021;13(7):1547.
47. Li M, Wang X, Chen G, Liang J, Guo H, Song G, Bao Y. Trigeminal schwannoma: a single-center experience with 43 cases and review of literature. Br J Neurosurg. 2021;35(1):49–56.
48. Wang X, Bao Y, Chen G, Guo H, Li M, Liang J, Bai X, Ling F. Trigeminal schwannomas in middle fossa could breach into subdural space: report of 4 cases and review of literature. World Neurosurg. 2019;127:e534.

Rokaya H. Abdalridha, Maen Saris, Zinah A. Alaraji,
Ahmed Muthana, Mahmood H. AlObaidy, Oday Atallah,
and Asmaa H. AL-Sharee

Case 26: CSF Rhinorrhea

Case Scenario

A 2-year-old male with a history of facial trauma sustained 1 month ago presented with watery discharge from the nasal cavity (Fig. 3.1).

Imaging Description

MRI shows CSF leakage is seen as a cleft line (arrow) extending from the left frontal subarachnoid space into the ipsilateral anterior ethmoidal air cell through a bony defect of the left cribriform plate as a cystic lesion (line).

R. H. Abdalridha
College of Medicine, Babylon University, Babylon, Iraq

M. Saris
Dar Al-Shifaa Hospital, Kuwait, Kuwait

Z. A. Alaraji · M. H. AlObaidy
College of Medicine, Al-Nahrain University, Baghdad, Iraq

A. Muthana
College of Medicine, University of Baghdad, Baghdad, Iraq

O. Atallah
Hannover Medical School, Hannover, Germany

A. H. AL-Sharee (✉)
Department of Neuroradiology, Neurosurgery Teaching Hospital, Baghdad, Iraq

© The Author(s), under exclusive license to Springer Nature
Switzerland AG 2024
S. Hoz et al. (eds.), *Neuroradiology Board's Favorites*,
https://doi.org/10.1007/978-3-031-64261-6_3

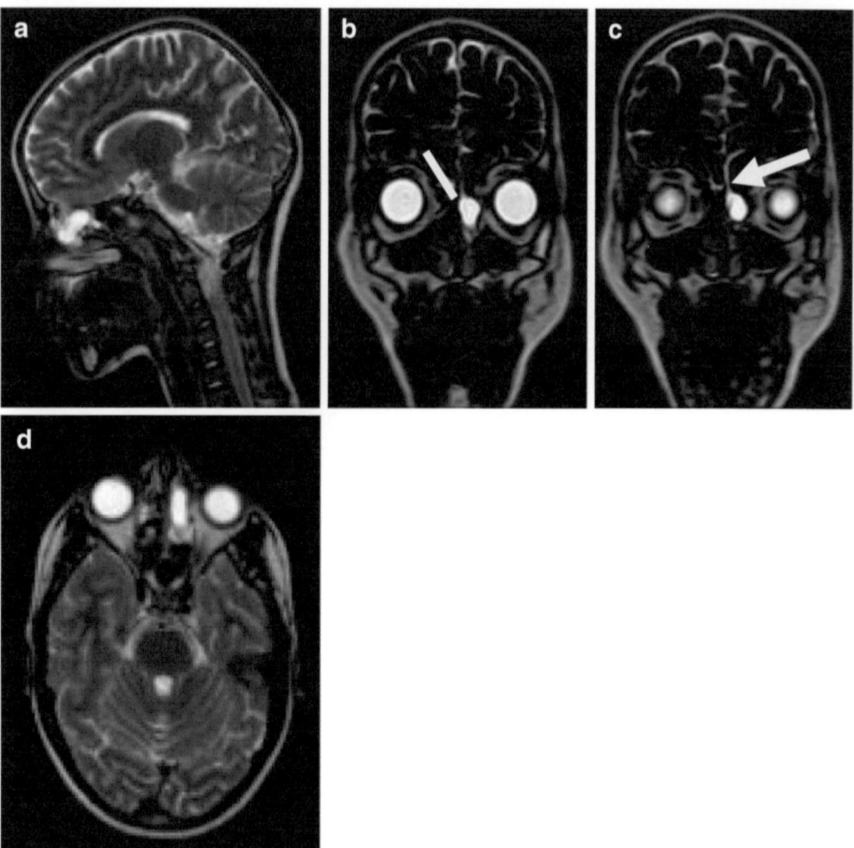

Fig. 3.1 Serial MRI images of the brain with the following: (**a**) sagittal T2, (**b**) coronal T2 3D imaging driven equilibrium radiofrequency reset pulse (DRIVE), (**c**) coronal T2 3D DRIVE, and (**d**) axial T2. (Figure courtesy of Dr. Samer Hoz)

CSF Rhinorrhea

CSF rhinorrhea refers to the leakage of CSF through a defect in the bony or dural base, passing through the paranasal sinuses into the nasal cavity and exiting through the anterior nares. The causes of CSF rhinorrhea can be categorized into congenital and acquired factors. Congenital factors include conditions like meningocele, nasal encephaloceles, and persistent cricopharyngeal canal. Acquired causes can be further divided into traumatic, iatrogenic, and non-traumatic factors. Closed head trauma with an anterior skull base fracture is the most common traumatic cause, disrupting anatomical barriers between the anterior cranial fossa and paranasal sinuses. Iatrogenic causes can result from various neurosurgical and otolaryngology procedures. Non-traumatic causes include conditions like pseudotumor cerebri with medial sphenoid meningocele formation and malignant tumors invading the skull base.

The cribriform plate is the most common site for spontaneous cases, as observed in this instance. CSF rhinorrhea can also manifest in the perisellar region and the lateral recess of the sphenoid sinus, which is situated within the sphenoid bone. Additionally, the temporal bone and middle ear are frequent sites, especially those involving the tegmen tympani and tegmen mastoideum.

Various imaging techniques can reveal radiographic features associated with CSF rhinorrhea. Plain CT scans allow the identification of significant bony defects, often accompanied by contrast streaks or pooling in the anterior nasal cavity, sinus cavity, and nasopharynx. CT cisternography is the preferred diagnostic method for detecting hidden sites of CSF leaks, involving the injection of contrast into the thecal sac. However, the success of this procedure relies heavily on proper patient positioning and timing. MRI, particularly 3D high-resolution T2W and T1W sequences, can also aid in CSF rhinorrhea diagnosis, with coronal sections offering improved visualization of bony defects. Nuclear medicine radionuclide studies have higher sensitivity to leak detection but offer poor anatomical resolution [1, 2].

Questions

1. **CSF rhinorrhea, the FALSE answer is:**
   A. CT cisternography is considered the diagnostic modality for identifying occult sites of CSF leaks.
   B. Nuclear medicine radionuclide studies have a higher sensitivity in detecting leaks, but they have poor anatomic resolution.
   C. Closed head trauma with middle base of skull fracture is the most common traumatic cause of CSF rhinorrhea.
   D. On plain CT scans, large osseous defects can be identified.
   E. The cribriform plate is the most frequent site for spontaneous cases of CSF rhinorrhea.
   The answer is C.
   Closed head trauma with anterior base of skull fracture is the most common traumatic cause of CSF rhinorrhea.

Case 27: Diffuse Axonal Injury (DAI)

Case Scenario

A 17-year-old male with a history of a motor vehicle accident a few weeks ago (Fig. 3.2).

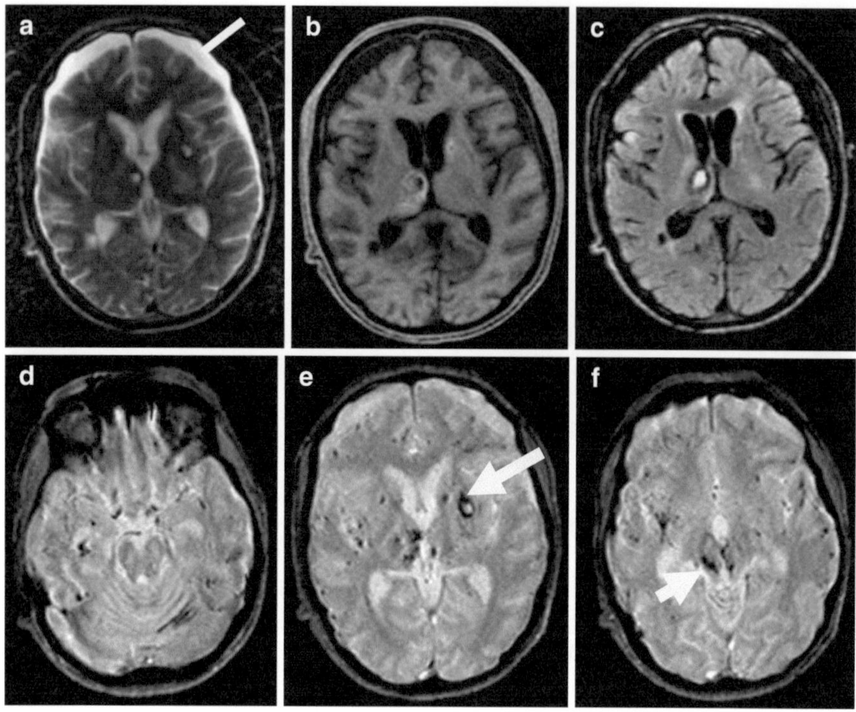

Fig. 3.2 Serial MRI images of the brain with the following: (**a**) axial T2, (**b**) axial T1, (**c**) axial FLAIR, (**d**) axial GRE, (**e**) axial GRE, and (**f**) axial GRE. (Figure courtesy of Dr. Samer Hoz)

Image Description

MRI demonstrates foci of high FLAIR signal with bilateral subdural hygroma, mainly in the frontal region (line). Multiple foci of susceptibility artifacts are observed in a typical distribution for DAI (arrows), i.e., in the gray-white matter junction, deep white matter, brainstem, cerebellum, and basal ganglia region, reflecting hemorrhagic DAI grade III.

Diffuse Axonal Injury

Diffuse axonal injury (DAI), also referred to as traumatic axonal injury, stands as one of the most severe forms of primary traumatic brain injury and plays a significant role in causing persistent vegetative states following severe head trauma. DAI typically arises in accidents involving rapid changes in motion, leading to shearing forces that particularly affect axons at the gray-white matter junction, as these areas have slightly different specific gravities.

Histological signs of axonal damage are found in various brain regions in different grades of DAI. Grade 1 exhibits axonal destruction in the white matter of the cerebral hemispheres, corpus callosum, brain stem, and, less commonly, the cerebellum. Grade 2 involves a localized lesion in the corpus callosum, while grade 3 adds an additional localized lesion in the dorsolateral quadrant of the rostral brainstem.

In patients with head injuries, a standard non-contrast brain CT scan is routine. However, this method may not detect subtle DAI, potentially leaving patients with unexplained neurological deficits despite seemingly normal CT scans. The CT findings depend on whether the lesions show hemorrhage. Hemorrhagic lesions appear hyperdense and are variable in size. In contrast, non-hemorrhagic lesions appear hypodense but become more apparent as edema develops around them, often accompanied by severe cerebral edema.

MRI is the preferred imaging technique for suspected DAI, even in cases with entirely normal CT scans. Special sequences like SWI or GRE in MRI, highly sensitive to paramagnetic blood products, can reveal small regions with susceptibility artifacts at the gray-white matter junction, in the corpus callosum, or in the brainstem. Some lesions may lack hemorrhage but will appear as areas with a high FLAIR signal. For additional shearing injuries not visible on T2/FLAIR or T2 GRE sequences, DWI becomes invaluable in their detection. Regions with restricted diffusion on DWI are typically considered to have suffered irreversible injuries [3, 4].

Differential Diagnosis

In patients who have experienced head trauma, the primary consideration is cortical contusion, which is generally a superficial injury affecting the cortex (as opposed to the gray-white matter junction) and is often associated with varying degrees of extra-axial hemorrhage. Another potential consideration is diffuse vascular injuries, which may include conditions like amyloid angiopathy, chronic hypertensive encephalopathy, and cavernoma type IV.

Questions

1. **DAI, the FALSE answer is:**
   A. In DAI, shearing forces have a predilection for gray-white matter junction axons.
   B. In grade I DAI, there is only histological evidence of axonal damage.
   C. With advanced imaging, a lack of evidence can definitively rule out DAI.
   D. CT scan is not sensitive to subtle DAI.
   E. CT scan findings depend on whether the lesions are overtly hemorrhagic.
   The answer is C.
   DAI is challenging to diagnose solely from imaging and even with advanced high-field strength scanners; the lack of evidence cannot definitively rule out axonal injury.

2. **MR techniques in DAI evaluation, the FALSE answer is:**
   A. T2 GRE/SWI sequences are useful for detecting microhemorrhages.

B. T2/FLAIR sequences are useful for identifying non-hemorrhagic lesions.
C. DWI sequence is of value in predicting outcomes in DAI.
D. MRS is not a reliable technique in patients with only histological evidence of injury.
E. Within 3 months, the majority of FLAIR changes resolve.

The answer is D.

In patients with grade I injury, MRS is a more beneficial technique than other MRI sequences.

Case 28: Traumatic Frontoethmoidal Encephalocele

Case Scenario

A 1-year-old male presented with a history of a fall from the third floor 2 months ago (Fig. 3.3).

Image Description

MRI reveals bony defects in the right cribriform plate with herniation of the right frontal lobe and meninges into the right ethmoidal air cells (the arrow).

Traumatic Frontoethmoidal Encephalocele

Encephalocele refers to the protrusion of brain tissue through a defect in the skull, and it can occur due to various factors, including congenital, traumatic, tumoral, or spontaneous causes. Acquired encephaloceles, most commonly resulting from trauma, can be triggered by iatrogenic factors or craniofacial injuries. In cases of craniofacial trauma, patients often experience symptoms such as CSF rhinorrhea within a few days to months after the injury. This condition is also associated with

Fig. 3.3 (a) Coronal T2 and (b) axial T2. (Figure courtesy of Dr. Samer Hoz)

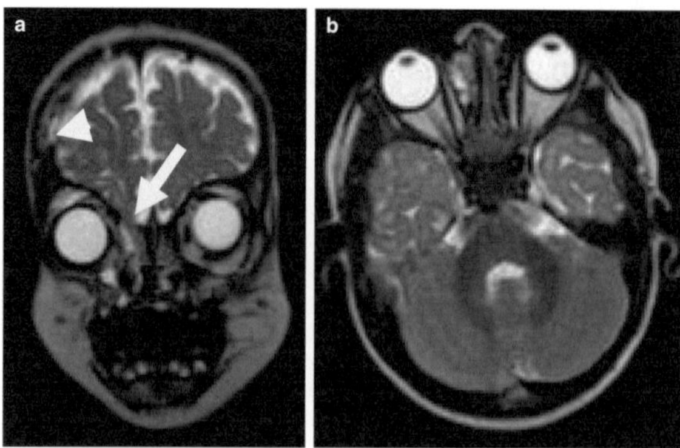

potentially life-threatening complications like meningitis, which may develop 3–8 months after the traumatic event.

Frontoethmoidal encephaloceles are classified into three subtypes based on the location of the skull defect: nasoethmoidal (the most common), nasofrontal, and naso-orbital encephaloceles.

Early detection of CSF leaks and encephaloceles is essential to minimize the risk of complications. CT scans can reveal fractures in the ethmoidal sinus wall and frontal sinus, with the cribriform plate being a common site. MRI, on the other hand, can detect the herniation of brain tissue within the frontal and ethmoidal sinuses, aiding in the diagnosis of encephaloceles [5, 6].

Questions

1. **Traumatic frontoethmoid encephalocele, the FALSE answer is:**
   A. The naso-orbital type is the most common.
   B. Can manifest as CSF rhinorrhea.
   C. Cross-sectional imaging plays a pivotal role in detection.
   D. Cribriform plate fracture is a common finding on CT.
   E. MRI demonstrates herniation of the brain tissue within the frontal and ethmoidal sinuses.

   The answer is A.

   The naso-ethmoidal type is the most common among frontoethmoidal encephaloceles, whereas the naso-orbital type is the least common.

References

1. Reddy M, Baugnon K. Imaging of cerebrospinal fluid rhinorrhea and otorrhea. Radiol Clin. 2017;55(1):167–87.
2. Hiremath SB, Gautam AA, Sasindran V, Therakathu J, Benjamin G. Cerebrospinal fluid rhinorrhea and otorrhea: a multimodality imaging approach. Diagn Interv Imaging. 2019;100(1):3–15.
3. Frati A, Cerretani D, Fiaschi AI, Frati P, Gatto V, La Russa R, Pesce A, Pinchi E, Santurro A, Fraschetti F, Fineschi V. Diffuse axonal injury and oxidative stress: a comprehensive review. Int J Mol Sci. 2017;18(12):2600.
4. Jang SH. Diagnostic problems in diffuse axonal injury. Diagnostics. 2020;10(2):117.
5. Orcajadas A, Palma A, Khalon BM. Frontoethmoidal encephalocele. Report of a case. Neurocirugía. 2019;30(2):94–9.
6. Cullu N, Deveer M, Karakas E, Karakas O, Bozkus F, Celik B. Traumatic fronto-ethmoidal encephalocele: a rare case. Eurasian J Med. 2015;47(1):69.

Vascular Diseases of the Brain

4

Mostafa H. Algabri, Maliya Delawan, Zainab Q. Saadi, Fatimah O. Ahmed, Ahmed Muthana, Mustafa Ismail, and Samer S. Hoz

Case 29: Cerebral Cavernous Venous Malformation (CCVM)

Case Scenario

A 20-year-old male presented with seizure (Fig. 4.1).

Image Description

MRI reveals a well-defined, spherical mass located in the left parasagittal region of the frontal lobe. The lesion appears as a mulberry-like structure with a hyperintense signal on T2, T1, and FLAIR sequences. There is a focal region of susceptibility-induced signal loss, which is well-seen in GRE sequences. After IV contrast administration, no enhancement is seen in the lesion. The histopathological examination confirmed that the mass is a cavernoma.

M. H. Algabri · Z. Q. Saadi · A. Muthana
College of Medicine, University of Baghdad, Baghdad, Iraq

M. Delawan
College of Medicine, Gulf Medical University, Ajman, United Arab Emirates

F. O. Ahmed
College of Medicine, University of Mustansiriyah, Baghdad, Iraq

M. Ismail
Neurosurgery Teaching Hospital, Baghdad, Iraq

S. S. Hoz (✉)
University of Pittsburgh Medical Center (UPMC), Pittsburgh, PA, USA

© The Author(s), under exclusive license to Springer Nature Switzerland AG 2024
S. Hoz et al. (eds.), *Neuroradiology Board's Favorites*,
https://doi.org/10.1007/978-3-031-64261-6_4

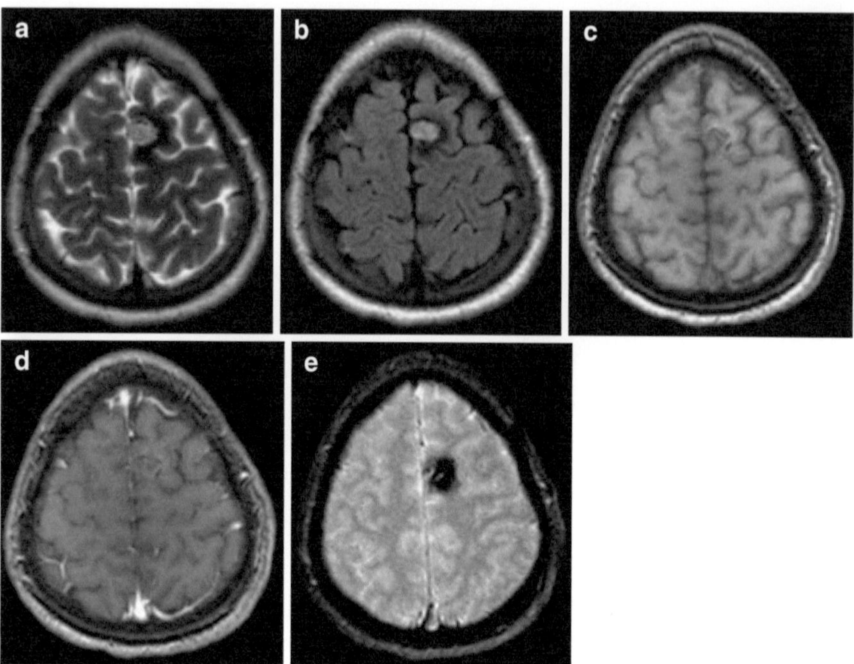

Fig. 4.1 Serial MRI images of the brain with the following: (**a**) axial T2, (**b**) axial FLAIR, (**c**) axial T1, (**d**) axial T1 C+, and (**e**) GRE. (Figure courtesy of Dr. Samer Hoz)

Cerebral Cavernous Venous Malformation (CCVM)

CCVMs, also known as cavernous hemangiomas or cavernomas, are the third most common cerebral vascular malformations following capillary telangiectasia and developing venous abnormalities. Symptomatic patients with cerebral cavernous malformations are typically between 40 and 60 years old. Most patients have single lesions, but multiple lesions can occur in familial multiple cavernous malformation syndrome. Additionally, these malformations are sometimes observed after cerebral irradiation, along with capillary telangiectasias.

Approximately 40% of cavernous malformations are discovered incidentally during neuroimaging. They are referred to as "mixed vascular malformations" when closely associated with a developmental venous abnormality (DVA). While they are most commonly located supratentorially (about 80% of the time), cerebral cavernous malformations can occur anywhere in the brain, including the brainstem. However, solitary lesions are more common, with up to one-third of patients having more than one cavernous malformation.

For the evaluation and monitoring of cerebral cavernous malformations, brain MRI with SWI or GRE sequences is recommended. CT scans may be used within the first week of symptom onset, and MRI can be used after 1 week to assess new symptoms that may suggest cerebral hemorrhage. Detecting these lesions on CT can be challenging unless they are large and may appear as rounded, well-defined structures with calcification speckles resembling blood products. They typically exhibit little to no enhancement.

MRI is the preferred imaging modality for detecting cavernous malformations. They have a characteristic appearance on MRI described as "popcorn-" or "berry-" shaped, often surrounded by a rim of signal loss due to hemosiderin. The signal intensity on T1 varies depending on the age of the blood products, and small fluid-fluid levels may be visible. The rim of the lesion appears hypointense on T2, with internal signal intensity variations based on the age of the blood products. Blood locules with fluid-fluid levels may be seen, and surrounding edema may be observed following a recent bleed. The GRE T2*/SWI sequence is particularly useful for detecting smaller lesions, especially in patients with familial or multiple cavernous malformations. Typically, there is no enhancement on contrast-enhanced T1 images, although exceptions may exist. The Zabramski classification categorizes cavernous malformations into four types based on their MRI characteristics, considering T1 and T2 signal intensity patterns. Additionally, DSA can be used to differentiate cerebral cavernous malformations from arteriovenous malformations (AVMs) or developmental venous anomalies. Cavernous malformations do not exhibit arteriovenous shunting and are angiographically occult [1, 2].

Differential Diagnosis

When multiple cavernous malformations are present, a range of conditions that cause cerebral microhemorrhages should be considered in the differential diagnosis. These conditions may include cerebral amyloid angiopathy, characterized by numerous small foci of microhemorrhages, chronic hypertensive encephalopathy (which more commonly affects the basal ganglia), diffuse axonal injury (DAI), cerebral vasculitis, radiation-induced vasculopathy, hemorrhagic metastases, and Parry-Romberg syndrome. In cases of larger lesions, it is important to consider hemorrhagic cerebral metastases, primary brain tumors with hemorrhage (such as ependymoma or glioblastoma), or other lesions like old neurocysticercosis or infections (e.g., tuberculoma).

Questions

1. **CCVMs, the FALSE answer is:**
    A. CCVMs are less common than capillary telangiectasia and developing venous abnormalities.
    B. The majority of symptomatic patients with CCVMs are between the ages of 40 and 60.
    C. CCVMs are typically solitary lesions.
    D. The rim of the lesion appears hyperintense on T2.
    E. MRI is useful for detecting smaller lesions that may be missed by conventional spin echo sequences.

The answer is D.

The rim of the lesion appears hypointense on T2, with internal signal intensity variations based on the age of the blood products.

Case 30: Serpentine Middle Cerebral Artery (MCA) Aneurysm

Case Scenario

A 29-year-old male presented with severe headaches (Fig. 4.2).

Image Description

MRI shows a large saccular aneurysm of the right MCA. The aneurysm appears as a layered structure with alternating high and low signal intensities, with the high signal intensity layers predominantly visible on T1W imaging. The low-signal intensity layers are visible on T2 and the GRE sequences. The peripheral

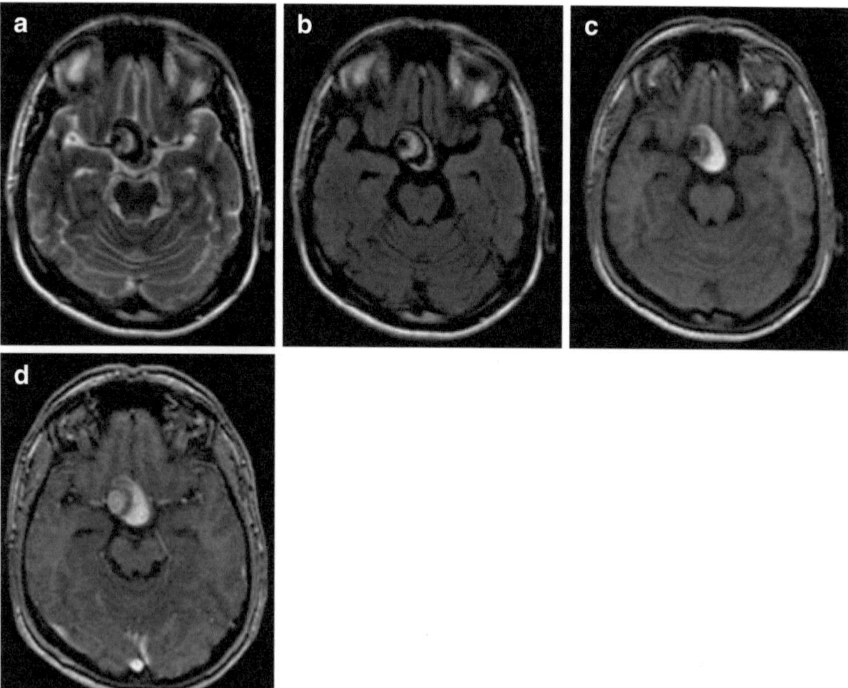

Fig. 4.2 Serial MRI of the brain with the following: (**a**) axial T2, (**b**) axial FLAIR, (**c**) axial T1, and (**d**) axial T1 C+. (Figure courtesy of Dr. Samer Hoz)

component of the aneurysm shows vivid enhancement following the administration of IV contrast. The aneurysm is causing a mass effect on the surrounding tissue.

Intracranial Serpentine Aneurysms

Serpentine aneurysms are a rare subtype characterized by their unique appearance. These aneurysms are typically large, measuring approximately 25 mm in size. They are partially thrombosed and are traversed by a patent serpiginous vascular channel within the aneurysm. Unlike giant aneurysms, this vascular channel exhibits distinct entry and exit points.

The MCA is involved in roughly half of the cases. CT can demonstrate the vascular channel within the aneurysm, along with the irregularly shaped, partially thrombosed giant aneurysm. MRI scans display variable signal intensities depending on the age of the underlying blood products, typically appearing hyperintense on T1-weighted scans and hypointense on T2-weighted scans within the partially thrombosed aneurysm. The serpentine intra-aneurysmal vascular channel's flow voids can be visualized using MRI angiographic sequences. Surrounding tissue may exhibit vasogenic edema, and susceptibility artifacts are commonly seen on T2* scans. MRA scans demonstrate the enhancing serpentine intra-aneurysmal vascular channel. DSA remains the most effective imaging modality for visualizing the inflow and outflow of the aneurysm, with the patent intra-aneurysmal tortuous vascular channel often resembling the shape of a pretzel [3, 4].

Questions

1. **Intracranial serpentine aneurysm, the FLASE answer is:**
   A. MRI scans show varying signal intensities due to the age of underlying blood products.
   B. It is comprised of a massive cerebral aneurysm (approximately 25 mm in size).
   C. The giant aneurysm is totally thrombosed in a majority of cases.
   D. It is traversed by a patent serpiginous intra-aneurysmal vascular channel.
   E. The MCA is involved in approximately half of the cases.
   The answer is C.
   The giant aneurysm is typically partially thrombosed.

Case 31: Inferior Medial Pontine Syndrome (Foville Syndrome)

Case Scenario

A 62-year-old male presented with left-sided weakness, diplopia, and right-sided facial palsy (Fig. 4.3).

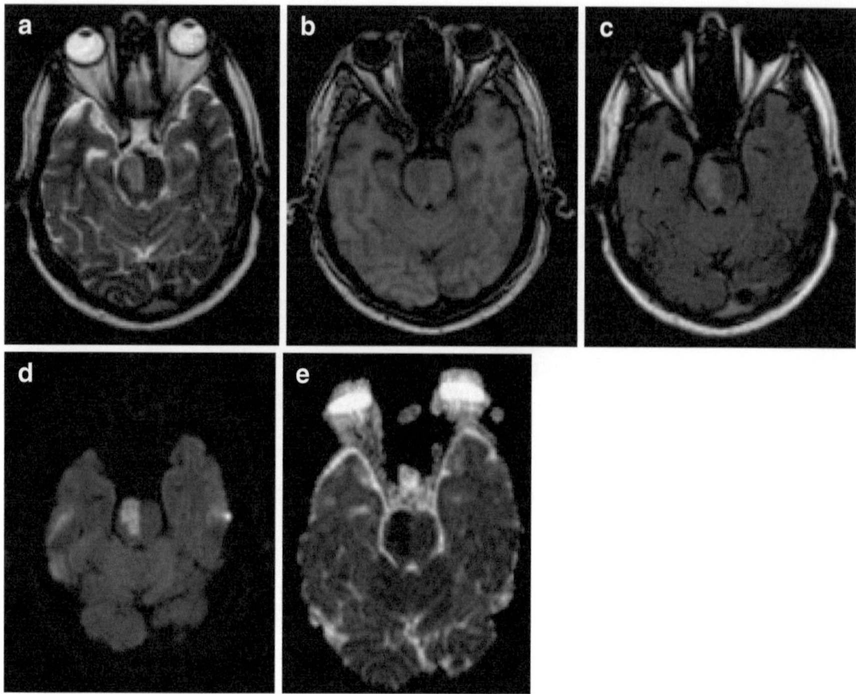

Fig. 4.3 Serial MRI of the brain with the following: (**a**) axial T2, (**b**) axial T1, (**c**) axial FLAIR, (**d**) axial DWI, and (**e**) axial ADC. (Figure courtesy of Dr. Samer Hoz)

Imaging Description

MRI shows an axial view of the brainstem, specifically the right medial inferior pons. The affected area is highlighted by an arrow pointing towards a wedge-shaped region. This region appears brighter on DWI with a low ADC signal, indicating restricted diffusion. The affected area also shows low T1 signal intensity and high T2/FLAIR signal intensity, suggesting water accumulation or edema. Additionally, there is a subtle T1 hypointensity visible in the same area.

Inferior Medial Pontine Syndrome

Foville syndrome, also known as inferior medial pontine syndrome, is a brainstem stroke syndrome resulting from the occlusion of the paramedian branches of the basilar artery, leading to infarction in the medial inferior pons. This infarction affects various structures within the brainstem, resulting in neurological deficits including contralateral hemiplegia or hemiparesis due to involvement of the cortico-spinal tract, contralateral loss of proprioception and vibration sense due to damage to the medial lemniscus, ipsilateral ataxia caused by impairment of the middle cerebellar peduncle, ipsilateral facial weakness due to damage to the CN VII nucleus, and lateral gaze paralysis with diplopia resulting from the involvement of the abducens nerve (CN VI) nucleus. In more severe cases, crossed hemihypesthesia and ipsilateral cerebellar signs may also be observed. This syndrome was initially described by the French physician Achille-Louis Foville in 1859.

CT scans are frequently used due to their accessibility and ability to identify hypodense areas associated with stroke. However, MRI remains the preferred diagnostic tool for ischemic strokes, especially for detecting acute ischemia using diffusion-weighted MRI. On MRI, Foville syndrome typically presents as a wedge-shaped region in the medial inferior pons with restricted diffusion, resulting in high DWI and low ADC signal. Additionally, it exhibits low T1 and high T2/FLAIR signal intensity, with subtle T1 hypointensity. Neurovascular imaging, such as CT or MR angiography, may be conducted in cases involving large artery obstruction, such as the basilar artery, potentially requiring mechanical thrombectomy. If embolism is suspected, further imaging studies like carotid Doppler or transthoracic echocardiography may be necessary to assess stenosis or the embolic source [5, 6].

Differential Diagnosis

The list of potential differential diagnoses encompasses various structural abnormalities that may affect the inferomedial region of the pons. These conditions comprise neoplastic growths, hemorrhages, inflammatory conditions, and vascular abnormalities like arteriovenous malformations (AVMs) or cerebral cavernous venous malformations (CCVMs).

Questions

1. **Inferior medial pontine syndrome, the FALSE answer is:**
   A. Caused by the blockage of the paramedian branches of the basilar artery.
   B. The infarction affects the corticospinal tract, causing contralateral hemiplegia/hemiparesis.
   C. The infarction affects the facial nerve nucleus, causing ipsilateral facial weakness.
   D. The clinical presentation includes gaze palsy away from the side of the lesion.
   E. The infarction affects the middle cerebellar peduncle, causing ipsilateral ataxia.
   The answer is D.
   The clinical presentation includes gaze palsy towards the side of the lesion.

Case 32: Basilar Artery and Anterior Communicating Artery Aneurysms

Case Scenario

A 55-year-old female presented with severe headaches (Fig. 4.4).

Image Description

MRI shows a sagittal view of the brain with two well-defined, round signal void lesions. The first lesion is located in the midline position, anterior to the midbrain, and appears to be causing a mild pressure effect. This lesion is consistent with a basilar tip aneurysm. The second lesion, located further inside the brain, is identified as an anterior communicating artery saccular aneurysm. Both lesions appear as dark areas on the image, indicating signal voids.

Basilar Artery and Anterior Communicating Artery Aneurysms

Basilar artery aneurysms, although less common than anterior circulation aneurysms, require careful evaluation due to their specific location, which is associated with a lower likelihood of rupture. These unruptured aneurysms in the basilar artery account for 3% of all intracranial aneurysms.

The anterior communicating artery originates from the ACA and plays a pivotal role in connecting the left and right sides of the anterior cerebral circulation. It spans approximately 4 mm in length and serves as a demarcation point between the A1 and A2 segments of the ACA. Studies utilizing angiography have reported a wide-ranging prevalence of saccular cerebral aneurysms in the asymptomatic general population, ranging from 0.2% to 8.9%. Among these aneurysms, approximately 90% are located in the anterior circulation, with the ACA-anterior communicating artery complex representing about 30–40% of them.

Radiographically, a basilar artery aneurysm may present as a rounded structure located anterior to the midbrain, often exhibiting a lobulated appearance. When a basilar artery aneurysm ruptures, the typical site of involvement is the interpeduncular cistern, but it can extend into the suprasellar cistern. CTA is preferred over plain CT scans in assessing the aneurysm and its relationship with other branches of the basilar artery. Basilar artery aneurysms can take on either a saccular or fusiform shape. When considering a potential intervention, angiography is preferred over CTA for a comprehensive evaluation of the basilar aneurysm's relationship with the branch vessels originating from the basilar artery, as this information is of utmost importance [7, 8].

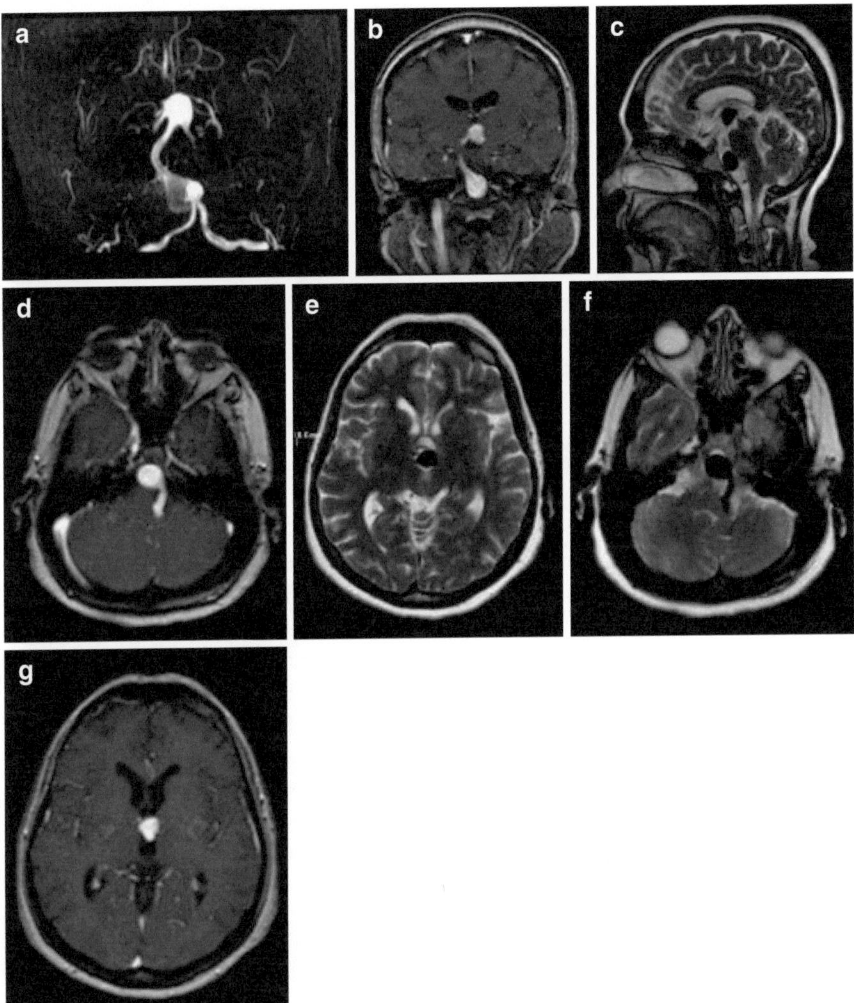

Fig. 4.4 Serial radiological of the brain with the following: (**a**) MRA, (**b**) coronal T1 C+, (**c**) sagittal T2, (**d**) axial T1 C+, (**e**) axial T2, (**f**) axial T2, and (**g**) axial T1 C+. (Figure courtesy of Dr. Samer Hoz)

Questions

1. **Basilar artery aneurysm, the FALSE answer is:**
   A. May present as a rounded structure anterior to the midbrain with a lobulated appearance.
   B. When a basilar artery aneurysm ruptures, the typical primary site of involvement is the suprasellar cistern.
   C. Can take on a saccular or fusiform shape.

 D. When considering a potential intervention, angiography is preferred over CTA.

 E. Are less frequent than aneurysms in the anterior circulation.

The answer is B.

When a basilar artery aneurysm ruptures, the typical primary site of involvement is the interpeduncular cistern, although it can extend into the suprasellar cistern.

Case 33: Cerebral Venous Thrombosis

Case 33.1: Deep Cerebral Vein Thrombosis

Case Scenario

A 31-year-old female presented with nausea, vomiting, and headache (Fig. 4.5).

Imaging Description

MRI demonstrates edema involving the bilateral thalami and basal ganglia, more on the right side. The lesion is hyper-intense on T2 and FLAIR, with some areas of hyper-intensity on T1 within the lesion reflecting hemorrhage, which is causing a mass effect within the deep gray nuclei and adjacent white matter. T2 GRE sequence demonstrates susceptibility-related signal loss consistent with hemorrhage or venous congestion, and DWI reveals focal restricted diffusion in the right thalamus with a small area of restriction at the site of the internal cerebral vein. There is no apparent enhancement with contrast, but a filling defect along the straight and internal cerebral vein can be seen in the sagittal view (arrow). These features are consistent with deep venous thrombosis of the deep internal cerebral veins, vein of Galen, and straight sinus.

Case 33.2: Cerebral Venous Sinus Thrombosis

Case Scenario

A 32-year-old female presented with right-sided weakness and dysphasia (Fig. 4.6).

Imaging Description

MRI demonstrates focal T1 and T2 hyperintensity in the left temporal lobe in a peripheral subcortical distribution, accompanied by surrounding edema. DWI demonstrates restricted diffusion, corresponding to an area of subacute hemorrhagic infarct. There is a subtle thin T2 hypointense rim and signal void on SWI (arrow) suggestive of hemosiderin deposition. The contrast study demonstrates an absence of flow in the left transverse and left sigmoid sinuses.

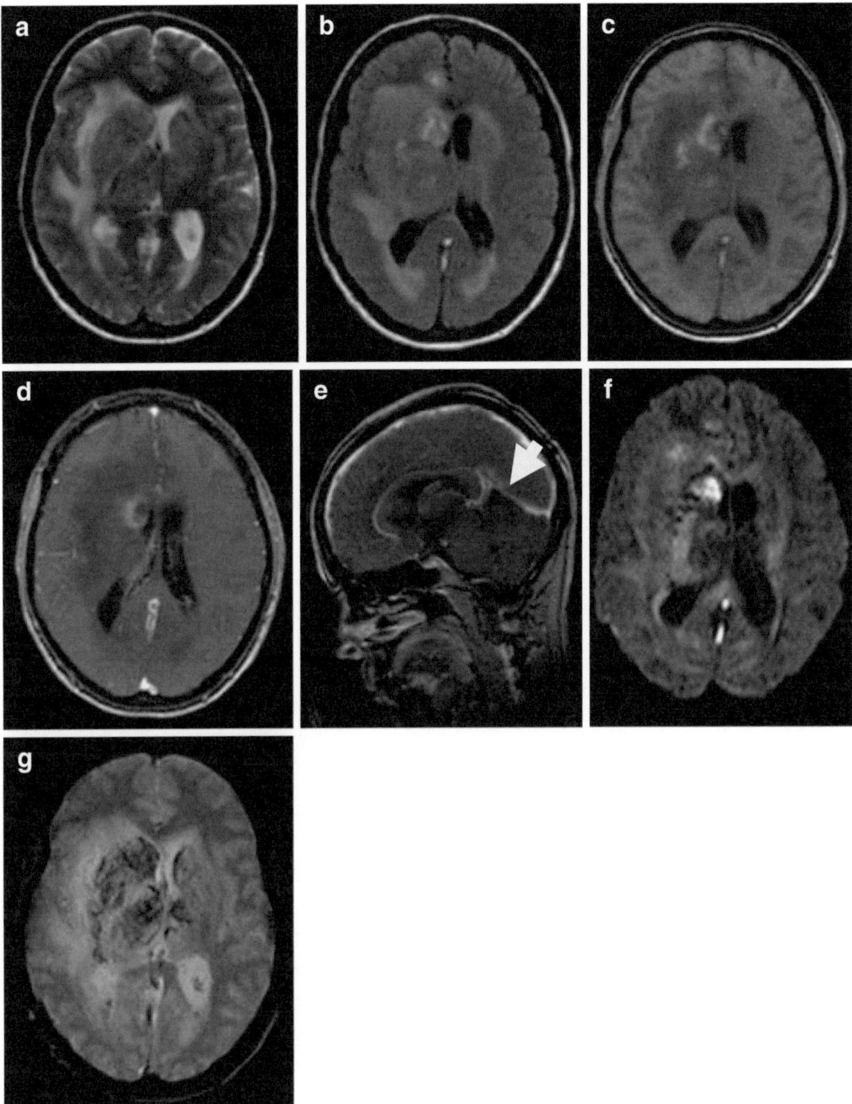

Fig. 4.5 Serial MRI images of the brain with the following: (**a**) axial T2, (**b**) axial FLAIR, (**c**) axial T1, (**d**) axial T1 C+, (**e**) sagittal T1 C+, (**f**) DWI, and (**g**) GRE. (Figure courtesy of Dr. Samer Hoz)

Deep Cerebral Vein Thrombosis

Deep cerebral vein thrombosis typically involves the internal cerebral veins and is frequently observed in conjunction with thrombosis of cortical veins or dural venous sinuses. The clinical presentation can vary based on the specific segment of the vein affected, and it can affect individuals of any age. However, it is more commonly

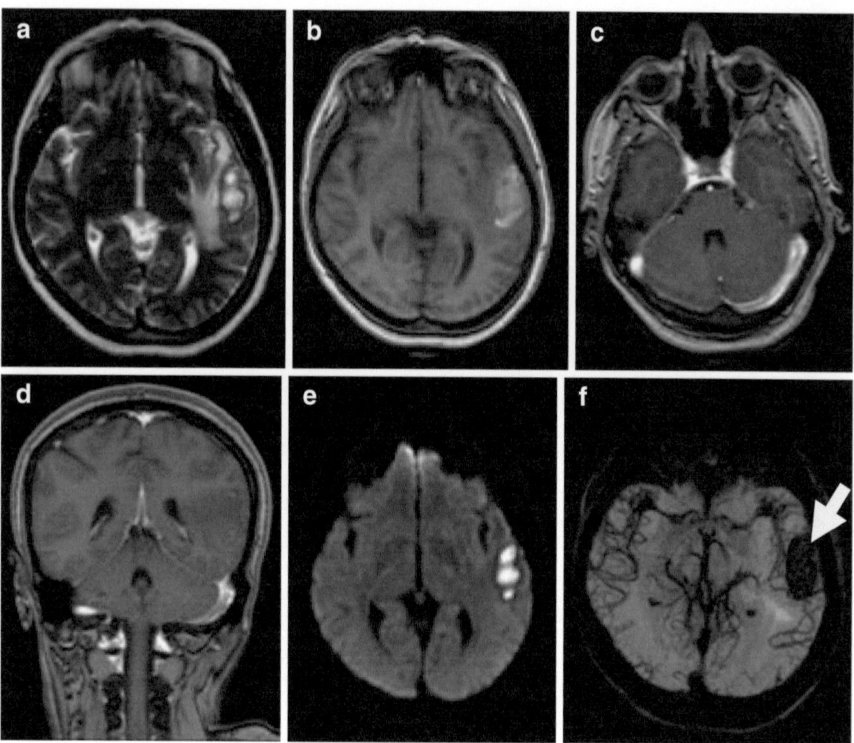

Fig. 4.6 Serial MRI images of the brain with the following: (**a**) axial T2, (**b**) axial T1, (**c**) axial T1 C+, (**d**) coronal T1 C+, (**e**) axial DWI, and (**f**) axial SWI. (Figure courtesy of Dr. Samer Hoz)

seen in elderly patients and females, particularly those in the peripartum period or those taking oral contraceptives.

Usually, this condition manifests with bilateral involvement of the thalami, which may appear edematous and could potentially exhibit signs of infarction and hemorrhage. However, there are instances of unilateral involvement, with the right side being more commonly affected than the left, though the reasons for this are not well understood. Due to the non-specific clinical symptoms associated with deep cerebral venous thrombosis, initial imaging typically involves a non-contrast CT scan, which may reveal features such as a dense clot sign, cortical edema, and peripheral hemorrhage. Additional information can be obtained from a CT venogram, which may show filling defects within the veins and sinuses.

MRI findings are influenced by the stage of hematoma development. In the hyper-acute stage (within the first day), the hematoma appears T1 iso-intense and T2 iso- to hyper-intense, with a high signal on DWI and low ADC values. In the acute stage (1–3 days), the clot remains iso- or hypo-intense on T1 but becomes hypo-intense on T2, with a low signal on DWI and reduced ADC values. During the early subacute phase (3–7 days), the signal gradually increases, becoming hyper-intense on T1 while remaining hypointense on T2. In the late subacute phase (7 to

14–28 days), the hematoma continues to appear hyper-intense on T1, and the signal on T2 gradually increases again. DWI signal becomes elevated. Chronic hematomas (beyond 14–28 days) exhibit peripheral hypo-intensity on both T1 and T2, with central regions appearing T1 iso-intense and T2 hyper-intense. ADC values become elevated.

MR venography can reveal absent flow and bilateral areas of infarction, such as in the thalami or parasagittal cortex. Additionally, signs of thrombosis, such as the dense clot sign or the cord sign, may be detectable, along with features like cortical hemorrhage and edema [9, 10].

Questions

1. **Deep cerebral vein thrombosis, the FALSE answer is:**
   A. In the hyper-acute stage, the hematoma can exhibit T2 hyperintensity.
   B. In the acute phase, the hematoma appears hypointense on T2.
   C. In the early subacute phase, the hematoma gradually becomes hypointense on T1.
   D. In the late subacute phase, the hematoma gradually becomes hyperintense on T2.
   E. Chronic hematomas are T1 isointense and T2 hyperintense in the center.
   The answer is C.
   In the early subacute phase, the hematoma gradually becomes hyperintense on T1.

Case 34: Superficial Cerebral Vein Thrombosis

Cortical Vein Thrombosis: Infarction of the Vein of Labbé

Case Scenario
A 35-year-old female presented with seizures (Fig. 4.7).

Imaging Description
There is a small abnormal hemorrhagic lesion involving the right cortical and subcortical areas of the temporal lobe. The lesion is hyperintense on T1, T2, and FLAIR. It is surrounded by edema and appears as a signal void on T2 FFE GRE, with no apparent enhancement. Magnetic resonance venography shows two small, rounded filling defects in the right transverse sinus (arrow), and the right vein of Labbe cannot be visualized. These features are suggestive of a hemorrhagic venous infarction secondary to cortical vein thrombosis.

Cortical Vein Thrombosis
Cortical vein thrombosis, also referred to as superficial cerebral vein thrombosis, involves the superficial cerebral veins apart from the dural sinus and is frequently associated with dural venous sinus or deep cerebral vein thrombosis. Cortical

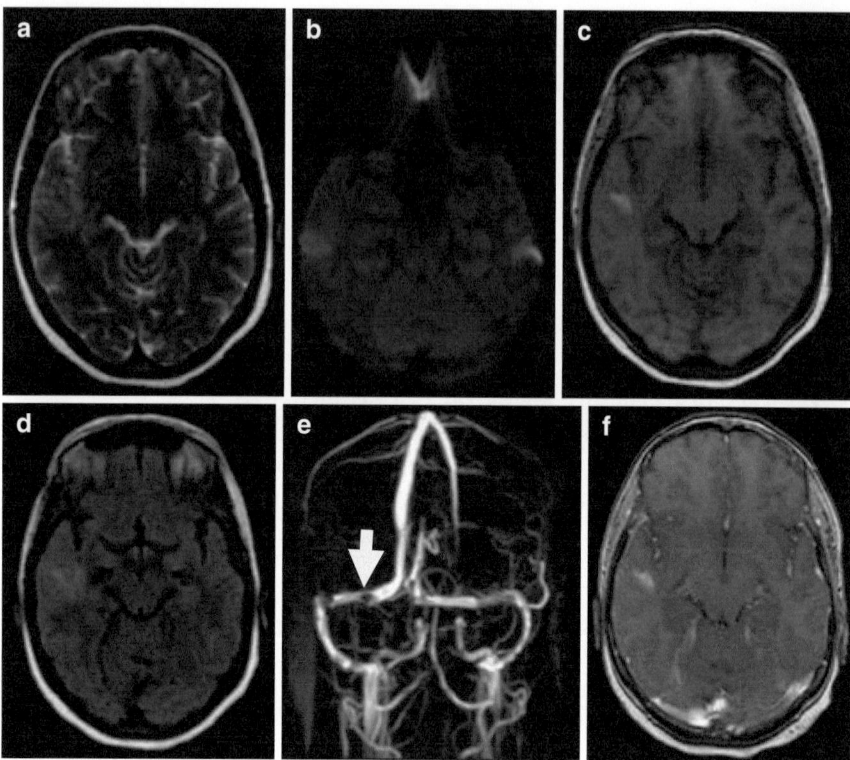

Fig. 4.7 Serial MRI images of the brain with the following: (**a**) axial T2, (**b**) axial DWI, (**c**) axial T1, (**d**) axial FLAIR, (**e**) MRV C+, and (**f**) axial T1 C+. (Figure courtesy of Dr. Samer Hoz)

cerebral veins that can be affected include the superficial middle cerebral vein, inferior anastomotic vein (a.k.a. the vein of Labbé), superior anastomotic vein (a.k.a. the vein of Trolard), and various small cortical veins. Clinical manifestations depend on the specific venous segment affected, the presence of collateral outflow, and coexisting deep cerebral vein thrombosis. Symptoms can range from being asymptomatic to seizures, coma, and in severe cases, death.

Due to the non-specific symptoms observed during presentation, the initial imaging modality is typically an unenhanced CT scan. While findings are often normal, potential indicators may include the cord sign, dense vein sign, hemorrhage, and cortical edema. Contrast-enhanced CT and Multiplanar Reformation (MPR) CT venography may reveal a filling defect within a superficial vein or an adjacent venous sinus. Additionally, focal gyral enhancement may be evident. MRI is more sensitive in cases where large cortical veins are occluded, as identifying occlusions in small cortical veins can be challenging.

The vein of Labbé, the largest superficial vein on the lateral surface of the brain, courses posterior-inferiorly from the Sylvian fissure, connecting to the superficial middle cerebral vein and extending to the anterolateral part of the transverse sinus.

Its presence varies widely, with reported frequencies ranging from 25% to 97% across different publications and imaging techniques. Approximately 60% of cases exhibit the vein in the mid-temporal region, while 30% are found in the posterior temporal region, and 10% are located in the anterior temporal region. The vein's anatomy also exhibits variation, ranging from a single dominant channel to branching channels and even venous lakes. The vein of Labbé serves to collect tributaries from the medial, anteroinferior, and posteroinferior temporal lobes, as well as draining the adjacent brain region. During temporal lobectomy for intractable epilepsy, preserving the vein of Labbé is crucial, often necessitating the retention of some adjacent cortical tissue, particularly when the vein traverses the anterior temporal region [11, 12].

Questions

1. **Infarction of the vein of Labbé, the FALSE answer is:**
   A. The vein of Labbé connects the superficial middle cerebral vein to the transverse sinus.
   B. Unenhanced CT is usually normal in cortical vein thrombosis.
   C. The vein of Labbé is found in the anterior temporal region in the majority of cases.
   D. Effort should be taken to preserve the vein of Labbé during temporal lobectomy.
   E. The cord sign may be seen on unenhanced CT in cortical vein thrombosis.
   The answer is C.
   The vein of Labbé is found in the mid-temporal region in the majority of cases.

Case 35: Moyamoya Disease

Case Scenario

A 12-year-old female with a history of fever and chest infection presented with seizure and ischemic-like symptoms (Fig. 4.8).

Imaging Description

There are multiple bilateral abnormal lesions seen in the frontal lobe involving subcortical and periventricular deep white matter and extending to the centrum semiovali (more on the right side). The lesions are hyper-intense on DWI due to cytotoxic edema with restricted diffusion, hyper-intense on T2 and FLAIR, and hypo-intense on T1 with mild mass effect, with no enhancement seen in the large lesion on the right. CTA shows multiple small stenoses at the origin of MCA, more on the right side (arrow).

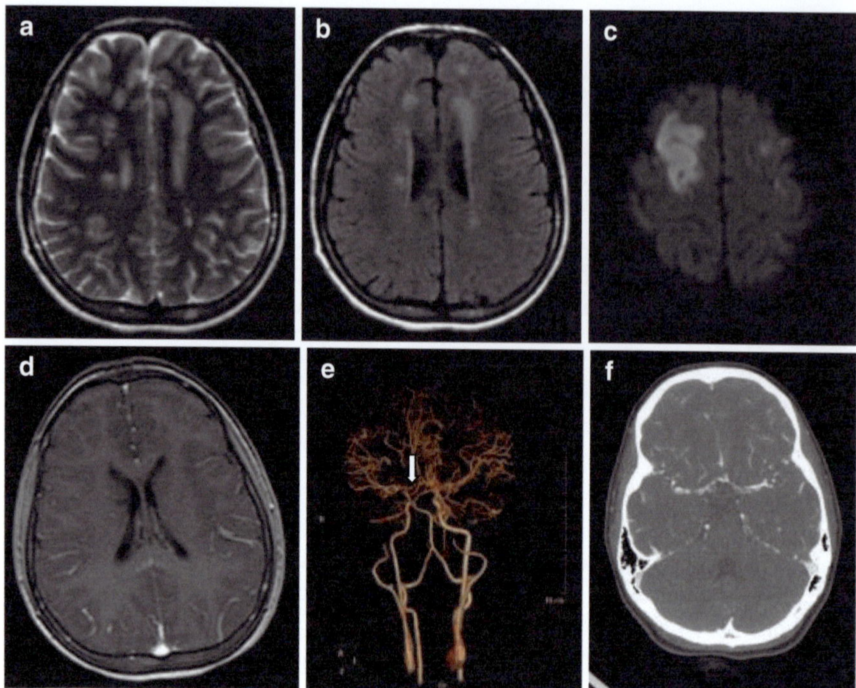

Fig. 4.8 Serial MRI images of the brain with the following: (**a**) axial T2, (**b**) axial FLAIR, (**c**) axial DWI, (**d**) axial T1 C+, (**e**) CTA, and (**f**) axial CTA. (Figure courtesy of Dr. Samer Hoz)

Moyamoya Disease

Moyamoya disease is a progressive vaso-occlusive condition affecting the circle of Willis and terminal internal carotid artery (ICA), characterized by non-atherosclerotic and non-inflammatory pathology. Its exact cause remains unknown, but it is observed to have a familial component in approximately 7–10% of cases, with a higher prevalence among individuals in Japan. While there are other medical conditions with underlying atherosclerotic or inflammatory mechanisms that may resemble moyamoya disease, they are categorized as moyamoya syndrome, not moyamoya disease.

The disease demonstrates a bimodal age distribution, with peaks occurring in early childhood around the age of four and then again in middle age, typically between 30 and 40 years old. While most cases involve bilateral distal ICA and circle of Willis, up to 18% may manifest with unilateral involvement. On direct angiography, the hallmark sign is the "puff of smoke," representing an abnormal network of proliferating vessels. This characteristic finding is not always visible on CTA and MRA. Over 50% of cases exhibit involvement of the posterior cerebral arteries, and patients commonly display generalized cerebral atrophy and watershed infarcts. Older adults and individuals with normalized perfusion through collateral circulation, such as those who have undergone long-term post-bypass surgery, are at a higher risk of cerebral hemorrhage. Several sources can contribute to collateral circulation in moyamoya disease, including:

1. Abnormal moyamoya vessels (such as lenticulostriate, thalamoperforating, leptomeningeal, and dural arteries), which appear as tortuous flow voids on T1 and T2 images.
2. Vessels less affected by moyamoya disease, particularly the posterior cerebral arteries, can give rise to pial collaterals, visible as the "ivy sign." This sign presents as high serpentine sulcal FLAIR signal intensity due to slow flow and a high signal on post-contrast T1 MRI.
3. Multiple microbleeds and prominent deep medullary veins, seen as the "brush sign" on SWI, are also indicative of collateral circulation.
4. Dural branches, such as those from the middle meningeal artery.

MRI aids in the diagnosis by revealing arterial stenosis (best visualized with MRA) and enlarged perforators, which are most evident on T2. Vessel wall imaging is particularly useful in differentiating moyamoya disease from moyamoya syndrome with atherosclerotic or inflammatory causes. In moyamoya disease, vessel wall imaging typically displays concentric luminal narrowing with a homogeneous T2 signal and minimal or no enhancement. This is distinct from atherosclerosis, where outward remodeling of vessels occurs, resulting in a greater area of the outer wall in the occluded segment compared to the normal proximal segment [13, 14].

Differential Diagnosis

The possible diagnoses to consider include the effects of cranial radiotherapy, atherosclerosis, phakomatoses like NF1 and TS, infections such as tuberculous meningitis, bacterial leptomeningitis, and post-varicella vasculitis, CNS vasculitides, connective tissue disorders like systemic lupus erythematosus, as well as hematological disorders including sickle cell disease, antiphospholipid syndrome, and aplastic anemia.

The Suzuki staging system, initially described by Suzuki and Takaku in 1969, remains in use today. This system is divided into six stages and traditionally relies on findings from conventional angiography for staging, although recent efforts have explored the application of MRA findings in staging. In children, the Suzuki stage tends to correspond with the extent of collateralization, although this correlation is not as consistent in adults. Most patients progress through one or more of the Suzuki stages, albeit at varying rates, with children often experiencing more rapid progression compared to adults.

Questions

1. **Moyamoya disease, the FALSE answer is:**
   A. MRI demonstrates arterial stenosis, best visualized on MRA.
   B. Vessel wall imaging helps differentiate moyamoya disease from moyamoya syndrome caused by atherosclerosis or inflammation.
   C. Outward remodeling of vessels is observed in moyamoya disease, similar to atherosclerosis.
   D. The Suzuki staging system, based on angiography findings, is commonly used for moyamoya disease.
   E. Children tend to progress through the Suzuki stages at a faster rate compared to adults.
   The answer is C.

   Unlike in cases of atherosclerosis, there is no outward remodeling of the vessels, where the area of the outer wall of the occluded segment is greater than that of the normal proximal segment.

Case 36: Anterior Inferior Cerebellar Artery (AICA) Loop Type I to II

Case Presentation

A 57-year-old female presented with left-sided tinnitus (Fig. 4.9).

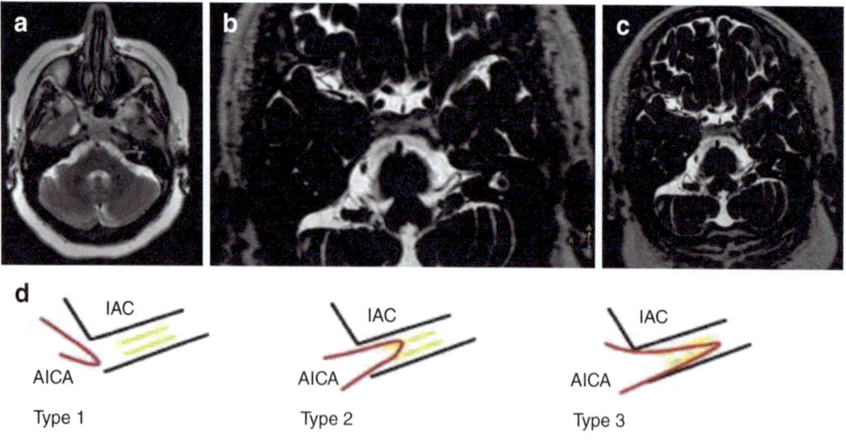

Fig. 4.9 Cranial MRI images with the following: (**a**) axial T2, (**b, c**) axial T2 drive, and (**d**) diagram. (Figure courtesy of Dr. Samer Hoz)

Imaging Description

The left anterior inferior cerebellar artery has a vascular loop that is located close to the left IAC (but does not extend beyond 50% of its length) and compresses the bundles of the CN VII and CN VIII, as seen on the axial T2 and 3D-FIESTA MRI images.

AICA Loop

An AICA loop refers to an alteration in the course of the anterior inferior cerebellar artery (AICA), where it passes through the internal auditory canal (IAC) and loops over the cranial nerves CN VII and CN VIII. This condition is associated with symptoms related to hearing, balance, and facial function due to vascular compression.

Chavda proposed a simplified method for classifying AICA loops:

- Type I: Situated within the cerebellopontine angle (CPA) without crossing over into the IAC.
- Type II: Extending into the IAC but not extending beyond 50% of its length.
- Type III: Extending into the IAC and extending beyond 50% of its length.

A prospective study involving individuals with tinnitus did not find evidence supporting the idea that vascular compression by the AICA is a causative factor for tinnitus development. Furthermore, other research studies did not reveal any association between the Chavda categorization and symptoms related to the ear and nervous system [15, 16].

Questions

1. **AICA loop, the FALSE answer is:**
   A. The AICA loop is a deviation in the route of the AICA that crosses through the IAC and wraps over the CNVII and CNVIII.
   B. Chavda provided a straightforward approach for categorizing AICA loops, including Type I, Type II, and Type III.
   C. Type II AICA loops extend into the IAC but not beyond 50% of its length.
   D. The vascular compression from AICA is the cause for developing tinnitus.
   E. The Chavda categorization of AICA loops is based on the presence and severity of otoneurologic symptoms.
   The answer is D.
   A prospective study of patients with tinnitus failed to establish vascular compression from AICA as a cause for developing tinnitus.

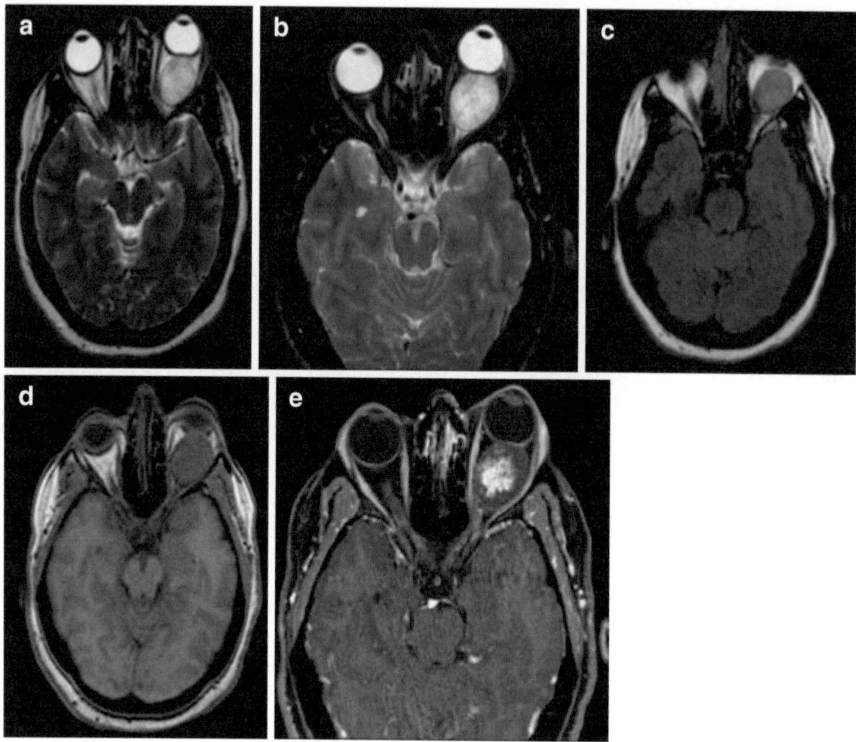

Fig. 4.10 Cranial axial MRI images with the following: (**a**) axial T2, (**b**) axial STIR, (**c**) axial FLAIR, (**d**) axial T1, and (**e**) axial T1 C+. (Figure courtesy of Dr. Samer Hoz)

Case 37: Orbital Cavernous Venous Malformation

Case Presentation

A 44-year-old female presented with left sided proptosis (Fig. 4.10).

Imaging Description

MRI reveals a well-defined oval-shaped mass in the left side of intraconal compartment, located lateral to the optic nerve. The ocular globe is displaced anteriorly by the mass, and the optic nerve is compressed and displaced medially. The signal intensity on T1 is intermediate, while on T2 and FLAIR it is high. Post-contrast sequences reveal increasing enhancement. Histopathological examination confirmed the diagnosis of cavernous hemangioma.

Cavernous Venous Malformations of the Orbit

Also referred to as cavernous hemangiomas, cavernous venous malformations are the most common vascular conditions affecting the orbits in adulthood, constituting 5–7% of all orbital tumors. Nevertheless, there is an ongoing debate regarding whether these lesions should be categorized as tumors. They typically become apparent between the ages of 30 and 50, and there appears to be a higher incidence in women than in men.

While cavernous venous malformations can occur anywhere in the body, including the orbit, over 80% of them are located within the intraconal region, most frequently on the lateral aspect (as in this case). Despite their tendency to encircle the globe, they typically exhibit a circular or oval cross-sectional shape. They do not distort the globe itself; instead, the soft consistency causes a deformation of these lesions. In the case of large lesions, they may be associated with the enlargement of the bony margins of the orbit. Ultrasound examination reveals a retrobulbar lesion with a smooth contour and uniform, moderate-to-high internal echogenicity, without detectable blood flow on Doppler scans.

On CT, cavernous hemangiomas appear as small, well-defined, round or oval masses with soft tissue density. As they increase in size, their soft composition can lead to distortion. They are slightly hypoattenuating compared to muscle and exhibit gradual and incomplete enhancement after contrast administration. Typically, the orbital apex remains unaffected. Sclerosing hemangiomas may demonstrate calcifications.

The morphological presentation on MRI closely resembles that on CT, with varying signal intensities as follows:

- T1 appears similar in signal intensity to muscle, and hyperintense regions may be evident if there are areas associated with thrombosis.
- T2 exhibits higher signal intensity compared to muscle and may display low-intensity septations and a low-intensity pseudocapsule.
- With T1 post-contrast (Gd) imaging, there is irregular, slow, gradual enhancement, followed by a delayed washout pattern.

Hemangiomas do not show up clearly on angiograms because their enhancement occurs very gradually [17, 18].

Differential Diagnosis

The potential diagnoses differ depending on the location, but they primarily consist of orbital vascular abnormalities and a few non-vascular cancers. The following conditions are part of the potential diagnoses for the more widespread intra-conal variety:

1. Optic nerve meningioma
2. Orbital schwannoma
3. Hemangiopericytoma sclerosing hemangioma (a variant rather than a distinct entity)
4. Orbital metastases
5. Orbital fibrous histiocytoma
6. Orbital lymphoma
7. Orbital venous varix
8. Capillary hemangioma of orbit

The potential diagnoses also include the following if the condition is extraconal:

1. Lacrimal gland tumors
2. Schwannoma
3. Orbital metastases
4. Orbital fibrous histiocytoma
5. Orbital lymphoma
6. Orbital venous varix

Questions

1. **Cavernous hemangiomas, the FALSE answer is:**
   A. They are circular or oval in the cross section.
   B. There is no flow seen on Doppler.
   C. Hypointense areas in T1 MRI may indicate thrombosis.
   D. Contributes to 5–7% of all orbital tumors.
   E. Differential diagnosis varies according to the site of the lesion.
   The answer is C.
   In T1 MRI, hyperintense areas may be seen if there are areas affected by thrombosis.
2. **Cavernous hemangiomas, the FALSE answer is:**
   A. Cavernous venous malformations are the most frequent vascular diseases of the orbits in adulthood.
   B. Women are more likely to develop cavernous venous malformations than men.
   C. More than 80% of cavernous venous malformations are found in the intraconal area of the orbit.
   D. Cavernous venous malformations do not deform the globe due to their soft consistency.
   E. Cavernous venous malformations show regular, moderate-to-high inner echogenicity on ultrasound.
   The answer is E.
   Cavernous venous malformations do not show flow on Doppler scanning, which means they do not exhibit echogenicity on ultrasound.

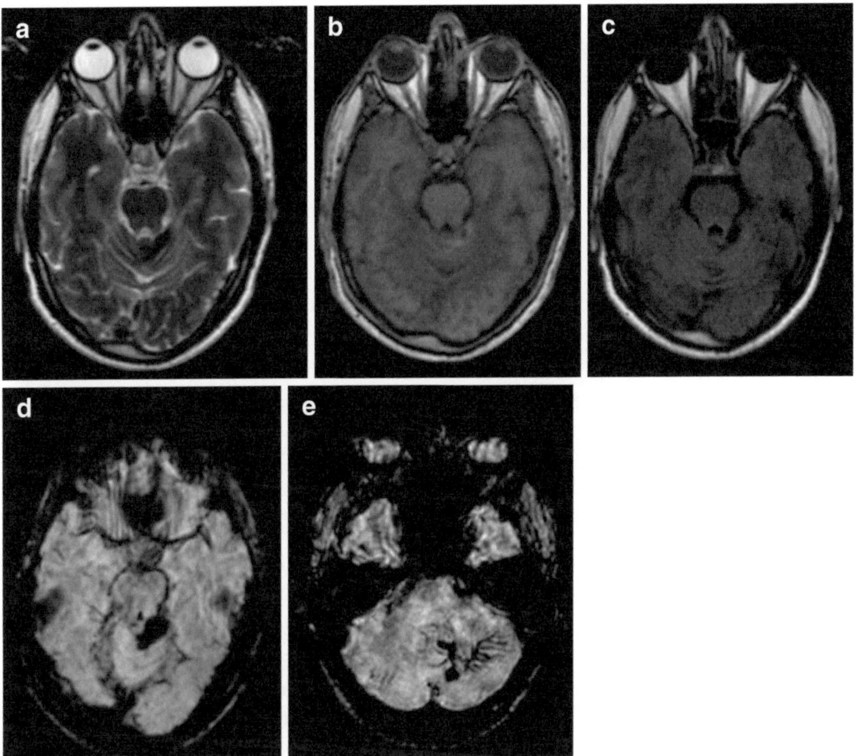

Fig. 4.11 Cranial axial MRI shows the following: (**a**) axial T2, (**b**) axial T1, (**c**) axial FLIAR, (**d**) GRE, and (**e**) axial GER. (Figure courtesy of Dr. Samer Hoz)

Case 38: Developmental Venous Anomaly (DVA) with Cavernoma

Case Scenario

A 34-year-old male presented with severe occipital headache (Fig. 4.11).

Imaging Description

MRI shows that on the left side of the cerebellum, there is an abnormal tangle of small, curvilinear, brilliantly branching blood capillaries, with a prominent draining vein pointing in the direction of the arrow (caput medusa sign). The surrounding white matter has a high signal area on T1 and a low signal area on T2 and FLAIR due to intracerebellar hemorrhage, with a small area of gliosis adjacent to them. There is minimal mass effect and peripheral blooming artifact on FFE GRE sequence. These MRI features are consistent with a DVA in the left cerebellum associated with hemorrhagic cavernoma.

Developmental Venous Anomaly (DVA)

DVA, also referred to as cerebral venous angioma, is a congenital abnormality affecting the cerebral veins. Prior to the advent of cross-sectional imaging, these abnormalities were believed to be rare. However, they are now recognized as the most common type of cerebral vascular malformation, accounting for approximately 55% of all such anomalies.

DVA is characterized by a distinctive appearance known as the "caput medusa head sign," which involves smaller veins converging into a larger collecting vein that subsequently drains into either a deep ependymal vein or a dural sinus. This arrangement has also been likened to the appearance of a palm tree. Isolated DVAs rarely result in bleeding, and the presence of intraparenchymal hemorrhage often suggests the coexistence of an associated cavernoma. Solitary DVAs seldom bleed, and detection on many MRI sequences may be subtle. The most sensitive sequences for identifying developmental venous abnormalities are T1-weighted with contrast enhancement and SWI.

DVAs are most commonly located in the frontoparietal region (36–64%), where they typically drain toward the frontal horn of the lateral ventricle, as well as in the cerebellar hemisphere (14–27%), where they drain toward the fourth ventricle. However, DVAs can be found in various locations and may have both deep and superficial drainage patterns. With the exception of cases related to blue rubber bleb nevus syndrome, DVAs are typically solitary (75%). Approximately 20% of cases involve mixed vascular malformations, often in conjunction with cavernous malformations. Instances of cortical dysplasia and venous abnormalities affecting the head and neck are rare.

On CT and MRI scans, developmental venous abnormalities appear as a cluster of vessels converging into a central vein. In larger DVAs, the draining vein may be visible on non-contrast CT scans and becomes more distinct with the administration of contrast, often appearing as a linear or curved enhancing structure. In some cases (up to 9.6%), dystrophic calcifications may be observed, most commonly in the thalami and basal ganglia, resulting in unilateral calcification.

Angiography reveals the characteristic "caput medusa" appearance, where dilated medullary veins converge into an enlarged transcortical or subependymal collector vein, which is only evident in the venous phase. The arterial phase typically appears normal, and there is no shunting observed. While developmental venous abnormalities can often be identified on various MRI sequences, they may be subtle and are best visualized on post-contrast T1-weighted sequences and SWI. SWI, in particular, offers enhanced sensitivity when coexisting cavernous hemangiomas are present. Additionally, around 10% of cases exhibit a strong T2/FLAIR signal in the adjacent white matter, though the cause remains unclear and could be attributed to gliosis, edema, or leukoaraiosis. SWI is the preferred sequence for detecting venous structures within these anomalies compared to standard T2-weighted imaging.

It is worth noting that isolated developmental venous abnormalities have a relatively low risk of complications (0.15% annually). This is primarily due to the spontaneous thrombosis of the collecting vein, which can lead to venous infarction and bleeding [19, 20].

Differential Diagnosis

The usual presentation is quite typical, and the diagnosis is straightforward. However, there are instances where the imaging appearance can deviate from the norm or be complicated by concurrent medical conditions (e.g., hemorrhage). In such situations, it is essential to take into account the following possible alternative diagnoses:

1. AVM
2. Dural sinus thrombosis or dural arteriovenous fistula with collateral transparenchymal drainage
3. Leptomeningeal angiomatosis
4. Demyelination—may also have enlarged medullary vein

Questions

1. **DVA, the FALSE answer is:**
   A. Occurs most frequently in the frontoparietal region.
   B. Characterized by a group of dilated medullary veins converging into a widened transcortical or subependymal collector vein.
   C. Has a relatively low complication rate.
   D. Best visualized on T2 MRI.
   E. Are typically solitary.
   The answer is D.
   Developmental venous abnormalities are best visualized on T1 C+ and SWI.
2. **DVA, the FALSE answer is:**
   A. DVAs are the most prevalent cerebral vascular malformation, accounting for approximately 55% of all such lesions.
   B. DVAs exhibit the caput medusae head sign.
   C. Solitary DVAs rarely bleed.
   D. The second most common location is the cerebellar hemisphere.
   E. The presence of cortical dysplasia and venous abnormalities of the head and neck is common.
   The answer is E.
   The cortical dysplasia and venous abnormalities of the head and neck are rare.

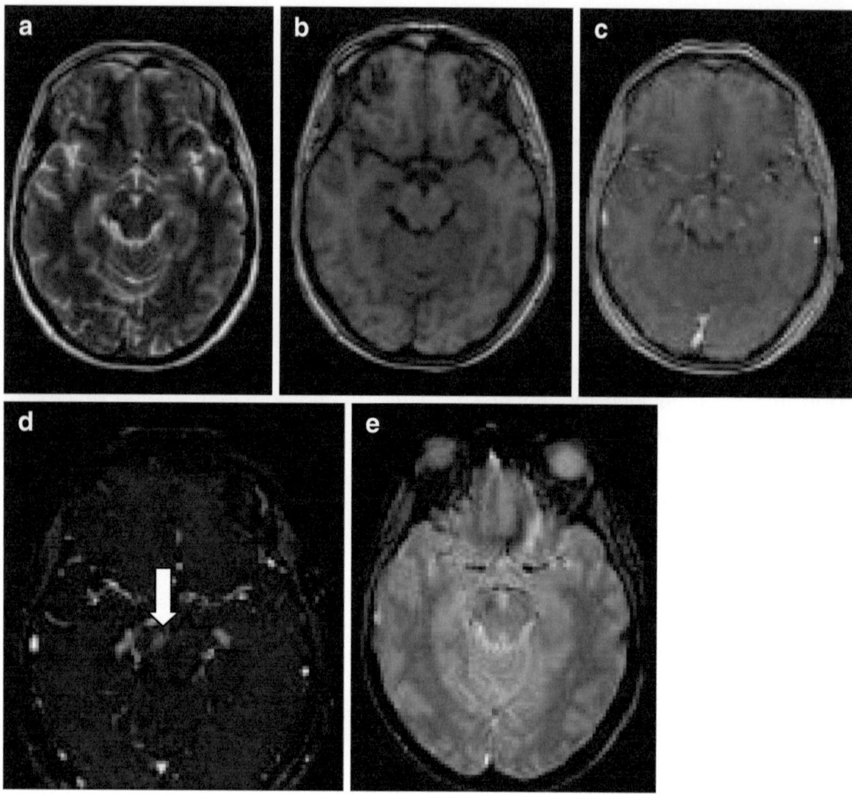

Fig. 4.12 Serial MRI images of the brain with the following: (**a**) axial T2, (**b**) axial T1, (**c**) axial T1 C+, (**d**) axial T1 C+, and (**e**) axial FFE GRE. (Figure courtesy of Dr. Samer Hoz)

Case 39: CNS Capillary Telangiectasias

Case Scenario

A brain lesion was discovered incidentally in a 38-year-old female on MRI (Fig. 4.12).

Imaging Description

The images show an axial view of a brain MRI. The abnormality is located in the anterior aspect of the midbrain, on the right side of the para midline, and does not appear to be associated with any mass effect. The abnormality appears slightly brighter than the surrounding tissue on T2W and has the same intensity as the surrounding tissue on T1W. Following administration of contrast, there is an ill-defined blush of contrast in the area (arrow), and SWI demonstrates signal dropout.

CNS Capillary Telangiectasias

CNS capillary telangiectasias are small vascular lesions within the brain that typically remain asymptomatic. They are commonly discovered during the initial MRI scans of middle-aged and older individuals and account for approximately 20% of cerebrovascular malformations identified in autopsy studies. These lesions consist of dilated capillaries intertwined with normal brain tissue and lack the presence of vascular smooth muscle or elastic fiber lining. This distinguishes them from cavernous venous malformations, which lack normal brain tissue. Capillary telangiectasias are most frequently located in the pons, cerebellum, and spinal cord and may have an association with hemorrhagic telangiectasia.

In particular, capillary telangiectasias are frequently encountered in the brainstem, notably within the pons. They can occur as solitary or multiple lesions, but they are typically not visible on CT or DSA imaging and are only detectable through MRI. When observed on MRI, they manifest as small, subtle lesions without significant mass effects. On T1 imaging, they usually appear isointense to hypointense in comparison to brain tissue, while on T2/FLAIR imaging, they exhibit iso- or slightly hyperintense signals. Occasionally, a blooming artifact may be present, which is believed to be related to deoxyhemoglobin from sluggish blood flow rather than hemorrhage. SWI, being more sensitive than T2 MRI, demonstrates low signals in all cases. If the lesion is large enough, T1 C+ (Gd) imaging may reveal stippled enhancement and branching or linear draining veins. Differential diagnoses include diffuse glioma, metastasis, resolving infarction, demyelination, cerebritis, cavernous malformation, arteriovenous malformation (AVM), and developmental venous anomaly (DVA) [21, 22].

Questions

1. **CNS capillary telangiectasias, the FALSE answer is:**
   A. Make up about 20% of cerebrovascular malformations found in autopsy studies.
   B. Capillary telangiectasias are commonly found in the brainstem, particularly in the pons.
   C. When viewed on MRI, they appear as small, subtle lesions without any significant mass effect.
   D. On T1 imaging, they are usually isointense to hyperintense compared to brain tissue, while on T2/FLAIR imaging, they appear iso- or slightly hypointense.
   E. SWI is more sensitive than T2 MRI and shows a low signal in all patients.
   The answer is D.
   On T1 imaging, they are usually isointense to hypointense compared to brain tissue, while on T2/FLAIR imaging, they appear iso- or slightly hyperintense.

References

1. Snellings DA, Hong CC, Ren AA, Lopez-Ramirez MA, Girard R, Srinath A, Marchuk DA, Ginsberg MH, Awad IA, Kahn ML. Cerebral cavernous malformation: from mechanism to therapy. Circ Res. 2021;129(1):195–215.
2. San Millán Ruíz D, Gailloud P. Cerebral developmental venous anomalies. Childs Nerv Syst. 2010;26:1395–406.
3. Xu K, Yu T, Guo Y, Yu J. Study and therapeutic progress on intracranial serpentine aneurysms. Int J Med Sci. 2016;13(6):432.
4. Christiano LD, Gupta G, Prestigiacomo CJ, Gandhi CD. Giant serpentine aneurysms. Neurosurg Focus. 2009;26(5):E5.
5. Khazaal O, Marquez DL, Naqvi IA. Foville syndrome.
6. Applebaum EL, Ferguson RL. The latero-medial inferior pontine syndrome. Ann Otol Rhinol Laryngol. 1975;84(3):379–83.
7. Kliś KM, Krzyżewski RM, Kwinta BM, Łasocha B, Brzegowy P, Stachura K, Popiela TJ, Borek R, Gąsowski J. Increased tortuosity of basilar artery might be associated with higher risk of aneurysm development. Eur Radiol. 2020;30:5625–32.
8. Zhang XJ, Gao BL, Hao WL, Wu SS, Zhang DH. Presence of anterior communicating artery aneurysm is associated with age, bifurcation angle, and vessel diameter. Stroke. 2018;49(2):341–7.
9. Oliveira IM, Duarte JÁ, Dalaqua M, Jarry VM, Pereira FV, Reis F. Cerebral venous thrombosis: imaging patterns. Radiol Bras. 2022;55:54–61.
10. Ulivi L, Squitieri M, Cohen H, Cowley P, Werring DJ. Cerebral venous thrombosis: a practical guide. Pract Neurol. 2020;20(5):356–67.
11. Moriles KE, Lockhart C. Isolated cortical venous thrombosis. In: StatPearls. StatPearls Publishing; 2023.
12. Ge S, Wen J, Kei PL. Cerebral venous thrombosis: a spectrum of imaging findings. Singapore Med J. 2021;62(12):630.
13. Li J, Jin M, Sun X, Li J, Liu Y, Xi Y, Wang Q, Zhao W, Huang Y. Imaging of moyamoya disease and moyamoya syndrome: current status. J Comput Assist Tomogr. 2019;43(2):257.
14. Lehman VT, Cogswell PM, Rinaldo L, Brinjikji W, Huston J, Klaas JP, Lanzino G. Contemporary and emerging magnetic resonance imaging methods for evaluation of moyamoya disease. Neurosurg Focus. 2019;47(6):E6.
15. Papadopoulou AM, Bakogiannis N, Sofokleous V, Skrapari I, Bakoyiannis C. The impact of vascular loops in the cerebellopontine angle on audio-vestibular symptoms: a systematic review. Audiol Neurotol. 2022;27(3):200–7.
16. Saraf A, Gupta N, Manhas M, Kalsotra P. Anterior inferior cerebellar artery loop and tinnitus—is there any association between them? Egypt J Otolaryngol. 2022;38(1):1–5.
17. Hongo H, Miyawaki S, Teranishi Y, Mitsui J, Katoh H, Komura D, Tsubota K, Matsukawa T, Watanabe M, Kurita M, Yoshimura J. Somatic GJA4 gain-of-function mutation in orbital cavernous venous malformations. Angiogenesis. 2023;26(1):37–52.
18. Zhang L, Li X, Tang F, Gan L, Wei X. Diagnostic imaging methods and comparative analysis of orbital cavernous hemangioma. Front Oncol. 2020;10:577452.
19. Althobaiti E, Felemban B, Abouissa A, Azmat Z, Bedair M. Developmental venous anomaly (DVA) mimicking thrombosed cerebral vein. Radiol Case Rep. 2019;14(6):778–81.
20. Idiculla PS, Gurala D, Philipose J, Rajdev K, Patibandla P. Cerebral cavernous malformations, developmental venous anomaly, and its coexistence: a review. Eur Neurol. 2020;83(4):360–8.
21. Larson AS, Flemming KD, Lanzino G, Brinjikji W. Brain capillary telangiectasias: from normal variants to disease. Acta Neurochir. 2020;162:1101–13.
22. Hart BL, Mabray MC, Morrison L, Whitehead KJ, Kim H. Systemic and CNS manifestations of inherited cerebrovascular malformations. Clin Imaging. 2021;75:55–66.

Infectious Diseases of the Brain

5

Ali Q. Al-asady, Khalid M. Alshuqayfi, Mahmood H. AlObaidy, Fatimah O. Ahmed, Osama S. Idris, Ahmed Muthana, and Asmaa H. AL-Sharee

Case 40: Encephalitis

Case Scenario

A 15-year-old male presented with fever, headaches, and a decreased level of consciousness (Fig. 5.1).

Imaging Description

In both T2W and FLAIR sequences, there is bilateral asymmetrical enlargement and hyper-intensity involving the underlying white matter of temporal lobes (arrow), cortex, the peri-sylvian region, and the inferior frontal lobe more on the left side. There is also restricted diffusion in DWI/ADC and leptomeningeal enhancement in T1W post-contrast. The findings on MRI suggest the diagnosis of herpes encephalitis.

A. Q. Al-asady · A. Muthana
College of Medicine, University of Baghdad, Baghdad, Iraq

K. M. Alshuqayfi
Faculty of Medicine, King Abdulaziz University, Jeddah, Saudi Arabia

M. H. AlObaidy
College of Medicine, Al-Nahrain University, Baghdad, Iraq

F. O. Ahmed
College of Medicine, University of Mustansiriyah, Baghdad, Iraq

O. S. Idris
College of Medicine, University of Mansoura, Mansoura, Egypt

A. H. AL-Sharee (✉)
Department of Neuroradiology, Neurosurgery Teaching Hospital, Baghdad, Iraq

© The Author(s), under exclusive license to Springer Nature Switzerland AG 2024
S. Hoz et al. (eds.), *Neuroradiology Board's Favorites*,
https://doi.org/10.1007/978-3-031-64261-6_5

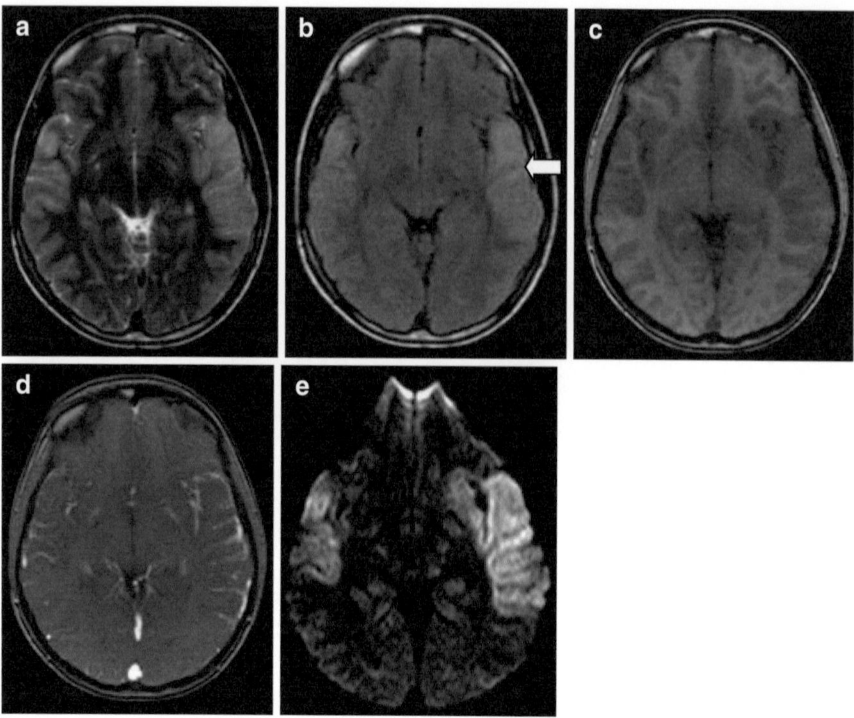

Fig. 5.1 Serial MRI images of the brain with the following: (**a**) Axial T2, (**b**) Axial FLAIR, (**c**) Axial T1, (**d**) Axial T1 C+, (**e**) Axial DWI. (Figure courtesy of Dr. Samer Hoz)

Herpes Simplex Encephalitis

The leading cause of fulminant necrotizing viral encephalitis is herpes simplex encephalitis (HSE). HSE encompasses two recognized subtypes, each differing in terms of demographics, virus type, and patterns of involvement. These subtypes are:

- Neonatal herpes encephalitis.
- Childhood and adult herpes encephalitis.

In cases of childhood and adult herpes encephalitis, approximately 90% are attributed to HSV-1, while the remaining 10% are caused by HSV-2. There is no specific predilection for age, gender, or season. In immunocompetent adult patients, the pattern of involvement typically exhibits bilateral asymmetrical lesions affecting the limbic system, medial temporal lobes, insular cortices, and inferolateral frontal lobes. Encephalitis can be distinguished from MCA infarction with the sparing of the basal ganglia. In contrast, children tend to display extra limbic involvement more frequently than adults. The parietal lobe is the most commonly affected region in children, with the basal ganglia usually spared. Over time, this involvement leads to volume loss and evident cystic encephalomalacia in affected areas.

Early diagnosis on CT scans can be challenging, and negative findings should not rule out the diagnosis. Initial features often include subtle low-density regions in the anterior and medial parts of the temporal lobe and the insular cortex (island of Reil). If the scan is performed later, these alterations may become more pronounced and may even progress to hemorrhage. During the first week of the disease, contrast enhancement is typically absent. Subtle, low-level enhancement may become visible in subsequent scans.

On MRI, affected areas exhibit consistent signal characteristics, regardless of their location. T1-weighted images may reveal generalized edema as a low signal in the affected region. In cases of subacute hemorrhage, hyper-intense signal patches may be observed on T1-weighted images. Enhancement with T1-weighted imaging typically becomes evident at a later stage of the disease, and the pattern of enhancement can vary, appearing as gyral enhancement, leptomeningeal enhancement, ring enhancement, or diffuse enhancement. Affected white matter and cortex may exhibit hyper-intensity on T2-weighted images, while hemorrhagic components may appear hypo-intense. DWI and ADC mapping are more sensitive than T2-weighted imaging. Restricted diffusion is common due to cytotoxic edema, although it is less intense than in cases of infarction. T2 shine-through resulting from vasogenic edema should also be considered. Gradient echo (GRE) or SWI may reveal blooming artifact if hemorrhage is present, although this is rare in neonates but common in older patients [1–3].

Differential Diagnosis

1. Limbic encephalitis.
2. Gliomatosis cerebri.
3. Status epilepticus.
4. MCA infarction: Typically involves the basal ganglia.
5. Trauma.
6. Viral encephalitides: It is challenging to differentiate between them based solely on clinical features. The definitive diagnosis depends on PCR testing.
7. Neurosyphilis.

Questions

1. **Encephalitis, the FALSE answer is:**
   A. During the first week of the disease, contrast enhancement is uncommon.
   B. Findings typically include hyperdensity in the anterior and medial regions of the temporal lobe and the island of Reil (insular cortex).
   C. If scanned later by CT, the alterations can be more noticeable and might potentially develop into hemorrhage.
   D. Early diagnosis is challenging by CT scan, and a "normal" scan should not dissuade from the diagnosis.

E. Immunocompromised individuals may experience more diffuse involvement.
   The answer is B.

   If present, findings typically include subtle low density in the anterior and medial regions of the temporal lobe and the island of Reil (insular cortex).

2. **Encephalitis, the FALSE answer is:**
   A. On T1, edema may be seen as an area with a low signal.
   B. T1 C+ (Gd) enhancement usually appears later in the disease progression, and the pattern of enhancement can vary.
   C. T2WI is more sensitive than DWI/ADC.
   D. Limited diffusion is common because of cytotoxic edema.
   E. Limited diffusion is less intense than in infarction.
      The answer is C.
      DWI/ADC is more sensitive than T2WI.

3. **Encephalitis, the FALSE answer is:**
   A. Encephalitis can be distinguished from MCA infarction with the sparing of the basal ganglia.
   B. Children are more likely than adults to have extralimbic involvement, which most frequently affects the parietal lobe while sparing the basal ganglia.
   C. One of the differential diagnoses includes limbic encephalitis.
   D. Affected white matter and hemorrhagic components may exhibit T2 hyper-intensity.
   E. Ninety percent of cases of childhood and adult herpes encephalitis are caused by HSV-1.
      The answer is D.
      Affected white matter and cortex may exhibit T2 hyper-intensity, and hemorrhagic components may exhibit hypo-intensity.

Case 41: Tuberculoma

Case 41.1

Case Scenario
A 10-year-old male presented with headache and vomiting (Fig. 5.2).

Imaging Description
MRI reveals well-defined, contrast-enhancing, circular ring lesions accompanied by a central contrast-enhancing nodule. The lesions are situated in the left cerebellum, vary in size, and induce prominent vasogenic edema, causing a significant mass effect on the brain stem and fourth ventricle. This results in transependymal edema and obstructive hydrocephalus. The lesion appears hypo-intense with a hyper-intense core on T2 and FLAIR imaging, with the center part of some lesions appearing restricted (arrow).

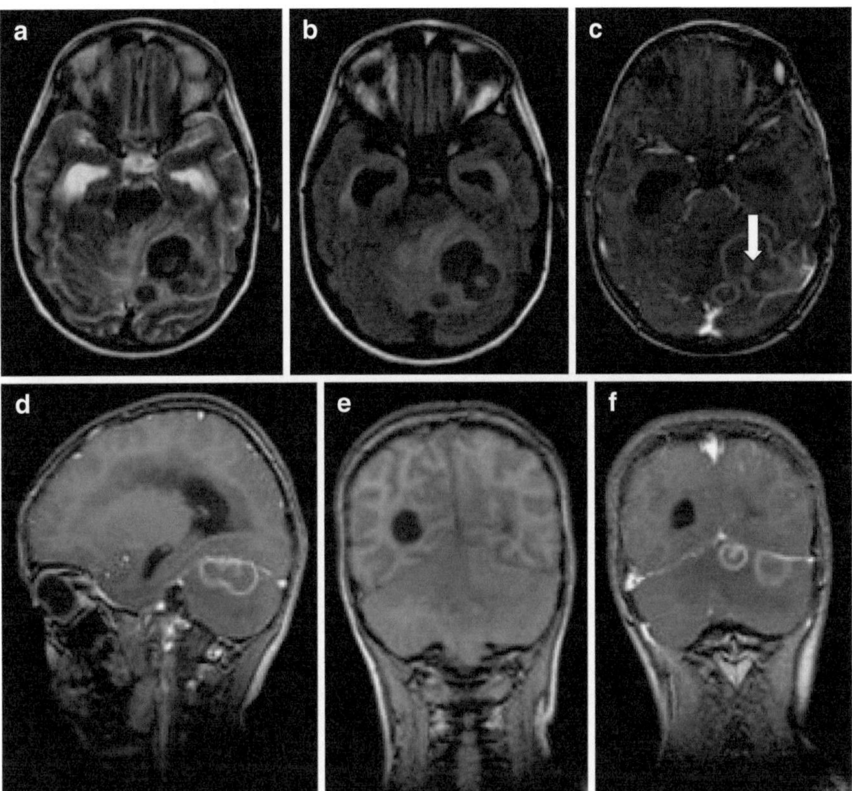

Fig. 5.2 Serial MRI images of the brain with the following: (**a**) Axial T2, (**b**) Axial FLAIR, (**c**) Axial TC+, (**d**) Sagittal T1 C+, (**e**) Coronal T1, (**f**) Coronal T1 C+. (Figure courtesy of Dr. Samer Hoz)

Case 41.2

Case Scenario
A 17-year-old female patient, previously treated for 3 weeks for a chest infection, presented with a 1-week history of recurrent convulsions (Fig. 5.3).

Imaging Description
The MRI sequences reveal multiple clusters of minor lesions intra-axially situated supratentorially in the left frontal precentral region. These lesions exhibit low signal on T1W, complex signal on T2W with patches of low/iso signal, surrounding vasogenic edema, and a signal void appearance in GRE. There is a peripheral cluster of ring enhancement of the tiny lesions (granulomas) on post-contrast enhancement sequences. A mass effect on the midline structures is noted with right side subfalcine herniation. An analysis of the histopathology confirmed the diagnosis of cerebral granuloma.

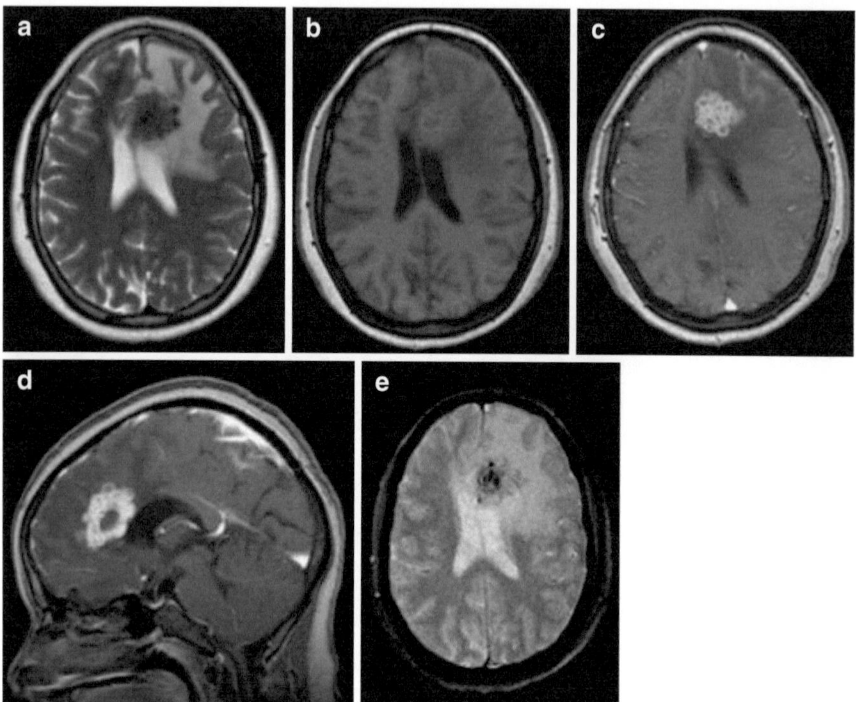

Fig. 5.3 Serial MRI images of the brain with the following: (**a**) Axial T2, (**b**) Axial T1, (**c**) Axial T C+, (**d**) Sagittal T1 C+, (**e**) Axial GRE. (Figure courtesy of Dr. Samer Hoz)

Intracranial Tuberculous Granulomas

In regions where tuberculosis (TB) is prevalent, intracranial tuberculous granulomas, also known as CNS tuberculomas, are frequently encountered. They can occur independently or concomitantly with tuberculous meningitis.

Tuberculomas can be histologically differentiated from tuberculous abscesses by the presence of a granulomatous reaction with caseous necrosis. In contrast, abscesses lack a granulomatous reaction and contain pus-filled cores. However, it is important to note that not all tuberculomas have a distinct granulomatous core, and some may undergo liquefaction.

On CT scans, tuberculomas may appear as lobulated or oval nodules accompanied by moderate to significant edema. Post-contrast enhancement typically exhibits either solid or ring-like patterns. Although the presence of centrally located calcification with a peripheral enhancing ring, known as the "target sign," has been described, it is not exclusive to TB. When calcification does occur (which is relatively rare), it tends to be more extensive than what is typically observed in neurocysticercosis.

MRI is the preferred imaging modality for identifying potential tuberculomas. The appearance can vary depending on the disease's stage, although it can often exhibit classic and highly diagnostic features such as central low T2 signal, surrounding contrast enhancement, and edema. In some cases, particularly when central liquefaction occurs, the imaging characteristics can closely resemble those of a tuberculous abscess, which shares similarities with pyogenic cerebral abscesses. Features include the following:

- Noncaseating granuloma: T1 C+ (Gd) typically shows homogeneous enhancement; T1 signal is iso- to hypo-intense; T2 signal is hyper-intense; FLAIR does not show suppression; and DWI/ADC do not indicate restricted diffusion.
- Caseating granuloma: On T1, it appears iso- to hypo-intense with a hyper-intense rim, while on T2, there is hyper-intensity representing gliosis with significant monocyte infiltration surrounded by vasogenic edema. T1 C+ (Gd) imaging can reveal homogeneous or ring-enhanced patterns. FLAIR does not suppress the signal, and there is no restricted diffusion on DWI/ADC.
- Caseating granuloma with central liquefaction: T1 signal is iso- to hypo-intense with a hyper-intense rim; T2 exhibits a hypo-intense rim with central hyper-intensity encircled by vasogenic edema. DWI/ADC may indicate variable diffusion restriction, FLAIR shows partial suppression, and T1 C+ (Gd) reveals ring enhancement.
- Calcified granuloma: It typically appears iso- to hypo-intense on T1, hypo-intense on T2, does not show FLAIR suppression, does not exhibit DWI/ADC restriction, and lacks enhancement on T1 C+ (Gd) imaging [4–6].

Differential Diagnosis

The primary differential diagnosis for tuberculomas mainly involves ring-enhanced lesions and includes:

1. Additional infections include bacterial cerebral abscesses, CNS cryptococcosis, neurocysticercosis, and cerebral toxoplasmosis.
2. Neurosarcoidosis.
3. Cerebral metastases.
4. CNS lymphoma.

Central iso-intensity or hypo-intensity on T2 can be a valuable distinguishing feature, as it is not typically found in most other causes.

Questions

1. **Intracranial tuberculous granulomas, the FALSE answer is:**
   A. A tuberculoma can be distinguished from a tuberculous abscess histologically by showing signs of granulomatous reaction with caseous necrosis.
   B. A centrally focused calcification with a peripheral enhancement ring ("target sign") may be seen but is not definitive for TB.
   C. In the rare instances where calcification does occur on a CT scan, it tends to be lower than the calcification observed in neurocysticercosis.
   D. On CT scan, post-contrast enhancement is either solid or ring-like.

   E. On CT, tuberculomas may manifest as lobulated or oval nodules with moderate to significant surrounding edema.

   The answer is C.

   When calcification appears on CT (which is rare), it is usually more pronounced than the calcification observed in cases of neurocysticercosis.

2. **Intracranial tuberculous granulomas, the FALSE answer is:**
   A. In some cases, when liquefactive necrosis occurs centrally, the imaging appearances are nearly identical to those of a tuberculous abscess.
   B. Noncaseating granulomas are iso- to hypo-intense on T1.
   C. Caseating granuloma with central liquefaction are iso- to hypo-intense with a hyper-intense rim on T1.
   D. On T2, caseating granulomas with central liquefaction have a hypo-intense rim with central hyper-intensity and are encircled by vasogenic edema.
   E. Caseating granulomas show restricted diffusion on DWI/ADC.

   The answer is E.

   Caseating granulomas do not show restricted diffusion on DWI/ADC.

3. **Intracranial tuberculous granulomas, the FALSE answer is:**
   A. When diagnosing a tuberculous abscess, detection of tuberculous organisms is required, despite the fact that they might not necessarily be identified in tuberculomas.
   B. Not all tuberculomas have a hard granulomatous core, and some may liquefy.
   C. Abscesses lack a granulomatous reaction and have pus-filled cores.
   D. Histologically, a tuberculoma cannot be distinguished from a tuberculous abscess.
   E. MRS shows a decrease in NAA/Cr, a mild decrease in NAA/Cho, and typically high levels of lipid-lactate peaks (86%).

   The answer is C.

   A tuberculoma can be distinguished from a tuberculous abscess histologically by signs of granulomatous reaction with caseous necrosis.

Case 42: Hydatid Cyst

Case 42.1: Intraconal and Cerebral Hydatid Cysts

Case Scenario

A 24-year-old female presented with right-sided proptosis, papilledema, loss of vision, and headache (Fig. 5.4).

Imaging Description

MRI reveals a multicystic lesion with a distinctly marked multivesicular appearance. There are numerous differently sized daughter cysts within the mother cyst. This lesion involves the right intraconal region, on both sides of the optic nerve. The optic nerve is compressed and the ocular globe is displaced anteriorly. Additionally, the lesion extends into the intraparenchyma at the right frontal area accompanying

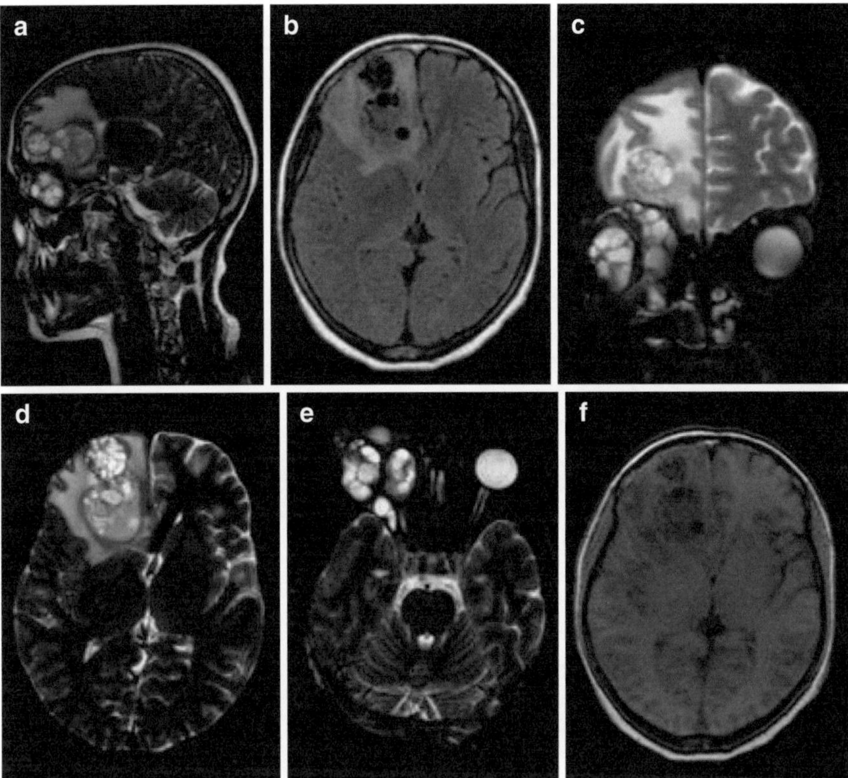

Fig. 5.4 Serial MRI images of the brain with the following: (**a**) Sagittal T2, (**b**) Axial FLAIR, (**c**) Coronal T2 STIR, (**d**) Axial T2 STIR, (**e**) Axial T2 STIR, (**f**) Axial T1. (Figure courtesy of Dr. Samer Hoz)

vasogenic edema and a noticeable mass effect. On imaging, the lesion appears as low signal on T1 and FLAIR and demonstrates a strong signal on T2, with no restriction observed in DWI. The diagnosis of hydatid cysts was confirmed through histopathological examination.

Case 42.2: Cerebral Hydatid Cyst

Case Scenario
A 20-year-old male presented with headache (Fig. 5.5).

Imaging Description
Deep white matter was affected by two sizable, clearly defined subcortical and deep white matter cyst mass lesions, with a massive mass effect in the right frontal area. No obvious intracystic features are present. No enhancement or perifocal edema is seen.

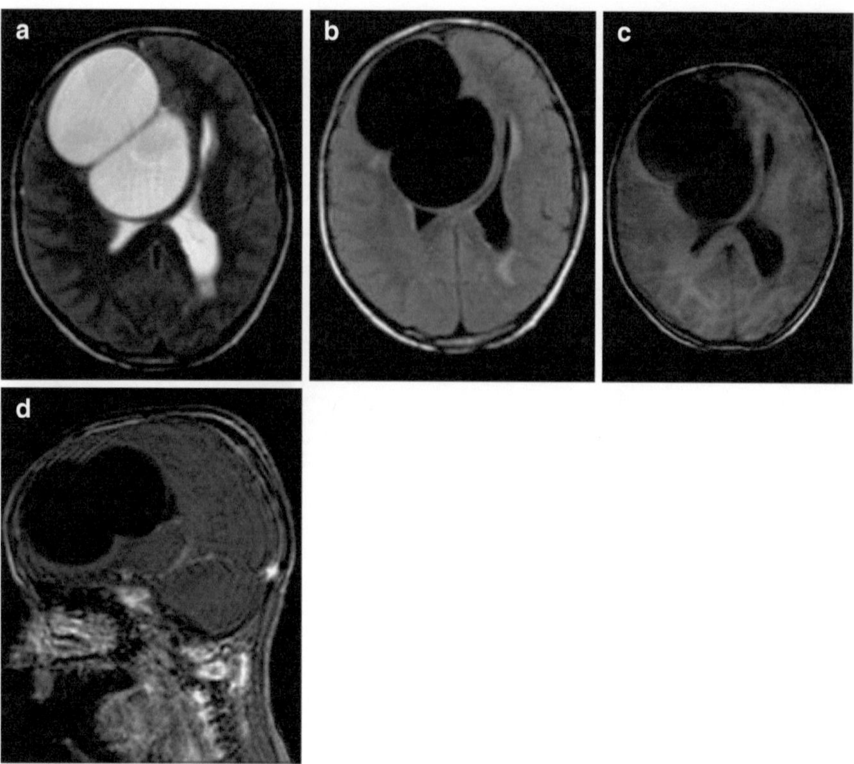

Fig. 5.5 Serial MRI images of the brain with the following: (**a**) Axial T2, (**b**) Axial FLAIR, (**c**) Axial T1, (**d**) Sagittal T1 C+. (Figure courtesy of Dr. Samer Hoz)

Cerebral Hydatid Disease (Neurohydatidosis)

Cerebral hydatid disease is most commonly caused by *Echinococcus granulosus* and less commonly by *E. alveolaris*. In humans, the larval stage triggers hydatid disease. Intracranial hydatid disease is exceedingly rare, accounting for only 1–2% of all cases of cystic echinococcosis. The majority of cerebral hydatid cysts are located in supratentorial structures within the vascular territory of the MCA. They can be categorized as follows:

1. Primary hydatid cysts originate from direct larval invasion that is filtered through the lung and liver before reaching the brain. They are typically solitary but can also occur in multiple numbers, and they have the capacity to reproduce.
2. Secondary hydatid cysts, on the other hand, are typically multiple, infertile, and lack a brood capsule. They develop when primary hydatid cysts rupture in other organs and then spread to the brain through embolization.

The cyst can be divided into four distinct types based on its morphology:

- Type I: Simple cyst with no internal architecture
- Type II: Cyst with daughter cyst(s) and matrix
- Type IIa: Round daughter cysts at the periphery
- Type IIb: Larger, irregularly shaped daughter cysts occupying almost the entire volume of the mother cyst
- Type IIc: Oval masses with scattered calcifications and occasional daughter cysts
- Type III: Calcified cyst (dead cyst)
- Type IV: Complicated cyst, e.g., ruptured cyst

MRI typically reveals a distinct, well-defined, round cystic lesion located within the MCA's territory. This lesion does not show enhancement with contrast. In all MRI pulse sequences, the cyst fluid appears similar to CSF in terms of signal intensity. There is no evidence of calcification, and there is no surrounding edema. The presence of perilesional edema usually suggests complications like secondary infection or rupture [7–9].

Differential Diagnosis
1. Pyogenic or fungal abscess.
2. Primary brain tumor, e.g., cystic astrocytoma.
3. Cystic granulomas.
4. Arachnoid cyst.
5. Porencephalic cyst.

Questions

1. **Cerebral hydatid disease, the FALSE answer is:**
   A. In humans, hydatid disease is brought on by the larval stage.
   B. Primary hydatid cysts develop as a result of a direct larval invasion that is controlled and filtered via the lung and liver before reaching the brain.
   C. The majority of cerebral hydatid cysts are found in the infratentorial region.
   D. *Echinococcus granulosus*, or less frequently *E. alveolaris*, is the culprit.
   E. Only less than 2% of all cases of cystic echinococcosis are diagnosed as intracranial hydatid disease.
   The answer is C.
   The majority of cerebral hydatid cysts are found in supratentorial structures within the region supplied by the MCA.
2. **Cerebral hydatid disease, the FALSE answer is:**
   A. A simple cyst with no internal architecture is classified as a type I cyst.
   B. Differential diagnoses include pyogenic or fungal abscess.
   C. Appears as well-defined, spherical, non-enhancing intra-axial cystic lesions on MRI.
   D. Primary hydatid cysts are usually multiple and infertile.
   E. Ruptured cysts are classified as type IV cysts.
   The answer is D.

Primary hydatid cysts are typically solitary and fertile.

Case 43: Rhombencephalitis

Case Scenario

A 19-year-old male with a history of cerebral palsy from birth presented with vomiting and severe headache (Fig. 5.6).

Imaging Description

In both T2W and FLAIR segments, there is symmetrical bilateral swelling and hyper-intensity affecting the thalami, brainstem, and cerebellum, restricted diffusion in DWI/ADC, and leptomeningeal enhancement in T1WI following contrast. These findings suggest an inflammatory process and are likely indicative of rhomboencephalitis.

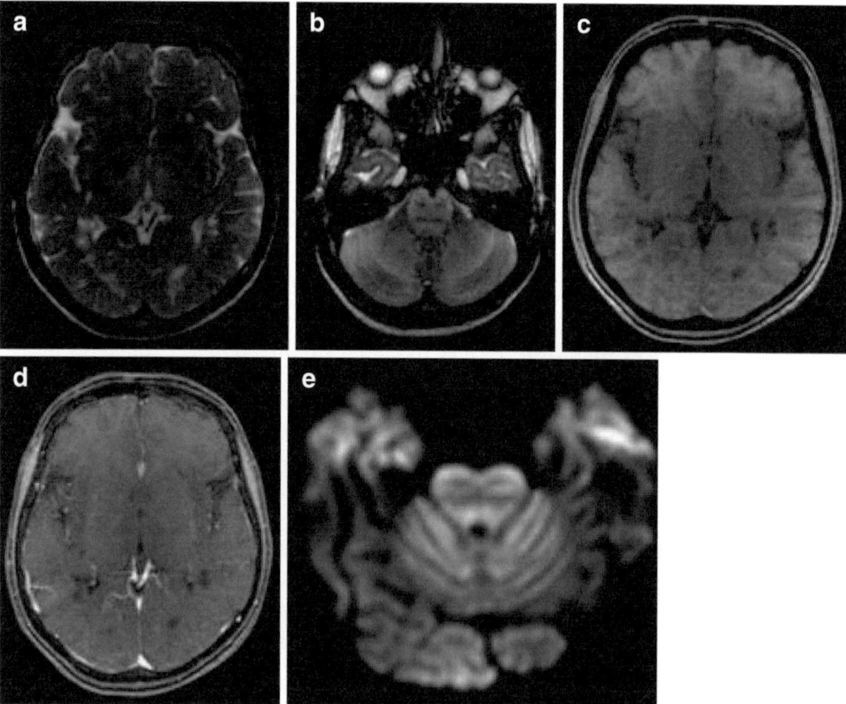

Fig. 5.6 Serial MRI images of the brain with the following: (**a**) Axial T2, (**b**) Axial T2, (**c**) Axial T1, (**d**) Axial T1 C+, (**e**) Axial DWI. (Figure courtesy of Dr. Samer Hoz)

Rhombencephalitis

The term "rhombencephalitis" pertains to inflammatory conditions affecting both the cerebellum and brainstem. These conditions have various causes, including infections, autoimmune diseases, and paraneoplastic syndromes. Among these, *Listeria monocytogenes* is the most common causative agent. Rhombencephalitis carries a significant risk of morbidity and mortality, with an estimated mortality rate ranging from 10% to 15%.

Types:

1. Infectious: *Listeria monocytogenes* is the most common bacterial cause of infectious rhombencephalitis. This is followed by *M. tuberculosis*, enterovirus 71, and herpes simplex virus.
2. Autoimmune: Systemic lupus erythematosus, multiple sclerosis, and Behçet disease.
3. Paraneoplastic syndromes: are associated with antibodies and are predominantly linked to small-cell lung cancer.
4. Malignancy: Lymphoma (rare).

Visualizing the brainstem in CT scans is hindered by beam-hardening artifacts. Therefore, when evaluating individuals suspected of having pathology in this area, the preferred imaging technique is contrast-enhanced MRI. In cases of rhombencephalitis caused by infectious agents, specific MRI findings typically include the following: on T1 images with contrast (Gd), there may be linear cranial nerve enhancement, ring enhancement in the presence of an abscess, and a heterogeneous appearance if inflammation is widespread. On T1 images, the lesions usually appear hypo- or iso-intense, while on T2/FLAIR images, they are hyper-intense. Additionally, on DWI, the lesions exhibit hyper-intensity, while on ADC images, they appear hypo-intense [10–12].

Differential Diagnosis

1. Brainstem tumor.
2. Chronic lymphocytic inflammation with pontine perivascular enhancement: This condition is responsive to steroids.

MRS can be utilized to differentiate between tumors and abscesses.

Questions

1. **Rhombencephalitis, the FALSE answer is:**
   A. T1WI with contrast may demonstrate linear cranial nerve enhancement.
   B. Lesions are hyper- or iso-intense on T1.
   C. Lesions are hypo-intense on ADC.

D. Lesions are hyper-intense on T2/FLAIR.

E. Lesions are hyper-intense on DWI.

The answer is B.

Lesions are hypo- or iso-intense on T1.

2. **Rhombencephalitis, the FALSE answer is:**

A. The term "rhombencephalitis" refers to inflammatory conditions that affect the cerebellum and brainstem.

B. *M. tuberculosis* is the most common causative agent.

C. The visualization of the brainstem on CT scans is limited by beam-hardening artifacts.

D. MRS enables the differentiation of tumors from abscesses.

E. The mortality rate is estimated to be between 10% and 15%.

The answer is B.

Listeria monocytogenes is the most common causative agent.

Case 44: Cerebral Abscess

Case Scenario

A 27-year-old male presented with seizures (Fig. 5.7).

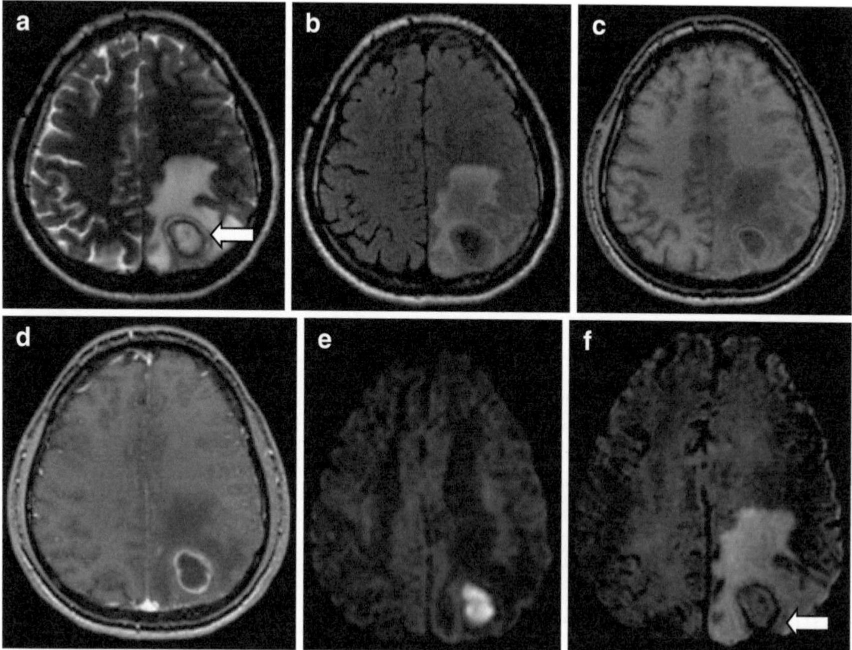

Fig. 5.7 Serial MRI images of the brain with the following: (**a**) Axial T2, (**b**) Axial FLAIR, (**c**) Axial T1, (**d**) Axial T1 C+, (**e**) Axial DWI, (**f**) Axial GRE. (Figure courtesy of Dr. Samer Hoz)

Imaging Description

MRI reveals a well-defined, rounded mass lesion in the left occipital lobe with a low signal on T1, ring enhancement, and a thin wall showing a dual rim sign on T2W/ SWI (arrow). There is a high signal in the center on DWI and a low signal on ADC. Perilesional swelling is extremely severe.

Cerebral Abscess

A cerebral abscess refers to a localized area of necrosis that can be life-threatening and requires swift medical attention. It typically originates from an initial infection in the brain parenchyma, referred to as cerebritis. In the past, the most common cause was direct extension from infections in the sinuses or scalp. However, more recently, hematological spread has become more prevalent. A very small number of cases result from surgical procedures or direct trauma. The most commonly encountered organism responsible for cerebral abscesses is *Streptococcus* spp., accounting for 35–50% of cases.

Cerebritis follows four distinct stages, each characterized by unique imaging and histopathological features:

1. Initial cerebritis: This stage represents a focal infection without the presence of a capsule or pus formation. It can either resolve on its own or progress into a full-fledged abscess and typically occurs within two to three days.
2. Late cerebritis: Following the initial infection, there is a phase known as late cerebritis, which typically occurs around one week after the onset of infection.
3. Early abscess or encapsulation: This stage may develop approximately 10 days after the initial infection, marked by the formation of an encapsulated region within the brain.
4. Late abscess or encapsulation: The late abscess or encapsulation stage can occur more than 14 days after the initial infection, characterized by the presence of a well-defined capsule around the abscessed area.

Infective endocarditis, intravenous drug use, lung infections like lung abscess, bronchiectasis, empyema, pneumonia, and sinonasal infections such as dental abscess are all risk factors associated with cerebral abscess.

Both CT and MRI can reveal similar characteristics, but MRI is more adept at distinguishing cerebral abscesses from other ring-enhancing lesions. For CT imaging, it is recommended to acquire both pre-contrast and post-contrast scans in patients suspected of having intraparenchymal sepsis, unless further MRI evaluation is planned regardless of CT results.

In most cases, cerebral abscesses appear with an outer hypodense and an inner hyperdense rim (double rim sign) but can also appear with an iso- or hyperdense ring of tissue, usually of uniform thickness. Other typical features include minimal attenuation in the center (indicating fluid or pus), low density in the surrounding area (signifying vasogenic edema), and potential findings such as ventriculitis, obstructive hydrocephalus, and ependymal enhancement, especially when intraventricular dissemination has occurred.

MRI is generally more sensitive compared to CT. While peripherally enhancing lesions may not be specific for abscess on imaging, DWI revealing restricted diffusion in the central region is essential for suggesting the diagnosis of a cerebral abscess. On T1-weighted MRI, cerebral abscesses typically appear hypo-intense in the center as well as the periphery (due to vasogenic edema). Ventriculitis and ring enhancement are also common findings, along with possible hydrocephalus. On T2-weighted and FLAIR MRI sequences, cerebral abscesses display central and peripheral hyper-intensity and an abscess capsule with a thin, intermediate-to-low signal rim. DWI and ADC maps indicate restricted diffusion within the central area of the abscess, further supporting the diagnosis [13].

Differential Diagnosis

1. High-grade glioma (e.g., GBM) or metastasis: The inner wall of abscesses typically appears smoother. The presence of satellite lesions suggests infection, and hypo-intensity in the capsule may be observed in abscesses. High-grade gliomas show elevated relative cerebral blood volume (rCBV), while abscesses exhibit decreased rCBV. Similar to abscesses, the cystic component does not display restricted diffusion, and there is no dual rim sign.
2. Subacute infarction, subacute bleeding, or contusion.
3. Demyelinating lesion.
4. Radiation necrosis: When a lesion displays both ring enhancement and centrally restricted diffusion, the list of potential differential diagnoses is highly restricted, with cerebral abscess being the most likely diagnosis. However, the radiation necrosis should also be included in the list of differentials.
5. Cerebral metastases, particularly necrotic adenocarcinoma.

Questions

1. **Brain abscess, the FALSE answer is:**
   A. CT shows a double rim sign characterized by an outer hypodense and an inner hyperdense rim.
   B. Is preceded by cerebritis.
   C. *Streptococcus* spp. is the most common causative agent.
   D. Pre- and post-contrast CT scans should be acquired in patients with suspected intraparenchymal sepsis, unless it is intended to move on to an MRI regardless of the CT results.
   E. Stage three of cerebral infection (late abscess or encapsulation) occurs >14 days after infection.
   The answer is E.

Stage three of cerebral infection (early abscess/encapsulation) occurs approximately 10 days after the infection. Stage four (late abscess/encapsulation) may manifest later, typically more than 14 days after the initial infection.

2. **Brain abscess, the FALSE answer is:**
   A. DWI demonstrating central diffusion restriction is essential for suggesting the diagnosis of a cerebral abscess over other peripherally enhancing lesions.
   B. T1WI demonstrates low intensity in the center as well as the periphery (due to vasogenic edema).
   C. T2/FLAIR demonstrates central hypo-intensity.
   D. SWI demonstrates a hypo-intense rim.
   E. On MRS, there are increased peaks for lipids/lactate, succinate, acetate, and amino acids.

   The answer is C.

   T2/FLAIR demonstrates central hyper-intensity (hypo-intense to CSF, does not attenuate on FLAIR).

Case 45: Gradenigo Syndrome (Petrous Apicitis)

Case Scenario

A 63-year-old male presented with right ear discharge, hearing loss, tinnitus, right-sided sixth cranial nerve palsy and pain (Fig. 5.8).

Imaging Description

The MRI findings reveal extensive homogeneous enhancement in the right pterygoid muscles and right retropharyngeal and parapharyngeal spaces, indicating an inflammatory process that has extended to the Meckel's cave and cavernous sinus. Additionally, there is an abnormal signal involving the petrous region with associated enhancement, suggestive of osteomyelitis. The lesion appears hypo-intense in T1 and hyper-intense in T2 and FLAIR sequences. Otitis media, mastoiditis, and pansinusitis are present on the right side.

Gradenigo Syndrome

The syndrome typically arises due to petrous apicitis, a condition caused by the spread of chronic suppurative otitis media to the petrous apex of the temporal bone. The petrous apex is in close proximity to the Dorello's canal, which houses the abducens nerve, and Meckel's cave, where the trigeminal ganglion is located. Consequently, inflammation in the extradural space resulting from petrous apicitis can affect these neighboring structures, leading to the characteristic symptoms of Gradenigo syndrome. Common pathogens associated with this condition include

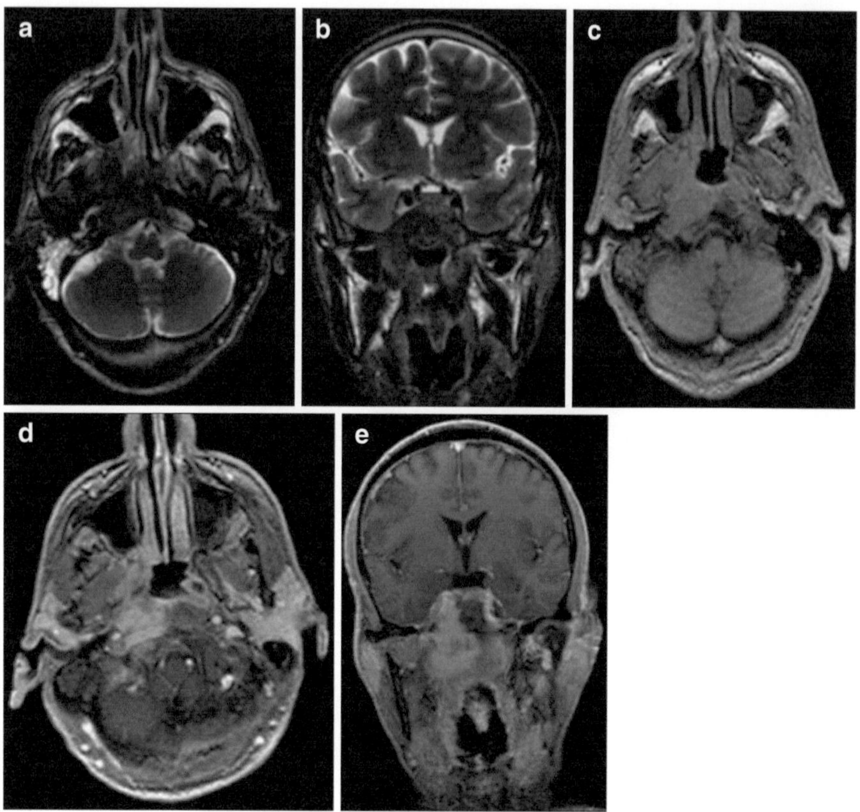

Fig. 5.8 Serial MRI images of the brain with the following: (**a**) Axial T2, (**b**) Coronal T2, (**c**) Axial T1, (**d**) Axial T1 C+, (**e**) Coronal T1 C+. (Figure courtesy of Dr. Samer Hoz)

Enterococcus spp. and *Pseudomonas* spp. Gradenigo syndrome was first identified by Italian otolaryngologist Giuseppe Conte Gradenigo in 1907.

The classic triad of symptoms for this condition includes:

1. Chronic discharge from the ear, suppurative otitis media, and earache.
2. Paralysis of the abducens nerve due to its involvement as it passes through the Dorello's canal.
3. Pain is experienced in the areas of the forehead and upper jaw, which corresponds to the cutaneous distribution of the frontal and maxillary branches of the trigeminal nerve, resulting from the extension of inflammation into Meckel's cave.

Patients also commonly experience persistent ear discharge and pain in the areas served by the ophthalmic and maxillary branches of the trigeminal nerve. Although the incidence of this condition has significantly decreased with the introduction of antibiotics, sporadic cases still occur, often diagnosed late due to the subtle nature of the symptoms.

On CT scans, potential findings include the presence of an intracranial abscess, opacification of the petromastoid air cells, and bone destruction. The preferred imaging modality for assessing temporal bone changes is CT, which can reveal erosive lysis with indistinct and irregular borders in cases of petrous apicitis. Contrast-enhanced scans may demonstrate peripheral contrast enhancement, dural thickening, and enhancement. However, MRI remains the most effective tool for detection.

MRI can also help differentiate between neoplastic and infectious processes and is superior to CT in depicting the inflammatory changes associated with petrous apicitis. The imaging typically reveals a fluid signal within the petrous apex, often with peripheral contrast enhancement, which likely reflects a combination of residual mucosa, granulation tissue, and adjacent dura and periosteum. Consequently, T1-weighted images show a low to intermediate fluid signal, T2-weighted images exhibit a hyper-intense fluid signal, and T1 post-contrast images display peripheral contrast enhancement. It is crucial to assess for dural thickening in Meckel's cave and the cavernous sinus, as well as any signs of cavernous sinus thrombosis [14, 15].

Differential Diagnosis

1. Cholesterol granuloma.
2. Congenital cholesteatoma.
3. Glomus jugulare.
4. Petrous apical mucocoele: uncommon.
5. Tumors at the base of the skull include chordoma, chondrosarcoma, and metastases.

Questions

1. **Gradenigo syndrome, the FALSE answer is:**
   A. Significantly decreased since the advent of antibiotics.
   B. T1 post-contrast images display peripheral contrast enhancement.
   C. Contrast-enhanced scans may show peripheral contrast enhancement as well as dural thickening and enhancement.
   D. The standard technique for assessing bony changes in the temporal bone is CT.
   E. CT findings could include the presence of an intracranial abscess and petromastoid air cell opacification, typically with bone preservation.
   The answer is E.
   Findings on CT include the presence of an intracranial abscess, opacification of the petromastoid air cells, and bone destruction.
2. **Gradenigo syndrome, the FALSE answer is:**
   A. T1WI shows a low to intermediate fluid signal.
   B. It is important to look for thickening of the dura in Meckel's cave and the cavernous sinus as well as any signs of cavernous sinus thrombosis.

C. MRI is superior to CT in showing the inflammatory changes of petrous apicitis.
D. Imaging typically reveals a fluid signal within the petrous apex that is accompanied by contrast enhancement in the center.
E. T2WI demonstrates a hyper-intense fluid signal.

The answer is D.

Imaging typically reveals a fluid signal within the petrous apex that is accompanied by peripheral contrast enhancement.

Case 46: Pott Puffy Tumor of the Frontal Sinus

Case Scenario

A 35-year-old male presented with a left frontal swelling with neurological deficits (Fig. 5.9).

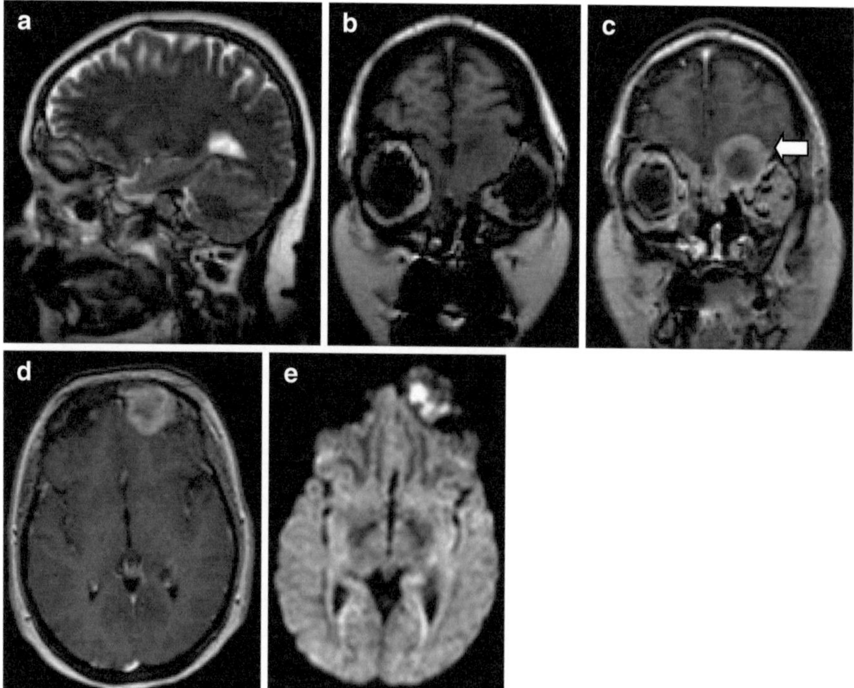

Fig. 5.9 Serial MRI images of the brain with the following: (**a**) Sagittal T2, (**b**) Coronal T1, (**c**) Coronal T1 C+, (**d**) Axial T1 C+, (**e**) Axial DWI. (Figure courtesy of Dr. Samer Hoz)

Imaging Description

MRI reveals expansive fluid filling the frontal sinuses and extending into the ethmoidal sinuses. Superior intracranial extension into the extra-axial compartment is also seen, exerting a mass effect on the frontal lobes bilaterally, particularly on the left side (arrow). An infection, evident as a large abscess with restricted diffusion on DWI, involves the frontal sinuses, with extension into the ethmoidal sinuses and left extraconal region. There is noticeable bone erosion between the anterior cranial foci and the frontal sinus, along with erosion and destruction of the nasal septum. This form of aggressive infection may be caused by a fungus. No evidence suggests intra-axial involvement. Additionally, mucosal thickening is observed in the maxillary antra.

Pott Puffy Tumor

This condition is a complication of acute sinusitis, which is typically non-neoplastic. It is characterized by osteomyelitis, a subperiosteal abscess, and a subgaleal collection. It is most commonly associated with frontal sinusitis but can also develop secondary to mastoid pathology. Less commonly, it can result from craniotomy, intranasal cocaine and methamphetamine use, or trauma. A subperiosteal abscess forms when the infection penetrates the wall of the blocked, infected sinus. This can lead to serious complications like intracranial extension, epidural abscess, subdural empyema, meningitis, and even the development of cerebral abscess. Another potential consequence is the thrombosis of the dural sinus, a condition initially described by Sir Percivall Pott. The term "puffy tumor" in the name refers to the typical forehead swelling caused by the subgaleal collection. This condition can affect individuals of any age but is more common in adolescents. Fortunately, with the increased availability of antibiotics, it has become less common.

CT demonstrates an opacified frontal sinus along with swelling and inflammation of the overlying scalp. The bone algorithm can show a defect in the anterior wall of the sinus. Contrast imaging can reveal a localized abscess and help delineate any intracranial complications.

MRI is particularly useful for detecting subtle intracranial involvement. MRI with contrast shows early linear enhancement of the dura mater, as well as an extra-axial fluid collection, a region of cerebritis, or the development of a localized cerebral abscess. If there is an organized fluid collection in the scalp, peripheral or rim enhancement may be noticeable on contrast imaging [16, 17].

Differential Diagnosis

1. Unilateral non-Hodgkin lymphoma.
2. Destructive metastasis to frontal sinus.

Questions

1. **Pott puffy tumor of the frontal sinus, the FALSE answer is:**
   A. Is characterized by osteomyelitis, a subperiosteal abscess, and a subgaleal collection.
   B. Its prevalence has decreased since antibiotics have been made more readily available.
   C. Adolescents are more likely to be affected than adults.
   D. CT typically reveals an opacified frontal sinus with extension to the sphenoid sinus.
   E. The pathology involves a non-neoplastic process.
   The answer is D.

   CT typically reveals an opacified frontal sinus along with swelling of the overlying scalp.

Case 47: Meningitis Secondary to Mastoiditis and Pan Sinusitis Resulting in Death

Case Scenario

A 55-year-old male presented with vomiting, headache, and neck stiffness (Fig. 5.10).

Imaging Description

MRI shows the leptomeningeal structures (arrow) exhibit contrast enhancement with diffuse and extensive thickening. Complete opacification is observed in the middle ear and mastoid air cells, particularly more pronounced on the right side. Additionally, there is mucosal thickening in the maxillary antrum and involvement of the ethmoidal air cells.

Meningitis

Leptomeningitis, which is more commonly referred to as meningitis, is characterized by inflammation of the subarachnoid space with the involvement of the arachnoid mater and pia mater. This inflammation can be triggered by infectious or noninfectious factors. *Haemophilus influenzae*, *Streptococcus pneumoniae*, and *Neisseria meningitides* are the most common causes of bacterial meningitis. Over recent decades, the median age at which bacterial meningitis is diagnosed has increased due to childhood immunization. However, this reduction in incidence is not observed in newborns under two months of age. In adults, bacterial meningitis is often associated with chronic and immunocompromising conditions. This

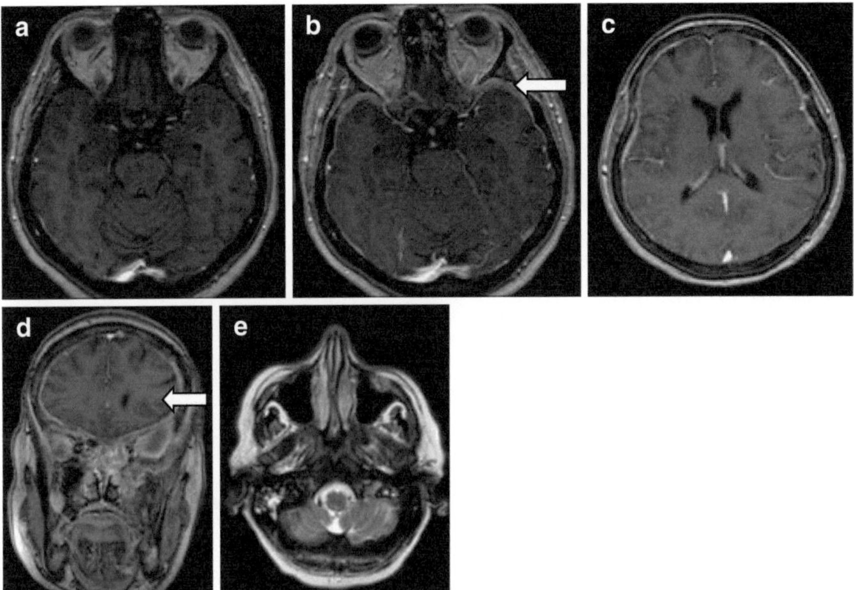

Fig. 5.10 Serial MRI images of the brain with the following: (**a**) Axial T1 3D FFE, (**b**) Axial T1 3D FFE C+, (**c**) Axial T C+, (**d**) Coronal T1 C+, (**e**) Axial T2. (Figure courtesy of Dr. Samer Hoz)

includes conditions such as elderly age (over 65 years), splenectomy or hyposplenic states, alcoholism, HIV/AIDS, diabetes mellitus, cancer, anatomical defects (related to recurrent meningitis), and organ transplant recipients.

Bacteria can enter the CNS through various routes, including direct implantation, transmission from a local septic condition (e.g., sinusitis), infection related to a foreign body (e.g., shunting catheter), or hematogenous dissemination. The imaging findings related to the causative pathogen are typically nonspecific. However, imaging results are valuable for detecting abnormalities and distinguishing them from other non-infectious causes. In general, cross-sectional imaging is not highly sensitive or specific for diagnosing meningitis.

One potential MRI finding is an increased FLAIR signal compared to normal cortex, although this is not unique to meningitis. Post-contrast MRI may reveal thin, linear enhancements in the leptomeninges on T1-weighted images, but this is only present in about 50% of cases. Specifically, smooth or linear enhancement is more typical in cases of acute lymphocytic viral meningitis and acute pyogenic (bacterial) meningitis. If a more nodular and thick contrast enhancement pattern is observed, especially involving the basal cisterns, it may indicate conditions like leptomeningeal carcinomatosis or granulomatous disease. Other MRI findings might include limited diffusion in cerebral sulci compared to the normal cortex.

The mnemonic "HACTIVE" can be used to recall some of the complications associated with meningitis:

- H: Hydrocephalus.
- A: Abscess.
- C: Cerebritis/cranial nerve lesion.
- T: Thrombosis.
- I: Infarct.
- V: Ventriculitis/vasculopathy.
- E: Extra-axial collection: empyema and hygroma [18, 19].

Differential Diagnosis

1. Fungal meningitis: usually exhibits thicker, lumpy, or nodular enhancement.
2. Viral meningitis.

Questions

1. **Meningitis, the FALSE answer is:**
   A. Leptomeningitis is characterized by inflammation of the subarachnoid space with the involvement of the arachnoid mater and pia mater.
   B. Associated with arterial narrowing or occlusion on MRA.
   C. The most frequent positive findings on post-contrast MRI are thin, linear leptomeningeal enhancements.
   D. Smooth or linear enhancement is more typical with acute lymphocytic viral meningitis and acute pyogenic (bacterial) meningitis.
   E. The majority of cases are caused by *Staphylococcus aureus*.
   The answer is E.
   The majority of cases are caused by *Haemophilus influenzae*, *Streptococcus pneumoniae*, and *Neisseria meningitides*.

References

1. Eran A, Hodes A, Izbudak I. Bilateral temporal lobe disease: looking beyond herpes encephalitis. Insights Imaging. 2016;7:265–74.
2. Bradshaw MJ, Venkatesan A. Herpes simplex virus-1 encephalitis in adults: pathophysiology, diagnosis, and management. Neurotherapeutics. 2016;13:493–508.
3. Kumar R. Understanding and managing acute encephalitis. F1000Res. 2020;9:F1000 Faculty Rev-60.
4. Mohindra S, Savardekar A, Gupta R, Tripathi M, Rane S. Tuberculous brain abscesses in immunocompetent patients: a decade long experience with nine patients. Neurol India. 2016;64(1):66.
5. Zahrou F, Elallouchi Y, Ghannane H, Benali SA, Aniba K. Diagnosis and management of intracranial tuberculomas: about 2 cases and a review of the literature. Pan Afr Med J. 2019;34(1):23.

 6. Perez-Malagon CD, Barrera-Rodriguez R, Lopez-Gonzalez MA, Alva-Lopez LF, Alva-Lopez L. Diagnostic and neurological overview of brain tuberculomas: a review of literature. Cureus. 2021;13(12):e20133.
 7. Saro A, Selim SM. Intracranial hydatid cysts: a report on 16 cases. Menoufia Med J. 2021;34(2):709.
 8. Senapati S, Parida D, Pattajoshi A, Gouda A, Patnaik A. Primary hydatid cyst of brain: two cases report. Asian J Neurosurg. 2015;10(02):175–6.
 9. Padayachy LC, Ozek MM. Hydatid disease of the brain and spine. Childs Nerv Syst. 2023;39(3):751–8.
10. Jeanneret V, Winkel D, Risman A, Shi H, Gombolay G. Post-infectious rhombencephalitis after coronavirus-19 infection: a case report and literature review. J Neuroimmunol. 2021;357:577623.
11. Wei P, Bao R, Fan Y. Brainstem encephalitis caused by Listeria monocytogenes. Pathogens. 2020;9(9):715.
12. Silva TL, Corbiceiro WC, Corrêa DG. Rhombencephalitis caused by Cytomegalovirus. Can J Neurol Sci. 2023;50:905–6.
13. Tung GA, Evangelista P, Rogg JM, Duncan JA III. Diffusion-weighted MR imaging of rim-enhancing brain masses: is markedly decreased water diffusion specific for brain abscess? Am J Roentgenol. 2001;177(3):709–12.
14. Taklalsingh N, Falcone F, Velayudhan V. Gradenigo's syndrome in a patient with chronic suppurative otitis media, petrous apicitis, and meningitis. Am J Case Rep. 2017;18:1039.
15. Athapathu AS, Bandara ER, Aruppala AA, Chandrapala KM, Mettananda S. A child with Gradenigo syndrome presenting with meningism: a case report. BMC Pediatr. 2019;19(1):350.
16. Hasan I, Smith SF, Hammond-Kenny A. Potts puffy tumour: a rare but important diagnosis. J Surg Case Rep. 2019;2019(4):rjz099.
17. Sideris G, Delides A, Proikas K, Papadimitriou N. Pott puffy tumor in adults: the timing of surgical intervention. Cureus. 2020;12(11):e11781.
18. Kocsis B, Tiszlavicz Z, Jakab G, Brassay R, Orbán M, Sárkány Á, Szabó D. Case report of Actinomyces turicensis meningitis as a complication of purulent mastoiditis. BMC Infect Dis. 2018;18:1–4.
19. Barry C, Rahmani G, Bergin D. Pneumocephalus and meningitis as complications of mastoiditis. Case Rep Radiol. 2019;2019:7876494.

Miscellaneous Pathologies of the Brain

6

Hasan M. Jabbar, Mohammed E. Al-Hamadani,
Fatimah O. Ahmed, Mahmood F. Alzaidy,
Hussein M. Hasan, Ahmed Muthana,
and Asmaa H. AL-Sharee

Case 48: Wilson Disease

Case Scenario

A 14-year-old male presented with dysarthria, choreoathetosis, ataxia, and Kayser-Fleischer rings in the cornea (Fig. 6.1).

Imaging Description

MRI images in Fig. 6.1 show bilateral and symmetrical abnormal signal intensity evident in the putamen, caudate nuclei, and thalami, which appears as hyper-intense in T2 and FLAIR sequences, hypo-intense in T1, and shows no restriction in DWI. Additionally, mild cortical atrophy is observed. No abnormal enhancement was detected on T1 with contrast.

H. M. Jabbar
College of Medicine, University of Misan, Misan, Iraq

M. E. Al-Hamadani · M. F. Alzaidy · H. M. Hasan · A. Muthana
College of Medicine, University of Baghdad, Baghdad, Iraq

F. O. Ahmed
College of Medicine, University of Mustansiriyah, Baghdad, Iraq

A. H. AL-Sharee (✉)
Department of Neuroradiology, Neurosurgery Teaching Hospital, Baghdad, Iraq

© The Author(s), under exclusive license to Springer Nature Switzerland AG 2024
S. Hoz et al. (eds.), *Neuroradiology Board's Favorites*,
https://doi.org/10.1007/978-3-031-64261-6_6

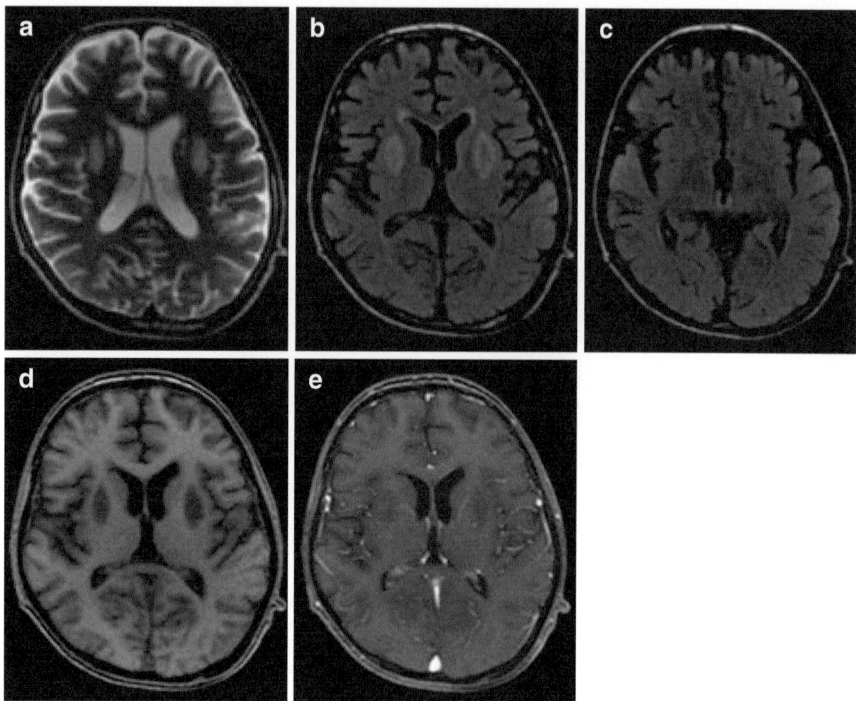

Fig. 6.1 Serial MRI images of the brain with the following: (**a**) Axial T2, (**b**) Axial FLAIR, (**c**) Axial FLAIR, (**d**) Axial T1, (**e**) Axial T1 C+. (Figure courtesy of Dr. Samer Hoz)

Wilson Disease

Wilson disease is a rare autosomal recessive disorder characterized by the abnormal accumulation of copper in various organ systems. Its prevalence is estimated to affect 1 in every 30,000–40,000 individuals, with about 1 in 90 people being carriers of a single copy of the mutated gene. The condition primarily manifests as early onset liver cirrhosis, accompanied by neurological symptoms, particularly those involving the basal ganglia and midbrain. Hence, it is also referred to as hepatolenticular degeneration.

Imaging features of Wilson disease can vary depending on several factors, including whether the disease is being treated, the degree of liver dysfunction, and other individual factors. The basal ganglia is the most commonly affected region, especially the putamen, followed by the midbrain, pons, and thalamus. Typically, the distribution of these lesions is bilateral and symmetric.

On MRI, the most common abnormality observed is an abnormal T2 hyperintensity in the putamen. Additional regions of abnormal T2 signal often affect the deep gray nuclei. In cases where the midbrain tegmentum is involved, the "face of the giant panda" sign can be seen on axial images. Similarly, axial T2 MRI images of the pons may show a "miniature panda" sign, and when both signs are present simultaneously, it is referred to as the "double panda" sign. However, the signal on T1-weighted images can vary, depending on factors such as edema and hepatic impairment. Patients with neurological symptoms typically show reduced a T1 signal, whereas those with severe liver dysfunction may exhibit an increased T1 signal, especially in the globus pallidus, resembling the signal seen in acquired hepatocerebral degeneration due to manganese deposition. In some cases, diffusion restriction may be observed early in the course of the disease.

On CT scans, atrophic changes may be visible in the basal ganglia, cortical, and cerebellar regions. However, copper deposition does not lead to increased density on non-enhanced CT images, and post-contrast CT scans do not reveal enhancement of the lesions [1, 2].

Questions

1. **Wilson disease, the FALSE answer is:**
   A. The face of the giant panda sign can be seen with the involvement of the midbrain tegmentum.
   B. Putamen is the most commonly affected region.
   C. T1 axial imaging shows reduced signal intensity in patients with hepatic dysfunction and increased signal in patients with neurologic manifestations.
   D. Hyper-intense signals can be seen in T2 and FLAIR.
   E. Atrophic changes can be seen on CT.
   The answer is C.
   Reduced T1 signal is seen in patients presenting with neurologic manifestations. In contrast, patients with severe hepatic dysfunction show areas of increased T1 signal, especially in the globus pallidus.

Case 49: Myelin Oligodendrocyte Glycoprotein Antibody-Associated Disease (MOGAD)

Case 49.1

Case Scenario

A 7-year-old female presented with right-sided blindness and paraplegia (Fig. 6.2).

Imaging Description

MRI reveals mild enlargement in the retrobulbar intraorbital segment of the right optic nerve, which does not extend to the intracranial segment. The optic chiasm appears normal. In the post-contrast study, apparent enhancement is evident, suggesting optic neuritis. Additionally, in the spinal cord, a hyper-intense signal in T2 (arrow) is noted with a central medullary distribution and a central bright spot. Furthermore, T2 imaging reveals multiple scattered subcortical and deep white matter foci with increased signal intensity.

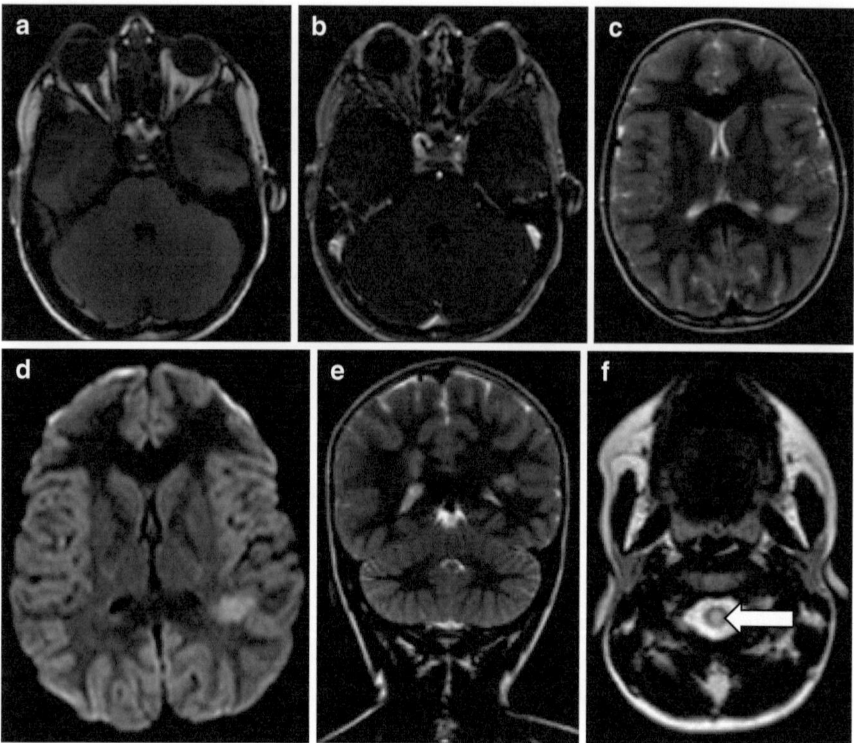

Fig. 6.2 Serial MRI images of the brain and spine with the following: (**a**) Axial T1, (**b**) Axial fat-saturated T1 C+, (**c**) Axial T2, (**d**) Axial DWI, (**e**) Coronal T2, (**f**) Axial T2. (Figure courtesy of Dr. Samer Hoz)

Case 49.2

Case Scenario

Follow-up of the same patient after 1 year of appropriate management (Fig. 6.3).

Imaging Description

After 1 year of management, significant clinical improvement is evident in the patient. Radiologically, a comparison between the previous images and the new ones reveals a normalization of the right optic nerve and a reduction in the size of deep white matter foci at left side. These changes are observed in both sagittal and axial images, and the subcortical abnormalities have disappeared. However, the hyper-intense signal in the upper cervical region of the spinal cord persists in the PSIR sequence (arrow), with a posterior medullary distribution.

MOGAD

MOGAD is a relatively newly recognized disease that encompasses a range of inflammatory demyelinating disorders. It is characterized by the presence of IgG

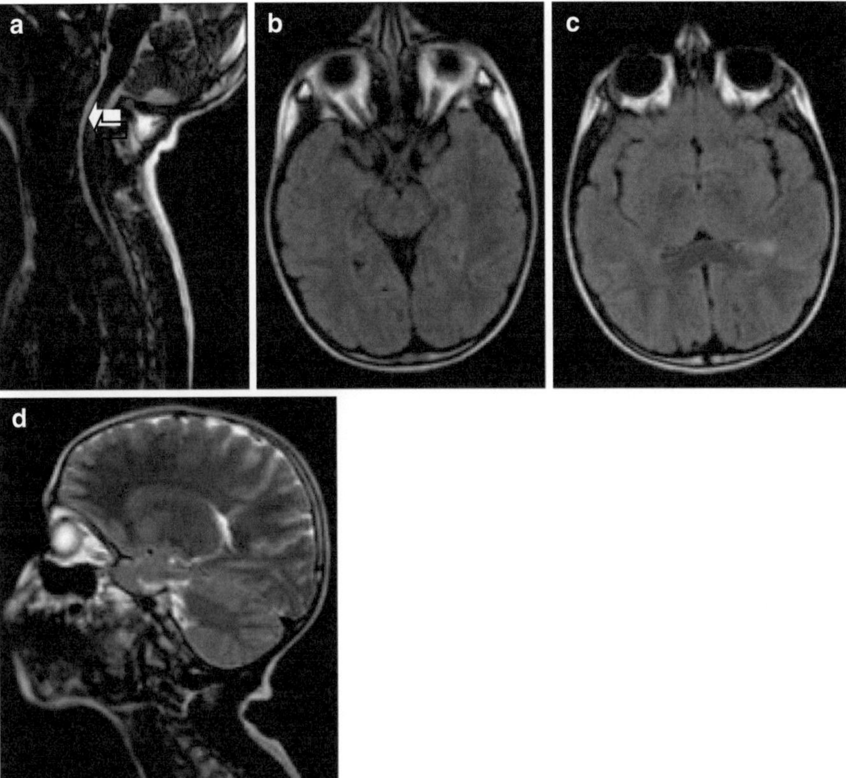

Fig. 6.3 Serial MRI images of the brain with the following: (**a**) Sagittal PSIR, (**b**) Axial FLAIR, (**c**) Axial FLAIR, (**d**) Sagittal T2. (Figure courtesy of Dr. Samer Hoz)

antibodies specific to the myelin oligodendrocyte glycoprotein (MOG). In recent times, MOGAD has been identified as a separate and distinct condition, although it shares some features with other disorders like ADEM, neuromyelitis optica spectrum disorder, and MS. Throughout the process of identifying this disease, various terms have been used, such as anti-MOG-associated encephalomyelitis, MOG-IgG-associated optic neuritis, encephalitis, and myelitis, and anti-MOG encephalitis, among others.

MOGAD is more commonly observed in young adults and children. In children, the presence of MOG IgG antibodies is more frequent than aquaporin 4 antibodies, and the presentation is similar to ADEM. However, adults often exhibit symptoms resembling neuromyelitis optica syndrome.

As mentioned earlier, MOGAD shares overlapping features with other acquired demyelinating conditions. Symptoms can be similar to those seen in this group, but there can be significant variability from one individual to another. Interestingly, approximately 50% of cases have a history of viral infection preceding the onset of symptoms.

On MRI, the presentation of MOGAD is diverse. As with clinical symptoms, there are no distinct imaging features that can definitively diagnose the condition. Instead, the imaging features often overlap with those of other demyelinating inflammatory conditions that affect the white matter of the central nervous system, such as ADEM, neuromyelitis optica spectrum disorder, and, to a lesser extent, MS. However, there have been some imaging characteristics that are more commonly associated with MOGAD and can raise suspicion of this diagnosis when observed [3–5].

Differential Diagnosis

1. Neuromyelitis optica spectrum disorder. Despite similarities, two aspects can be used to distinguish aquaporin-4-IgG-positive neuromyelitis optica spectrum disorder from MOGAD patients:
   (a) Optic neuritis in patients with MOGAD predominantly affects the anterior optic nerves, resulting in severe swelling, edema, and tortuosity. However, the involvement of the optic chiasm and tracts is generally spared.
   (b) In MOGAD, brain lesions are frequently observed in the thalami and pons, while involvement of the area postrema and medulla is infrequent.
2. MS. Some typical features encountered in MS are absent in MOGAD:
   (a) Dawson's fingers are absent in MOGAD.
   (b) Short segments of spinal involvement in MOGAD are less common than lengthy involvement (longitudinally extensive spinal cord lesions).
   (c) In MOGAD, when brain lesions are present, they exhibit an ADEM-like pattern characterized by larger size and a lower number of lesions.
   (d) Optic neuritis in MOGAD is typically bilateral and more distinct.
      Leptomeningeal enhancement is an uncommon finding in MS, and even when observed, it generally demonstrates subtle manifestations.

Questions

1. **MOGAD, the FALSE answer is:**
   A. Dawson's fingers are absent in MOGAD.
   B. The caudal portion of the spinal cord demonstrates preferential involvement.
   C. Short-segment lesions are more commonly seen in MOGAD.
   D. Leptomeningeal enhancement is a useful sign to distinguish MOGAD.
   E. Brain lesions in MOGAD can involve the gray matter and the brainstem.
      The answer is C.
      Short segments of spinal involvement in MOGAD are less common than lengthy involvement (longitudinally extensive spinal cord lesions).

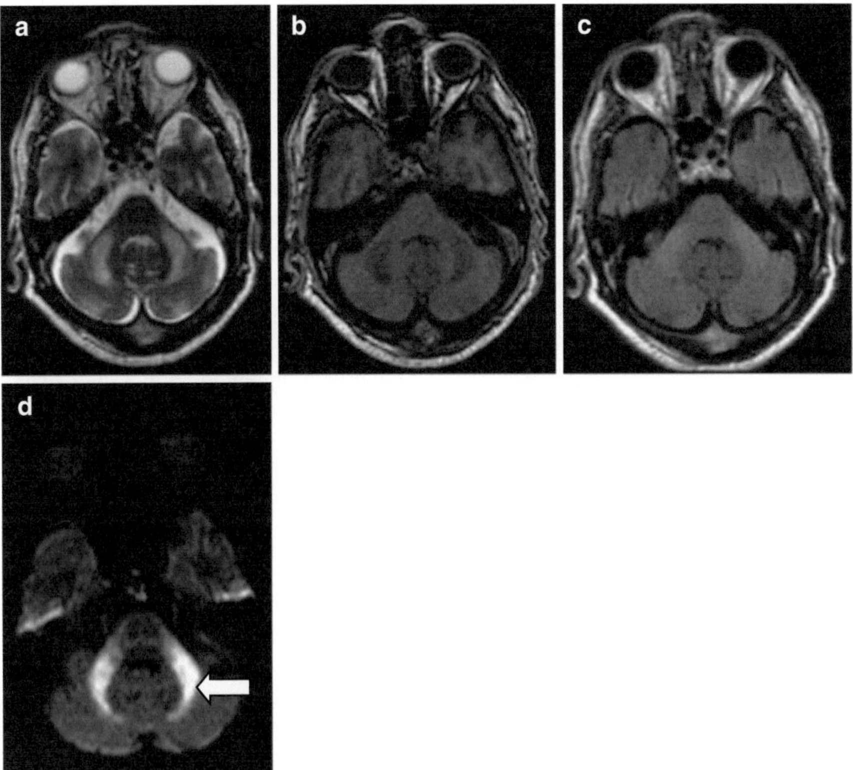

Fig. 6.4 Serial MRI images of the brain with the following: (**a**) Axial T2, (**b**) Axial T1, (**c**) Axial FLAIR, (**d**) Axial DWI. (Figure courtesy of Dr. Samer Hoz)

Case 50: Fragile X-Associated Tremor/Ataxia Syndrome (FXTAS)

Case scenario

A 59-year-old male presented with kinetic tremor and cerebellar ataxia (Fig. 6.4).

Imaging Description

There is T2 hyper-intense signal intensity in the middle cerebellar peduncles (MCP) with restriction in DWI (arrow). Additionally, there are findings of diffuse age-appropriate cerebellar and cerebral mild atrophy, characterized by proportionally dilated ventricles and cortical sulci.

FXTAS

FXTAS is a progressive neurodegenerative disorder that occurs in individuals carrying a premutation CGG repeat expansion (ranging from 55 to 200 repeats) within the 5′ untranslated region (5′UTR) of the fragile X mental retardation 1 (FMR1) gene. Interestingly, the FMR1 premutation tends to be more prevalent in males than females. Additionally, FXTAS exhibits age-dependent penetrance, affecting roughly 45% of male FMR1 premutation carriers over the age of 50 and approximately 15% of females. Consequently, the condition predominantly affects elderly males. In cases where an affected male has a daughter, she will inherit the premutation and pass it on to her offspring. This can lead to the full mutation and a confirmed diagnosis of fragile X syndrome in subsequent generations, as the premutation can expand during oogenesis, resulting in the full phenotype of fragile X syndrome.

The diagnosis of FXTAS requires a molecular assessment of the FMR1 gene mutation, specifically identifying a CGG repeat in the range of 55–200. However, the diagnostic criteria for FXTAS involve a combination of major and minor clinical and radiological features, along with the presence of characteristic ubiquitin-positive intranuclear inclusions, which are identified post-mortem. These various features allow patients to be categorized into three different levels of diagnostic certainty.

1. Definite: One major clinical and one major radiological; or one major clinical and intranuclear inclusions.
2. Probable: Two major clinical; or one minor clinical and one major radiological.
3. Possible: One major clinical and one minor radiological.

The primary clinical manifestations of FXTAS encompass intention tremor and cerebellar gait ataxia. Minor clinical features include parkinsonism, significant short-term memory impairment, executive function deficits, and peripheral neuropathy.

The updated diagnostic criteria for FXTAS include two major radiological features. The most notable of these is the MCP sign, characterized by increased signal intensity in the MCP due to spongiosis of the deep cerebellar white matter, typically observed on T2/FLAIR MRI scans. The MCP sign is detectable in 60% of affected males and 13% of affected females. However, the MCP sign is not exclusive to FXTAS; it can also be present in other conditions like multiple system atrophy of the cerebellar type, recessive ataxia, and acquired hepatocerebral degeneration. The second significant radiological feature is the hyper-intensity of the corpus callosum splenium on T2/FLAIR imaging, which occurs with similar frequency as the MCP sign and is more prevalent in males [6, 7].

Questions

1. **FXTAS, the FALSE answer is:**
   A. FXTAS predominantly affects elderly males.
   B. The MCP sign is specific to FXTAS.
   C. The white matter of the splenium can show T2/FLAIR hyper-intensities.
   D. Two major clinical features are required to diagnose a patient with probable FXTAS.
   E. Generalized brain atrophy can be seen in asymptomatic premutation carriers.
   The answer is B.

The MCP sign is not specific only to FXTAS, as this sign can be seen in other conditions such as multiple system atrophy of the cerebellar type, recessive ataxia, and acquired hepatocerebral degeneration.

Case 51: Optic Pathway Gliomas in NF1

Case Scenario

A 16-year-old female presented with a decrease in vision (Fig. 6.5).

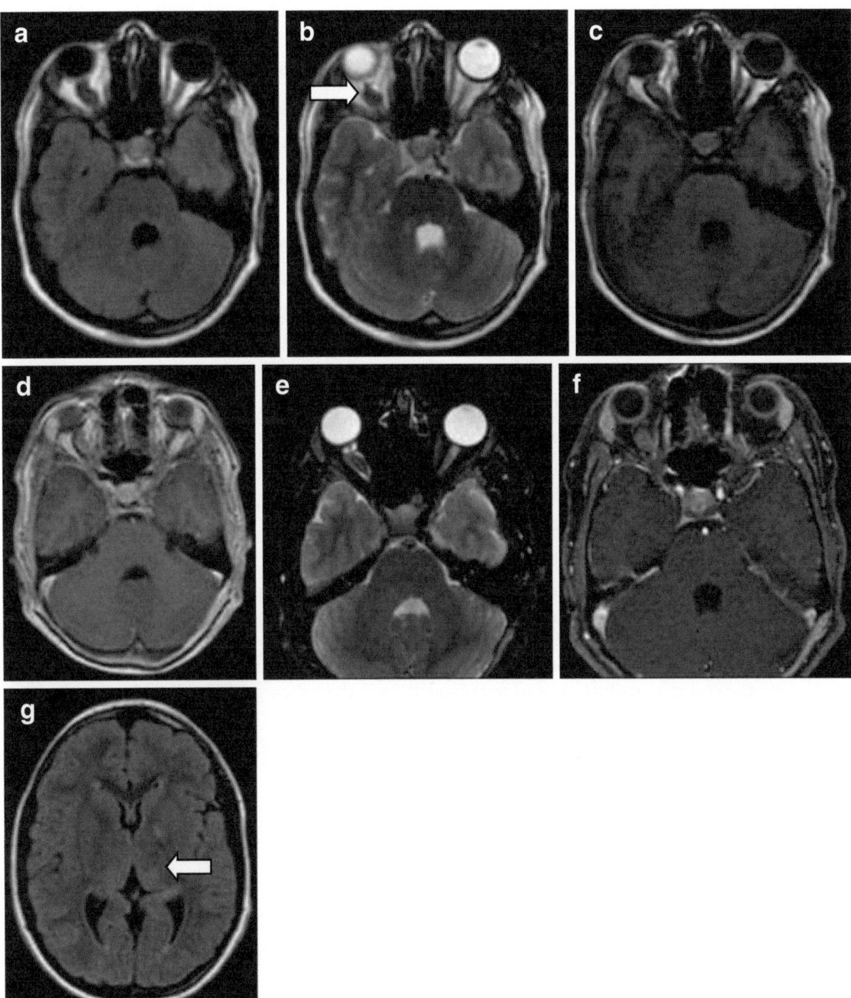

Fig. 6.5 Serial MRI images of the brain with the following: (**a**) Axial FLAIR, (**b**) Axial T2, (**c**) Axial T1, (**d**) Axial T1 C+, (**e**) Axial STIR, (**f**) Axial fat-saturated T1 C+, (**g**) Axial FLAIR. (Figure courtesy of Dr. Samer Hoz)

Imaging Description

MRI shows the optic apparatus in a child with NF1, with enlargement of the right optic nerve (arrow). Enlargement of the nerve is prominent within the orbit on the right side and extends back to the optic nerve canal, but does not appear to reach the chiasm. The affected optic nerve appears hypo-intense in T1 and iso-intense in T2, displaying homogeneous enhancement after a contrast study. Additionally, FLAIR images reveal hyper-intense foci on the left basal ganglia, particularly the globus pallidus (arrow), with no enhancement seen after contrast study. These features are consistent with a right optic nerve glioma with focal areas of signal intensity (FASI) in the basal ganglia.

Optic Pathway Gliomas in NF1

Neurofibromatosis type 1 (NF1), also known as Von Recklinghausen disease, is a complex hereditary neurocutaneous disorder with an estimated incidence of 1:2500–3000 live births. It stands as the most common condition of phacomatosis, arising from mutations in the NF1 tumor-suppressor gene located on chromosome 17q11.2. The disease is typically inherited through an autosomal dominant pattern in about half of the cases, while the other half results from de novo mutations. Additionally, NF1 displays a wide range of manifestations, with complete penetrance occurring by the age of 5. It is often categorized as a tumor-predisposing disease due to its association with elevated risks of various tumor types, including gliomas such as juvenile pilocytic astrocytoma (found in around 20% of patients), optic nerve glioma, diffuse brainstem glioma, spinal astrocytoma, and spinal pilocytic astrocytoma.

The clinical diagnosis of NF1 is typically established using the diagnostic criteria outlined in the National Institutes of Health Consensus Development Conference. Diagnosis is confirmed when two or more of the following manifestations are present:

1. Six or more cafe au lait spots (prepubertal >5 mm, postpubertal >15 mm in size) were evident during 1 year.
2. Two or more cutaneous neurofibromas or one plexiform neurofibroma.
3. Optic nerve glioma.
4. Skeletal dysplasia (such as sphenoid wing dysplasia or thinning of long bone cortex with or without pseudoarthrosis).
5. Two or more Lisch nodules (iris hamartomas).
6. Axillary or inguinal freckling (in places inaccessible to light).
7. A first-degree relative having NF1 with the above criteria.

A certain mnemonic that can help remember these features is CAFE SPOT.

NF1 typically manifests at a much younger age compared to NF2 and schwannomatosis. Around 50% of patients meet the diagnostic criteria for NF1 before reaching the age of 1, and nearly 97% do so by the time they are 8 years old. Moreover, cognitive dysfunction affects approximately 45% of patients, and roughly 1% experience hypertension resulting from renal artery stenosis.

The radiographic features include:

- FASI is typically found in the deep white matter, basal ganglia, and corpus callosum, and presents as regions with hyper-intensity on T2/FLAIR imaging, without any enhancement visible on contrast scans.
- Optic nerve glioma or optic pathway glioma which may manifest as enlarged optic foramen.
- Progressive sphenoid wing dysplasia.
- J-shaped sella.
- Lambdoid suture defects.
- Dural calcification at the vertex.
- Moyamoya phenomenon (rare).
- Buphthalmos [8, 9].

Questions

1. **Optic pathway glioma in NF1, the FALSE answer is:**
   A. A J-shaped sella can be observed in patients with NF1.
   B. NF1 is associated with high incidence rates of brain gliomas.
   C. FASI are one of the most common neuroimaging findings in patients with NF1.
   D. FASI appear as hypo-intense areas in T2/FLAIR MRI.
   E. Optic nerve glioma can manifest as an enlarged optic foramen.
   The answer is D.
   FASI appear as hyper-intense areas on T2/FLAIR MRI.

Case 52: Focal Areas of Signal Intensity (FASI)

Case Scenario

A 4-year-old male presented with cognitive dysfunction (Fig. 6.6).

Imaging Description

MRI revealed multiple hyper-intense foci in the basal ganglia (globus pallidus), thalami, brainstem (mainly in the pons), and the left side of the cerebellum on T2/FLAIR sequences. There is no evidence of atrophy or mass effect. No contrast enhancement is noted following the contrast study, and there are no restrictions observed in DWI.

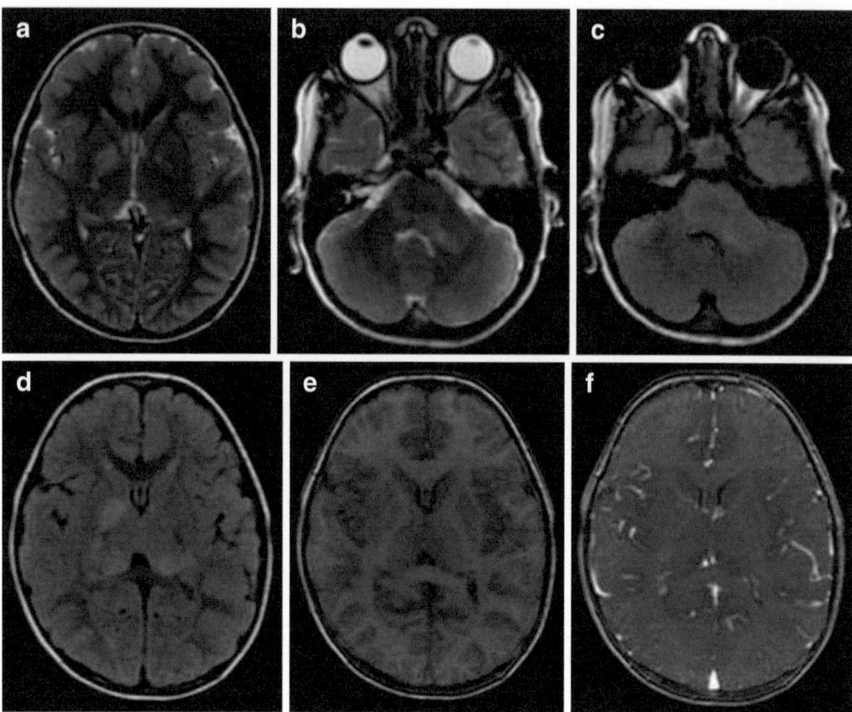

Fig. 6.6 Serial MRI images of the brain with the following: (**a**) Axial T2, (**b**) Axial T2, (**c**) Axial FLAIR, (**d**) Axial FLAIR, (**e**) Axial T1, (**f**) Axial fat-saturated T1 C+. (Figure courtesy of Dr. Samer Hoz)

Focal Areas of Signal Intensity

Focal areas of signal intensity (FASI), also known as unidentified bright objects or focal abnormal signal intensity, are regions of hyper-intensity that become more prominent in T2W MRI and are particularly visible in FLAIR images. These areas are typically found in various brain regions, including the basal ganglia (often the globus pallidus), thalamus, brainstem (pons), cerebellum, and subcortical white matter in children with NF1. In some cases, FASI may also be present in the thalami, and certain studies have linked this occurrence to cognitive changes. However, most studies have not found any significant impact on other brain regions. The exact nature of FASI remains unclear, but they are generally considered a benign phenomenon caused by increased fluid within intra-myelinic vacuoles. Nevertheless, they may also be associated with hyperplastic or dysplastic glial proliferation.

FASI are the most commonly observed neuroimaging findings in NF1 patients, with approximately 86% of children with NF1 having one or more FASI. In patients younger than 10 years, it is common to observe an increase in either the size or number of FASI lesions. However, if such an increase occurs beyond the age of 10, it may raise concerns about the possibility of a neoplasm.

On MRI, FASI lesions typically appear iso-intense to hyper-intense on T1W images and hyper-intense on T2W and FLAIR images. They do not exhibit contrast enhancement, although some studies have reported enhancing FASI, including cases where enhancement regressed gradually in late adolescence. Additionally, FASI lesions show restrictions on DWI. MRS results are typically normal and can be valuable in distinguishing FASI from tumors [10, 11].

Questions

1. **FASI, the FALSE answer is:**
   A. FASI are a common radiological finding in children with NF1.
   B. FASI have no mass effect on MRI.
   C. FASI are hyper-intense on T2WI/FLAIR.
   D. FASI are iso-intense to hypo-intense on T1WI.
   E. FASI lack of contrast enhancement on T1 C+ (Gd).
      The answer is D.
      FASI are iso-intense to hyper-intense on T1W.

Case 53: Cholesteatomas

Case Scenario

A 57-year-old male presented with conductive hearing loss (Fig. 6.7).

Imaging Description

MRI shows a well-defined mass lesion is identified in the left mastoid, measuring 23 × 14 mm. The lesion exhibits T1 hypo-intensity and T2 hyper-intensity. Notably, there is no enhancement observed following gadolinium administration. The lesion demonstrates marked diffusion restriction.

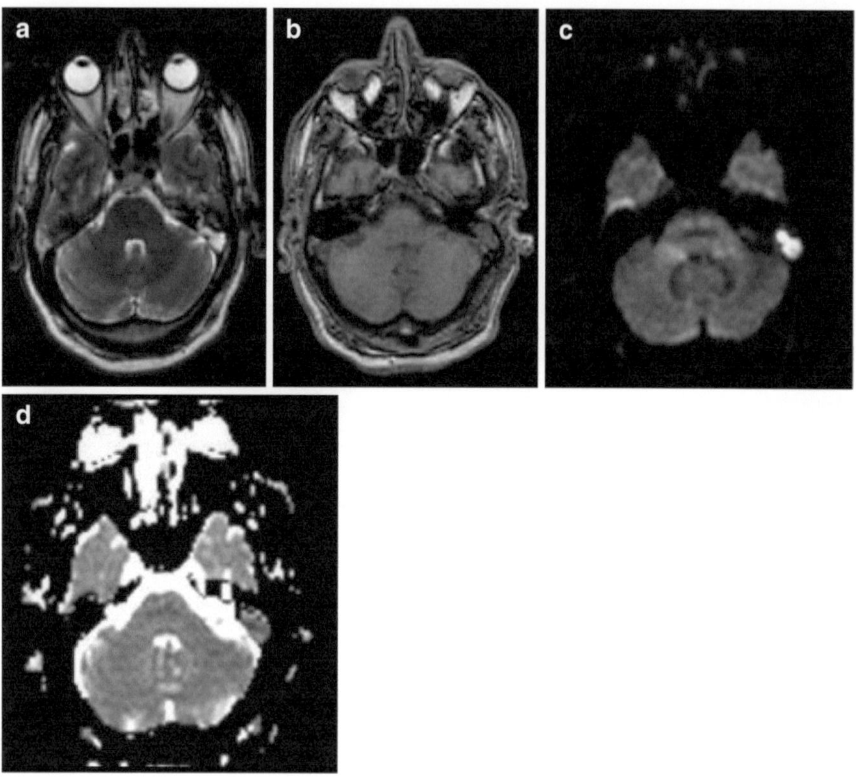

Fig. 6.7 Serial MRI images of the brain with the following: (**a**) Axial T2, (**b**) Axial T1, (**c**) Axial DWI, (**d**) Axial ADC. (Figure courtesy of Dr. Samer Hoz)

Cholesteatomas

Cholesteatoma is a benign cystic lesion that can affect various regions of the temporal bone, especially the middle ear cleft. Cholesteatomas share histological similarities with epidermoid cysts and are characterized by the accumulation of desquamated debris, surrounded by layers of keratinizing stratified squamous epithelium.

Cholesteatomas occurring in the middle ear and other pneumatized areas of the temporal bone are typically categorized into two main types: congenital, which account for a small 2%, and acquired cholesteatomas, which make up the remaining 98%. Acquired cholesteatomas are further classified into primary cases, which develop without a history of chronic otomastoiditis, and secondary cases, which are associated with chronic otomastoiditis. There is also a special group of cholesteatomas that includes external ear canal cholesteatomas, mural cholesteatomas, and petrous apex cholesteatomas.

High-resolution CT is the primary imaging method recommended for assessing suspected cases of cholesteatoma. High-resolution CT provides detailed information about the bone structure, which is crucial for preoperative planning. It helps pinpoint the exact location of the opacity, identify any associated bony and ossicular erosion, detect variant anatomical structures, and assess possible complications such as bony tegmen perforation, tegmen dehiscence, and labyrinthine fistulas. For instance, based on their location and associated findings, high-resolution CT can differentiate between pars flaccida cholesteatoma, which originates in the Prussak space and typically extends posteriorly, and pars tensa cholesteatoma, which originates in the posterior mesotympanum and tends to extend posteromedially. However, high-resolution CT may not reliably distinguish recurrent or residual cholesteatoma from fibrosis, granulation tissue, cholesterol granuloma, mucosal edema, or effusion, as these conditions can appear similarly dense on imaging.

Conventional MRI with DWI plays a critical role in identifying residual or recurrent cholesteatoma, particularly in patients with a history of prior surgery. To avoid false-positive diagnoses that could result from wax accumulation in the mastoid cavity, patients should prepare for the MRI by ensuring the external ear canal or postoperative cavity is clear. The MRI examination includes T2-weighted series in both coronal and axial sections, as well as non-echo planar DWI series with b-values of 0 and 1000. On T2W MRI, cholesteatomas appear hyper-intense compared to brain parenchyma. In DWI (b-value = 1000) images, they exhibit hyper-intensity, with low values on ADC maps due to restricted diffusion. The combination of these findings allows for the detection of residual or recurrent cholesteatomas, even as small as a few millimeters, with nearly 100% specificity [12, 13].

Differential Diagnosis

The primary conditions that cholesteatoma is often confused with are inflammation and cholesterol granuloma. Unlike cholesteatoma, these conditions typically exhibit a high signal on the ADC map in imaging. Other potential differential diagnoses include cerumen or ear wax, which can accumulate and produce similar imaging features, with the only distinction being their location in the external ear canal. Additionally, abscess formation in the middle ear is another consideration, but it presents with different clinical symptoms and signs.

Questions

1. **Cholesteatomas, the FALSE answer is:**
   A. High-resolution CT has a low specificity for detecting recurrent cholesteatomas.
   B. Cholesteatomas are hyper-intense on T2W MRI.
   C. Cholesteatomas show a low signal on ADC map.
   D. Cholesteatomas are hypo-intense on DWI MRI.
   E. DWI MRI shows high sensitivity and specificity in detecting cholesteatoma.
   The answer is D.
   Cholesteatomas appear hyper-intense on DWI.

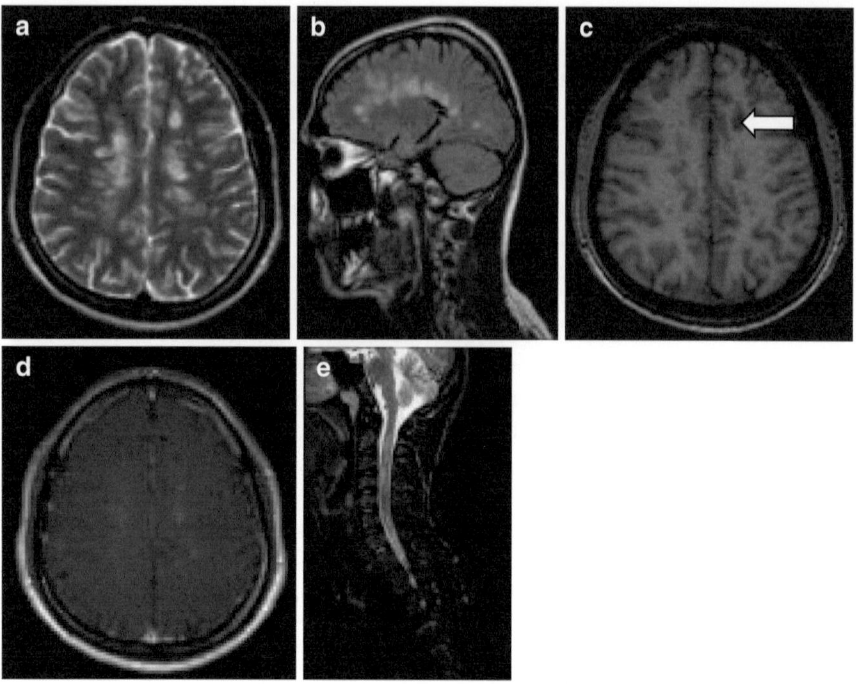

Fig. 6.8 Serial MRI images of the brain and spine with the following: (**a**) Axial T2, (**b**) Sagittal T2, (**c**) Axial T1, (**d**) MTC T1 C+, (**e**) Sagittal spine T2. (Figure courtesy of Dr. Samer Hoz)

Case 54: MS Brain and Spine

Case Scenario

A 23-year-old female presented with a history of bilateral visual loss 3 years ago, which she regained after undergoing treatment with steroids. Over the course of the past 3 years, she has experienced progressive weakness in her lower limbs (Fig. 6.8).

Imaging Description

The MRI reveals ovoid hyper-intensities on T2/FLAIR sequences arranged perpendicular to the body of the lateral ventricles and callososeptal interface, commonly referred to as Dawson's fingers. These structures exhibit low signals on T1WI, appearing as black holes (arrows). Post-gadolinium administration, there is evident ring enhancement. Similar lesions are also identified in the brainstem. Additionally, multiple demyelinating plaques in the spinal cord are most pronounced on T2 and STIR sequences.

Multiple Sclerosis

Multiple sclerosis (MS) is a relatively common, chronic demyelinating disease that affects the CNS. It ranks second only to trauma as the leading cause of neurological disability in young adults. The name itself suggests its characteristic feature, which is the presence of multiple lesions in various regions of the brain occurring at different times. MS typically presents between adolescence and the age of 60, with the highest incidence observed around the age of 35. It is more commonly diagnosed in females, with a gender ratio of 2:1, indicating a higher predisposition among women.

The diagnosis of MS, as defined by the McDonald diagnostic criteria for MS, relies on a combination of clinical findings and various investigations. These investigations encompass a typical medical history, the presence of specific markers in CSF, such as oligoclonal bands, immunoglobulin G levels in serum, abnormal results in visual evoked potential tests or MRI, and the absence of an alternative diagnosis.

On radiographic examination, MS lesions can appear anywhere within the CNS. They typically exhibit an oval shape with a distribution around veins in the perivenular area. CT imaging often lacks specificity, and significant changes may be observed on MRI even when CT scans appear normal. MRI has transformed the diagnosis and monitoring of MS patients. It not only confirms the diagnosis according to the McDonald criteria but also aids in tracking treatment responses and assessing disease patterns.

On MRI, various sequences provide valuable insights into MS lesions:

- T1-weighted images typically show lesions as iso- to hypo-intense areas, with some potentially appearing as small hypo-intense lesions at the callososeptal interface (referred to as Venus necklace). Advanced disease may show hyperintense lesions associated with brain atrophy.
- T2-weighted images usually depict hyper-intense lesions, with acute lesions often displaying surrounding edema.
- SWI often reveals the central vein sign, indicating that most plaques are perivenular in nature.
- FLAIR images typically show hyper-intense lesions. Early signs, such as the ependymal dot-dash sign, may be visible. Dawson's fingers, which are best seen on parasagittal images and represent lesions arranged radially outward, may also be observed.
- T1 post-contrast images show enhancement in active lesions, often with incomplete enhancement around the periphery, referred to as the open ring sign.
- DWI and ADC maps may depict active plaques with either high or low ADC values, indicating altered diffusion. These lesions often exhibit an open-ring morphology [14, 15].

Differential Diagnosis

For intracranial disease, the differential includes almost all other demyelinating diseases as well as:

1. CNS fungal infection (e.g., *Cryptococcus neoformans*): Patients tend to be immunocompromised.
2. Mucopolysaccharidosis (e.g., Hurler disease): Congenital and occurs in a younger age group.
3. Marchiafava-Bignami disease (for callosal lesions).
4. Susac syndrome.
5. CNS manifestations of primary antiphospholipid syndrome.
6. Chronic lymphocytic inflammation with pontine perivascular enhancement is responsive to steroids.

For spinal involvement, the following should be considered:

1. Transverse myelitis.
2. Infection.
3. Spinal cord tumors (e.g., astrocytomas).

Questions

1. **MS, the FALSE answer is:**
   A. T1 post-contrast shows complete enhancement around the periphery.
   B. Dawson's fingers are seen on FLAIR as lesions extending outward radially in a triangular configuration.
   C. Features that suggest progressive disease include large numerous plaques and hyper-intense T1 lesions.
   D. A very early sign in MS is the ependymal dot-dash sign.
   E. Dawson's fingers are best seen in the parasagittal plane.
   The answer is A.
   Active lesions show enhancement that is often incomplete around the periphery (known as open ring sign).

Case 55: Glomus Jugulare Paraganglioma

Case Scenario

A 43-year-old female presented with pulsatile tinnitus and hearing loss (Fig. 6.9).

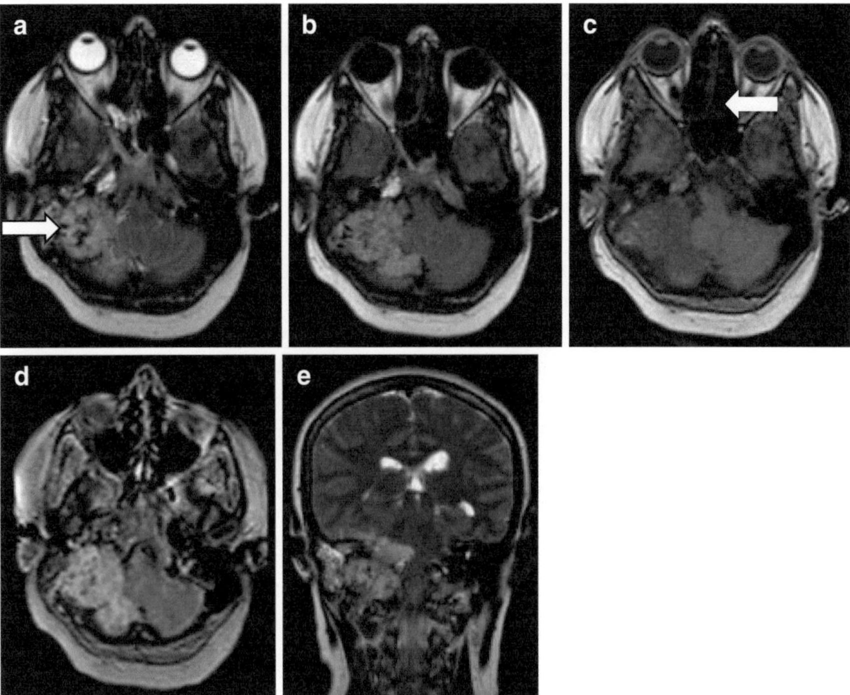

Fig. 6.9 Serial MRI images of the brain with the following: (**a**) Axial T2, (**b**) Axial FLAIR, (**c**) Axial T1, (**d**) Axial T1 C+, (**e**) Coronal T2. (Figure courtesy of Dr. Samer Hoz)

Imaging Description

The MRI reveals a large, lobulated, highly vascular mass lesion in the right jugular foramen, extending into the inner and middle ear. The lesion exhibits a heterogeneous signal and displays a classic "salt and pepper" appearance in both T1 and T2 sequences. The "salt" refers to areas representing hemorrhage, while the "pepper" signifies flow voids (arrow). In the coronal views, middle ear involvement is evident, causing a mass effect on the adjacent tissue.

Glomus Jugulare Paraganglioma

Glomus jugulare paraganglioma is a rare head and neck tumor primarily located within the jugular fossa. Despite its rarity, it is the most common tumor found in this area. These tumors typically affect adults, mostly in the age range of 40–60 years, and there is a slightly higher incidence in females.

Glomus jugulare tumors are categorized based on their location of origin within the jugular foramen rather than their anatomical source. They can originate from Jacobson's nerve, Arnold's nerve, or the jugular bulb. Although these tumors are rarely bilateral and do not often metastasize, they have a tendency to locally infiltrate nearby structures. In some cases, other tumors, like carotid body tumors, may coexist, and up to 10% of patients may have multiple lesions. These tumors typically grow in various directions, extending into the mastoid air cells, middle ear, and Eustachian tube.

Due to the tumor's proximity to bone, CT is the preferred imaging modality for assessing the bony margins of the tumor. These margins are typically irregularly eroded, resembling a moth-eaten pattern. As the tumor continues to grow, it may erode the jugular spine projecting from the jugular foramen and extend into the middle ear, as well as inferiorly into the infratemporal fossa. CT scans are also useful for evaluating the integrity of the ossicles and bony labyrinth, but radiologists may observe the Phelp sign, indicating erosion of the caroticojugular spine.

DSA reveals intense enhancement of the tumor, often associated with a prominent feeding vessel, most commonly the ascending pharyngeal vessel. Early draining veins may also be visible due to intra-tumoral shunting. DSA can play a role in preoperative embolization, typically performed 1–2 days before surgery. However, careful evaluation of feeding vessels is essential, and knowledge of the vascular anatomy in this region is crucial to prevent complications.

On MRI, glomus jugulare paragangliomas typically appear with low signal intensity on T1-weighted images and high signal intensity on T2-weighted images. T1 post-contrast images display marked and intense enhancement, often referred to as the "salt and pepper appearance." This appearance is visible on both T1 and T2 sequences, with the "salt" representing blood products from hemorrhage or slow flow and the "pepper" representing flow voids due to high vascularity. It is important to note that while this appearance is characteristic, it is not specific to glomus tumors and can sometimes be observed in other highly vascular lesions. Smaller glomus tumors may not exhibit this feature [16–19].

Differential Diagnosis

1. Jugular schwannoma: These tumors exhibit well-defined, smooth bony margins without internal flow voids. They tend to be less vascular when observed on angiography and do not display the characteristic "salt and pepper appearance." Additionally, indium-111 labeled octreotide scans typically yield negative results for these tumors.
2. Bony metastases: Tumors with high vascularity, such as renal cell carcinoma and thyroid cancer, can appear very similar on imaging.
3. Meningioma.
4. Normal anatomical variation: Asymmetry in jugular foraminal size and variations in the position of the jugular bulb can occur as part of normal anatomical diversity.
5. Thrombosis of jugular bulb.
6. Endolymphatic sac tumor: This type of tumor is typically located within the vestibular aqueduct rather than the jugular bulb.

Questions

1. **Glomus jugulare paraganglioma, the FALSE answer is:**
   A. Glomus jugulare tumors show local invasion but rarely metastasize.
   B. Nuclear medicine visualization with SPECT requires the tumor diameter to be larger than 1.5 cm.
   C. The Phelp sign on MRI is the presence of flow voids due to hypervascularity.
   D. The salt and pepper appearance on post-contrast MRI represents hemorrhage and flow voids respectively.
   E. CT is the optimal modality for visualizing bony margins, revealing moth-eaten pattern erosions extending from the middle ear to the infratemporal fossa.
   The answer is C.

   The Phelp sign can be seen on CT between the carotid canal and jugular fossa as erosion of the caroticojugular spine.

Case 56: Sphenoid Bone Fibrous Dysplasia

Case Scenario

A 12-year-old male presented with headache and visual disturbance (Fig. 6.10).

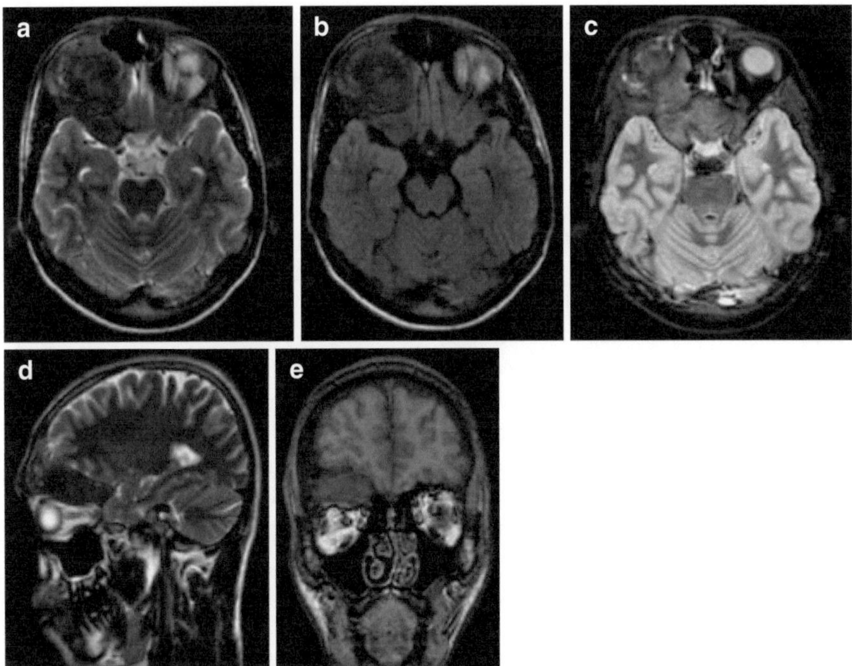

Fig. 6.10 Serial MRI images of the brain with the following: (**a**) Axial T2, (**b**) Axial T1, (**c**) Axial STIR, (**d**) Sagittal T2, (**e**) Coronal T1. (Figure courtesy of Dr. Samer Hoz)

Imaging Description

The MRI reveals characteristic features of fibrous dysplasia, manifesting as heterogeneous but predominantly low signal changes within the affected bones, with a more pronounced impact on the sphenoid sinus (more on the right side) and the anterior part of the frontal bone. There is bone expansion causing a regional mass effect on the ipsilateral frontal sinus and the right orbit, leading to obliteration of both sphenoid sinuses, indentation of the right optic canal, and subsequent compression of the intracanalicular part of the optic nerve. The diagnosis of fibrous dysplasia is confirmed through histopathological examination.

Fibrous Dysplasia

Fibrous dysplasia is a rare congenital, non-cancerous medullary disease characterized by the presence of a fibro-osseous process. It results in bone formation with a woven structure that can be multifocal, preventing the formation of mature lamellar bone. Fibrous dysplasia can affect any bone and can occur either in a single bone (monostotic) or involve multiple bones (polyostotic). In the WHO classification of soft tissue and bone tumors (5th edition), it is categorized as a benign bony neoplasm. Although uncommon, fibrous dysplasia can occur in individuals of all age groups, including children and adults. It is most commonly first diagnosed in children and young adults. The incidence is estimated to be around 5% of benign bone lesions, and gender does not appear to play a significant role in its occurrence.

Typical imaging features of fibrous dysplasia include an intramedullary, expansile lesion with well-defined borders. The lesion often maintains a smooth cortical contour, and endosteal scalloping may be present. On plain radiographs, fibrous dysplasia can manifest with three main imaging patterns: cystic/lucent, sclerotic, or mixed. In addition, the appearance of these lesions is usually smooth and homogeneous, with signs of endosteal scalloping and thinning of the cortex, despite the cortex appearing intact due to the expansive nature of the lesion. Other common features include a ground-glass appearance of the matrix, the absence of periosteal reaction, and the presence of a rind sign.

CT imaging is the preferred modality for evaluating fibrous dysplasia, especially in the context of craniofacial lesions. CT findings may include ground-glass opacities (56%), homogeneously sclerotic lesions (23%), or cystic areas (21%), with well-defined borders, bone expansion, and intact overlying bone. The lesions typically exhibit endosteal scalloping and can have attenuation values ranging from 60 to 140 Hounsfield units (HU). They often demonstrate contrast enhancement on post-contrast imaging.

MRI is less helpful in distinguishing fibrous dysplasia from other conditions because of the marked variability in the appearance of bone lesions, which can sometimes resemble tumors or more aggressive lesions. On MRI, lesions typically show intermediate to low heterogeneous signal on T1-weighted images, variable signal on T2-weighted images, and heterogeneous moderate to avid contrast enhancement on post-contrast T1-weighted images.

Nuclear medicine studies, such as Tc99 bone scans, may reveal increased tracer uptake in fibrous dysplasia lesions, indicating ongoing metabolic activity even in adulthood [20, 21].

Differential Diagnosis

1. Paget's disease: On histological examination, Paget's disease of bone exhibits a mosaic pattern of bone. Radiographically, it may display similarities with other conditions, but it typically affects a different demographic group.
2. Neurofibromatosis type 1: Osseous lesions are not a common feature of NF1. When they do occur, the vertebral column is the primary site affected, and additional characteristics of the disease are usually present.
3. Osteofibrous dysplasia: This condition is almost exclusively observed in the tibia, often leading to anterior bowing of the bone. The lesion originates in the cortex and is typically seen in children under the age of 10.
4. Adamantinoma: Approximately 80% of adamantinomas are located in the tibia and may appear very similar to other conditions, such as osteofibrous dysplasia.
5. Non-ossifying fibroma.
6. Simple bone cyst.
7. Giant cell tumor (GCT).
8. Enchondromatosis.
9. Hemangioma.

Questions

1. **Fibrous dysplasia, the FALSE answer is:**
   A. Appears as an intramedullary, expansile lesion with well-defined borders on plain radiographs.
   B. CT is the modality of choice showing ground glass opacities that are homogenously sclerotic.
   C. Endosteal scalloping is seen on plain radiographs and CT.
   D. Paget's disease may present similarly radiologically, but it affects distinct demographics.
   E. Is always monostotic.
      The answer is E.

      It can affect any bone and occur in only one bone (monostotic) or involve multiple bones (polyostotic).

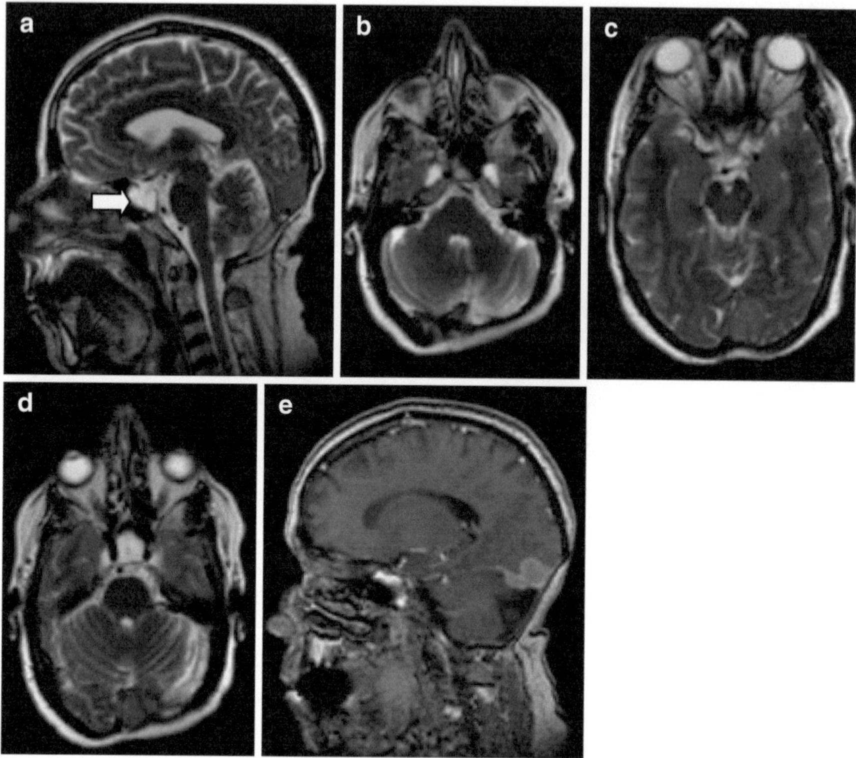

Fig. 6.11 Serial MRI images of the brain with the following: (**a**) Sagittal T2, (**b**) Axial T2, (**c**) Axial T2, (**d**) Axial T2, (**e**) Sagittal T1C +. (Figure courtesy of Dr. Samer Hoz)

Case 57: Idiopathic Intracranial Hypertension (IIH)

Case 57.1: Secondary IIH (Meningioma Invasion of the Cerebral Venous Sinus)

Case Scenario

A 50-year-old female with a history of operated brain tumor presented with headache, vomiting, and photophobia (Fig. 6.11).

Imaging Description

MRI reveals vertical tortuosity of the optic nerves, enlargement of the subarachnoid space around the optic nerves, and an enlarged and empty sella turcica (arrow) in axial T2, depicting the infundibulum sign. Additionally, Meckel's cave appears enlarged. There is a homogeneously enhancing mass lesion in the infra-tentorial regions that is attached to the tentorium cerebelli and displays a dural tail. The mass lesion appears to be invading the cerebral venous sinuses.

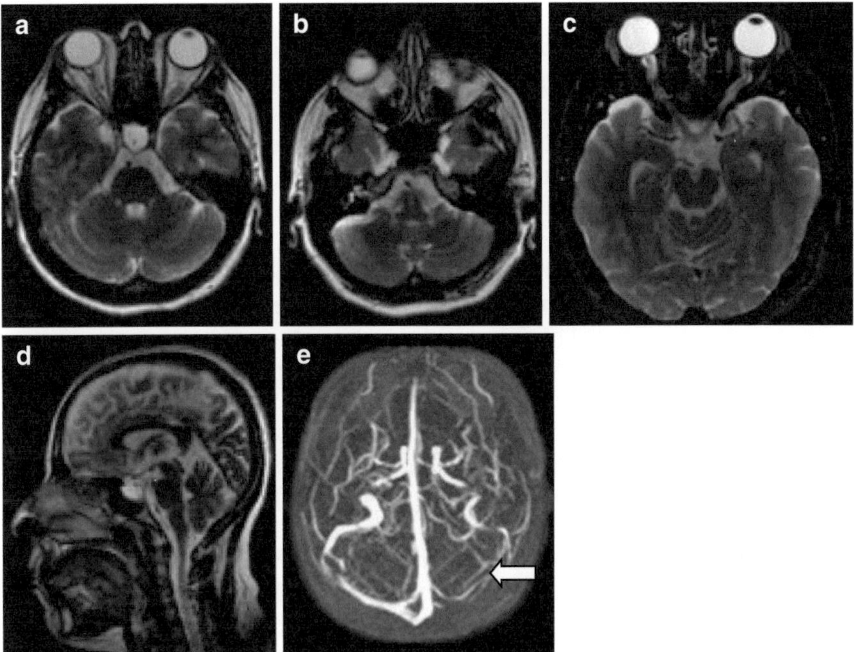

Fig. 6.12 Serial MRI images of the brain with the following: (**a**) Axial T2, (**b**) Axial T2, (**c**) Axial spectral presaturation with inversion recovery (SPIR), (**d**) Sagittal T2, (**e**) Magnetic resonance venography. (Figure courtesy of Dr. Samer Hoz)

Case 57.2: Primary IIH

Case Scenario

A 36-year-old female presented with headache, vomiting, and photophobia. Fundal examination revealed papilledema (Fig. 6.12).

Imaging Description

The MRI sequences reveal several findings: there is vertical tortuosity of the optic nerves, enlargement of the subarachnoid space around the optic nerves, and an empty sella turcica (seen as the infundibulum sign on axial T2). Additionally, Meckel's cave appears enlarged. The cerebral venous sinuses are patent, but there is evidence of smooth stenosis and attenuation, particularly in the lateral aspects of both transverse sinuses, more pronounced on the left side (arrow). There is no evidence of a mass lesion in the infra- or supra-tentorial regions.

IIH

Idiopathic Intracranial Hypertension (IIH), also known as pseudotumor cerebri, is a syndrome characterized by signs and symptoms of elevated ICP with an unknown cause. No hydrocephalus or mass lesion is identified as the underlying reason for this condition. The previously used term "benign intracranial hypertension" has fallen out of favor because some IIH patients can experience a rapidly deteriorating clinical condition that leads to permanent vision loss. Additionally, some experts prefer to revert to the older term "pseudotumor cerebri" because it was discovered that certain patients had IIH due to an identifiable secondary cause, such as venous stenosis. In such cases, the increased ICP is not considered idiopathic and is referred to as secondary intracranial hypertension.

Typically, IIH affects obese females of childbearing age who are otherwise healthy. However, the exact relationship between being female, obesity, and developing IIH is still not fully understood. IIH can also be found, to a lesser extent, in males who are typically older and less likely to be obese. In the pediatric population, IIH is rare and usually occurs after puberty, typically between the ages of 12 and 17.

In cases of suspected intracranial hypertension, it is essential to perform CT and MRI scans to rule out other conditions that could cause increased ICP, such as hydrocephalus, brain tumors, or dural venous sinus thrombosis. One notable distinction between IIH and normal-pressure hydrocephalus is that IIH patients do not exhibit hydrocephalus on CT or MRI scans. In children under 18 months of age, hydrocephalus might be observed because their skull sutures have not yet closed. The skull becomes fully ossified around the age of 3 years, after which ventricular expansion ceases.

When there is no apparent cause for intracranial hypertension, specific imaging features have been associated with IIH, including:

- Papilledema, characterized by flattening of the posterior sclera and intraocular protrusion of the optic nerve head (observed in approximately 80% of cases).
- Prominent subarachnoid space around the optic nerves (seen in about 45% of cases).
- Vertical tortuosity of the optic nerves (observed in around 40% of cases).
- Enhancement of the prelaminar (intraocular) optic nerves (seen in approximately 50% of cases).

In addition to these optic nerve-related features, IIH can also manifest with other imaging findings, such as:

- Enlarged arachnoid outpouchings due to prolonged increased ICP are often associated with findings like a partially empty sella (seen in approximately 70% of cases), prominent arachnoid pits, aberrant arachnoid granulations, prominent perivascular spaces, enlarged Meckel's cave, small meningoceles commonly found within the temporal bone and sphenoid wing, and enlargement of other CSF spaces around the oculomotor nerve in the lateral wall of the cavernous sinus.
- Bilateral venous sinus stenosis, especially affecting the lateral segment of the transverse sinus with no evidence of current or previous thrombosis, which is considered the most sensitive and specific finding.
- Slit-like ventricles, although this is less common compared to other findings.
- Acquired tonsillar ectopia, which can mimic Chiari I malformation when the tonsils descend below the level of the foramen magnum.
- Increased thickness of subcutaneous fat in the scalp and neck region [22, 23].

Differential Diagnosis

Other potential causes of intracranial hypertension and papilledema should be thoroughly investigated. Other conditions that could lead to venous obstruction include venous sinus thrombosis and occlusion in the neck's venous outflow, as they can present with similar intracranial findings as those seen in IIH. Additionally, there is a significant overlap in the demographic profile and clinical presentation of patients with prominent cerebellar tonsillar ectopia and those diagnosed with Chiari I malformation.

Questions

1. **IIH, the FALSE answer is:**
   A. The main role of MRI and CT scans is to exclude other causes of intracranial hypertension.
   B. Papilledema is the hallmark of IIH.
   C. Bilateral transverse sinus stenosis can be associated with IIH.
   D. Acquired tonsillar ectopia can be mistaken for Chiari I malformation.
   E. Slit-like ventricles are common in IIH.
   The answer is E.
   Slit-like ventricles are uncommon in IIH.

Case 58: ADEM

Case Scenario

A 14-year-old female presented with progressive right hemiparesis, which developed following a respiratory viral infection a week earlier (Fig. 6.13).

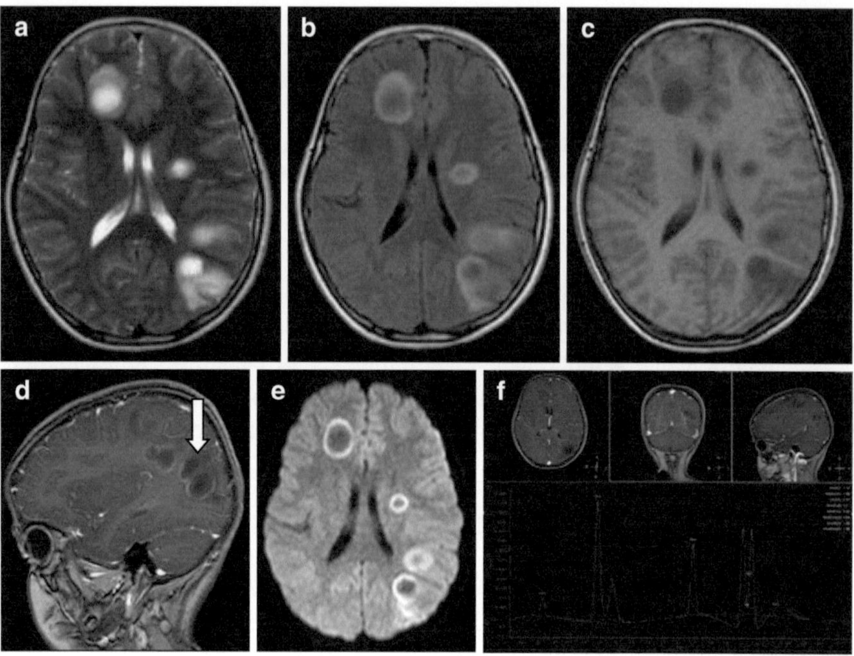

Fig. 6.13 Serial MRI images of the brain with the following: (**a**) Axial T2, (**b**) Axial FLAIR, (**c**) Axial T1, (**d**) Sagittal T1 C+, (**e**) Axial DWI, (**f**) MRS. (Figure courtesy of Dr. Samer Hoz)

Imaging Description

The MRI findings reveal bilateral asymmetrical multiple lesions, more pronounced on the left side. These lesions exhibit a well-defined white matter and subcortical involvement with a high T2/FLAIR signal, a low T1 signal surrounded by edema, and peripheral incomplete (open-ring) enhancement (arrow). The lesions show peripheral high signal intensity on DWI and peripheral low signal intensity on the ADC map, denoting true restricted diffusion along the C-shaped enhancing rim. MRS demonstrates increased Cho, decreased NAA, and increased lipids and lactate.

Acute Disseminated Encephalomyelitis (ADEM)

ADEM is an immune-mediated condition characterized by acute inflammation and multifocal demyelination of the white matter, typically occurring after recent viral infection (within 1–2 weeks) or vaccinations. Gray matter, particularly in the basal ganglia and, to a lesser extent, in the spinal cord, can also be affected. ADEM can develop at any age but primarily affects children and teenagers, typically those under 15 years old. Some studies have noted seasonal peaks in winter and spring, supporting the idea of an infectious cause. ADEM triggered by vaccination is relatively rare, accounting for less than 5% of cases. Unlike some other demyelinating disorders like MS or neuromyelitis optica, ADEM does not show a clear gender preference, although some studies have reported a slight male predominance in pediatric cases.

Radiographically, ADEM lesions can vary from small punctate spots to larger tumefactive regions, often with less mass effect than expected based on their size. These lesions are distributed bilaterally in the white matter of the brain, both supratentorial and infratentorial, but they tend to have an asymmetrical pattern. Unlike MS, involvement of the callososeptal interface is uncommon. Additionally, while cerebral cortex, subcortical gray matter (especially the thalami), and the brainstem can be affected, these areas are less commonly involved in ADEM but can help distinguish it from MS. In some cases, spinal cord lesions with variable enhancement can be seen, especially in about one-third of cases. However, even with these clinical and radiological differences, distinguishing ADEM from MS can still be challenging during the initial presentation.

On CT scans, ADEM lesions typically appear as poorly defined areas of low-density white matter and may exhibit ring enhancement.

MRI is the preferred imaging modality for ADEM, as it is more sensitive than CT in revealing characteristic demyelination lesions. On T2-weighted images, ADEM lesions present as regions of high signal intensity with surrounding edema, typically found in subcortical areas, while the thalami and brainstem can also be involved. On contrast-enhanced T1-weighted images, these lesions often show punctate, ring, or arc enhancement (referred to as the "open ring sign") along the leading edge of inflammation. However, the absence of enhancement does not rule out the diagnosis. DWI may reveal peripheral restriction of diffusion. Importantly, the central part of the lesion, although appearing hyper-intense on T2 and hypo-intense on T1, does not exhibit increased restriction on DWI (unlike cerebral abscess) because of the increase in extracellular water content in the demyelinated region. This can help differentiate ADEM from MS, as the normal-appearing brain in T2-weighted images has a normal magnetization transfer ratio and normal diffusivity, while both of these measures are decreased in MS [24, 25].

Differential Diagnosis

1. Susac syndrome.
2. MS.
3. Acute necrotizing encephalitis of childhood.
4. Acute hemorrhagic leukoencephalitis (Hurst disease).
5. Chronic lymphocytic inflammation with pontine perivascular enhancement responsive to steroids.

Questions

1. **ADEM, the FALSE answer is:**
   A. ADEM lesions appear as low-density regions within the white matter.
   B. ADEM lesions are hyper-intense on T2WI.
   C. Magnetization transfer ratio and diffusivity are increased in ADEM.
   D. ADEM lesions have bilateral and asymmetrical distribution.
   E. ADEM lesions can demonstrate an open ring sign on T1 C+ (Gd).

The answer is C.
Magnetization transfer ratio and diffusivity are normal in ADEM.

Case 59: Hypoxic-Ischemic Encephalopathy (HIE)

Case Scenario

A 14-year-old female, who attempted suicide by ingesting antihypertension medication, presented with seizures and coma (Fig. 6.14).

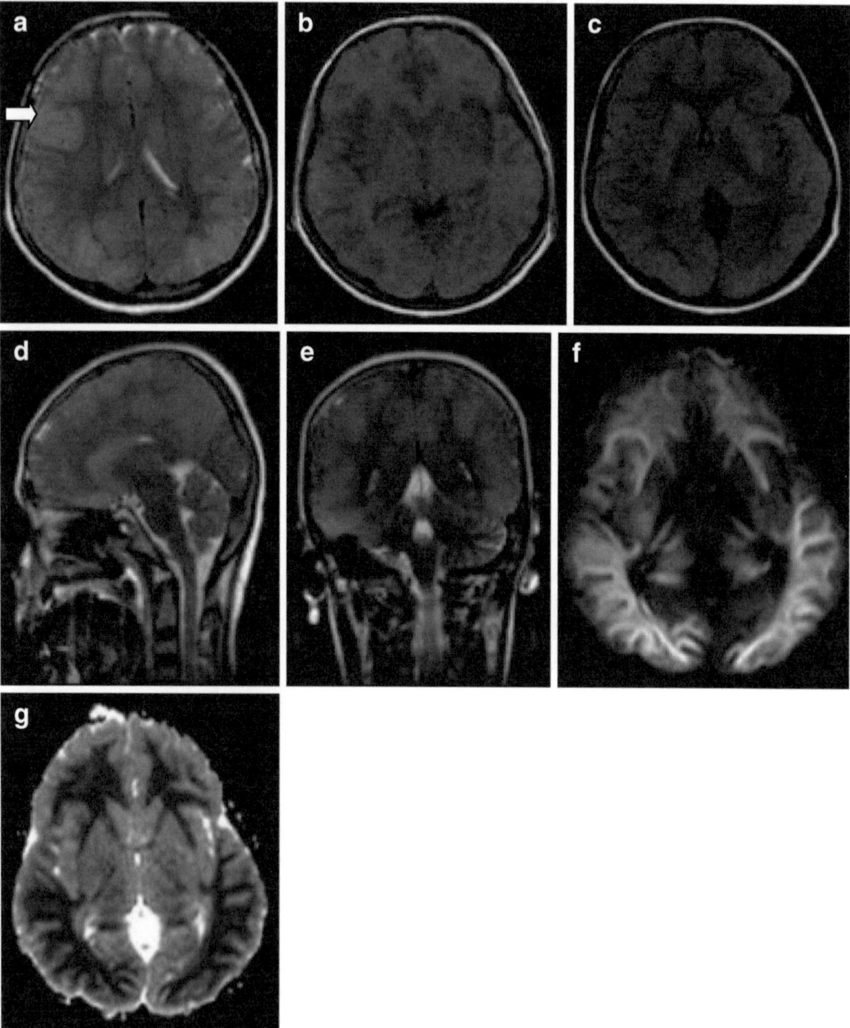

Fig. 6.14 Serial MRI images of the brain with the following: (**a**) Axial T2, (**b**) Axial T1, (**c**) Axial FLAIR, (**d**) Sagittal T2, (**e**) Coronal T2, (**f**) Axial DWI, (**g**) Axial ADC. (Figure courtesy of Dr. Samer Hoz)

Imaging Description

The MRI findings reveal diffuse cortical high-signal intensity and swelling of cerebral sulci on T2WI (arrow). There is effacement of the bilateral lateral ventricles and high-signal intensity in the cerebral and cerebellar hemispheres on DWI due to cytotoxic edema. There is no significant blooming observed on GRE.

Hypoxic-Ischemic Encephalopathy (HIE)

Hypoxic-ischemic encephalopathy (HIE), or global hypoxic-ischemic injury, is a condition that can occur across various age groups and often leads to severe neurological consequences. While it can affect adults and older children (excluding neonates), the underlying causes vary significantly between age groups. In older children, drowning and asphyxiation are common triggers, whereas in adults, it is more frequently associated with cardiac arrest or cerebrovascular disease, often resulting in secondary hypoxemia or hypoperfusion.

The patterns of severe global hypoxic-ischemic injury in adults exhibit substantial variability. While widespread involvement of the cortex and deep gray matter structures, particularly the basal ganglia, is common, several other patterns may also occur, such as:

1. Involvement of diffuse cortical and deep gray matter, including the perirolandic cortex (most prevalent).
2. Involvement of diffuse cortical and deep gray matter but sparing the perirolandic cortex.
3. Isolated perirolandic cortex damage affecting the primary motor/sensory areas and medial occipital lobes affecting the primary visual cortex.
4. Isolated damage to the basal ganglia.
5. Watershed distribution pattern.
6. Involvement of the diffuse white matter.
7. Damage to the cerebellum.

Due to the high metabolic demands of gray matter for oxygen and glucose, this type of tissue is particularly susceptible to hypoxic-ischemic injury. Gray matter contains the majority of dendrites housing postsynaptic glutamate receptors, making it vulnerable to glutamate excitotoxicity when nerve cells are excessively stimulated by glutamate.

In radiographic imaging, CT scans typically show diffuse edema, reduced cortical gray matter attenuation, and decreased bilateral basal ganglia attenuation. A reversal sign, possibly due to deep medullary vein distension and venous outflow obstruction, can be observed within 24 h. The white cerebellum sign may also be seen, characterized by diffuse edema and hypoattenuation of cerebral hemispheres. Later signs, such as linear hyperdensity and cortical enhancement, indicate cortical laminar necrosis. Pseudosubarachnoid hemorrhage may occur due to cerebral edema resulting in decreased parenchymal attenuation. Both the reversal sign and the white cerebellum sign indicate severe injury and are associated with a poor neurological prognosis.

MRI is a valuable imaging modality for HIE diagnosis, with diffusion-weighted MR imaging being the earliest to detect hypoxic-ischemic injury, typically becoming positive within the first few hours due to early cytotoxic edema. Restricted diffusion may be observed in various regions within the first 24 h, including the cerebellar hemispheres, basal ganglia, cerebral cortex (particularly the perirolandic and occipital cortices), thalami, brainstem, or hippocampi. Abnormalities in diffusion-weighted imaging often show pseudo-normalization by the end of the first week.

In younger patients, conventional T1-weighted and T2-weighted images frequently appear normal or display minor irregularities. During the early subacute period (24 h to 2 weeks), conventional T2-weighted images typically become positive, showing increased signal intensity and edema in the affected gray matter structures. T1 hyper-intensities, indicating cortical laminar necrosis, become evident after 2 weeks. This hyper-intense signal is attributed to denatured proteins accumulating in dying cells and can also be observed on FLAIR images within a few days [26–28].

Differential Diagnosis

Differential diagnosis includes molybdenum cofactor deficiency, an exceedingly rare and fatal genetic disorder.

Questions

1. **HIE, the FALSE answer is:**
   A. During the late subacute period, T2 imaging demonstrates reduced signal intensity and swelling of the injured white matter structures.
   B. The most common pattern of involvement is diffuse cortical and deep gray matter including the perirolandic cortex.
   C. T1 hyper-intensities appear due to denatured protein in dying cells.
   D. DWI is the imaging modality that demonstrates the earliest positive results following hypoxic-ischemic injury.
   E. Neurologic injury in HIE is caused by hypoxia or interruption of blood flow, usually from cardiac arrest or hanging.
   The answer is A.
   During the late subacute period, T2 demonstrates increased signal intensity and swelling of the injured gray matter structures.

2. **HIE, the FALSE answer is:**
   A. CT imaging may show high attenuation of the cerebral hemispheres relative to the cerebellum and brainstem.
   B. The linear hyperdensity outlining the cortex on CT imaging corresponds to cortical laminar necrosis.

C. The reversal sign is due to the distension of deep medullary veins secondary to partial obstruction of the venous outflow from elevated ICP caused by diffuse edema.

D. Pseudosubarachnoid hemorrhage seen on CT imaging may occur due to cerebral edema resulting in decreased parenchymal attenuation.

E. The predominant involvement of gray matter in ischemic-hypoxic injury is due to its high metabolic requirement of glucose and oxygen with a predominance of dendrites.

The answer is A.

CT imaging may show high attenuation of the cerebellum and brainstem relative to the cerebral hemispheres, known as the white cerebellum sign.

References

1. Zhong W, Huang Z, Tang X. A study of brain MRI characteristics and clinical features in 76 cases of Wilson's disease. J Clin Neurosci. 2019;59:167–74.
2. Yu XE, Gao S, Yang RM, Han YZ. MR imaging of the brain in neurologic Wilson disease. Am J Neuroradiol. 2019;40(1):178–83.
3. Carnero Contentti E, Okuda DT, Rojas JI, Chien C, Paul F, Alonso R. MRI to differentiate multiple sclerosis, neuromyelitis optica, and myelin oligodendrocyte glycoprotein antibody disease. J Neuroimaging. 2023;33(5):688–702.
4. Banwell B, Bennett JL, Marignier R, Kim HJ, Brilot F, Flanagan EP, Ramanathan S, Waters P, Tenembaum S, Graves JS, Chitnis T. Diagnosis of myelin oligodendrocyte glycoprotein antibody-associated disease: International MOGAD Panel proposed criteria. Lancet Neurol. 2023;22(3):268–82.
5. Deschamps R, Pique J, Ayrignac X, Collongues N, Audoin B, Zéphir H, Ciron J, Cohen M, Aboab J, Mathey G, Derache N. The long-term outcome of MOGAD: an observational national cohort study of 61 patients. Eur J Neurol. 2021;28(5):1659–64.
6. Cabal-Herrera AM, Tassanakijpanich N, Salcedo-Arellano MJ, Hagerman RJ. Fragile X-associated tremor/ataxia syndrome (FXTAS): pathophysiology and clinical implications. Int J Mol Sci. 2020;21(12):4391.
7. Hall DA, Birch RC, Anheim M, Jønch AE, Pintado E, O'Keefe J, Trollor JN, Stebbins GT, Hagerman RJ, Fahn S, Berry-Kravis E. Emerging topics in FXTAS. J Neurodev Disord. 2014;6:31.
8. Shofty B, Ben Sira L, Constantini S. Neurofibromatosis 1-associated optic pathway gliomas. Childs Nerv Syst. 2020;36:2351–61.
9. Friedrich RE, Nuding MA. Optic pathway glioma and cerebral focal abnormal signal intensity in patients with neurofibromatosis type 1: characteristics, treatment choices and follow-up in 134 affected individuals and a brief review of the literature. Anticancer Res. 2016;36(8):4095–121.
10. Salman MS, Hossain S, Alqublan L, Bunge M, Rozovsky K. Cerebellar radiological abnormalities in children with neurofibromatosis type 1: Part 1—Clinical and neuroimaging findings. Cerebellum Ataxias. 2018;5(1):14.
11. Alkan A, Sigirci A, Kutlu R, Ozcan H, Erdem G, Aslan M, Ates O, Yakinci C, Egri M. Neurofibromatosis type 1: diffusion weighted imaging findings of brain. Eur J Radiol. 2005;56(2):229–34.
12. Kuo CL, Shiao AS, Yung M, Sakagami M, Sudhoff H, Wang CH, Hsu CH, Lien CF. Updates and knowledge gaps in cholesteatoma research. Biomed Res Int. 2015;2015:854024.
13. Lingam RK, Connor SE, Casselman JW, Beale T. MRI in otology: applications in cholesteatoma and Ménière's disease. Clin Radiol. 2018;73(1):35–44.

14. Kearney H, Miller DH, Ciccarelli O. Spinal cord MRI in multiple sclerosis—diagnostic, prognostic and clinical value. Nat Rev Neurol. 2015;11(6):327–38.
15. Filippi M, Brück W, Chard D, Fazekas F, Geurts JJ, Enzinger C, Hametner S, Kuhlmann T, Preziosa P, Rovira À, Schmierer K. Association between pathological and MRI findings in multiple sclerosis. Lancet Neurol. 2019;18(2):198–210.
16. Lin EP, Chin BB, Fishbein L, Moritani T, Montoya SP, Ellika S, Newlands S. Head and neck paragangliomas: an update on the molecular classification, state-of-the-art imaging, and management recommendations. Radiol Imaging Cancer. 2022;4(3):e210088.
17. Alexopoulos G, Sappington J, Mercier P, Bucholz R, Coppens J. Glomus jugulare tumor presenting as mastoiditis in a patient with familial paraganglioma syndrome: a case report and review of the literature. Interdiscip Neurosurg. 2020 Jun;20:100657.
18. Noujaim SE, Pattekar MA, Cacciarelli A, Sanders WP, Wang AM. Paraganglioma of the temporal bone: role of magnetic resonance imaging versus computed tomography. Top Magn Reson Imaging. 2000;11(2):108–22.
19. Mafee MF, Raofi B, Kumar A, Muscato C. Glomus faciale, glomus jugulare, glomus tympanicum, glomus vagale, carotid body tumors, and simulating lesions: role of MR imaging. Radiol Clin North Am. 2000;38(5):1059–76.
20. Kinnunen AR, Sironen R, Sipola P. Magnetic resonance imaging characteristics in patients with histopathologically proven fibrous dysplasia—a systematic review. Skeletal Radiol. 2020;49:837–45.
21. Lisle DA, Monsour PA, Maskiell CD. Imaging of craniofacial fibrous dysplasia. J Med Imaging Radiat Oncol. 2008;52(4):325–32.
22. Bidot S, Saindane AM, Peragallo JH, Bruce BB, Newman NJ, Biousse V. Brain imaging in idiopathic intracranial hypertension. J Neuroophthalmol. 2015;35(4):400–11.
23. Boyter E. Idiopathic intracranial hypertension. JAAPA. 2019;32(5):30–5.
24. Pohl D, Alper G, Van Haren K, Kornberg AJ, Lucchinetti CF, Tenembaum S, Belman AL. Acute disseminated encephalomyelitis: updates on an inflammatory CNS syndrome. Neurology. 2016;87(9 Suppl 2):S38–45.
25. Garg RK. Acute disseminated encephalomyelitis. Postgrad Med J. 2003;79(927):11–7.
26. Douglas-Escobar M, Weiss MD. Hypoxic-ischemic encephalopathy: a review for the clinician. JAMA Pediatr. 2015;169(4):397–403.
27. Rutherford M, Malamateniou C, McGuinness A, Allsop J, Biarge MM, Counsell S. Magnetic resonance imaging in hypoxic-ischaemic encephalopathy. Early Hum Dev. 2010;86(6):351–60.
28. Huang BY, Castillo M. Hypoxic-ischemic brain injury: imaging findings from birth to adulthood. Radiographics. 2008;28(2):417–39.

Part II

Spine

Congenital Spinal Defects

7

Rokaya H. Abdalridha, Mahmood S. Bilal, Ahmed Y. Azzam,
Sara A. Mohammad, Muntadher H. Almufadhal,
Fatimah O. Ahmed, and Asmaa H. AL-Sharee

Case 60: Syringobulbia (SB)

Case Scenario

A 50-year-old female presented with progressive leg weakness and sensory changes (Fig. 7.1).

R. H. Abdalridha
College of Medicine, Babylon University, Babylon, Iraq

M. S. Bilal
College of Medicine, Al-Mustansiriyah University, Baghdad, Iraq

A. Y. Azzam
October 6 University Faculty of Medicine, Giza, Egypt

S. A. Mohammad · M. H. Almufadhal
College of Medicine, University of Baghdad, Baghdad, Iraq

F. O. Ahmed
College of Medicine, University of Mustansiriyah, Baghdad, Iraq

A. H. AL-Sharee (✉)
Department of Neuroradiology, Neurosurgery Teaching Hospital, Baghdad, Iraq

© The Author(s), under exclusive license to Springer Nature
Switzerland AG 2024
S. Hoz et al. (eds.), *Neuroradiology Board's Favorites*,
https://doi.org/10.1007/978-3-031-64261-6_7

193

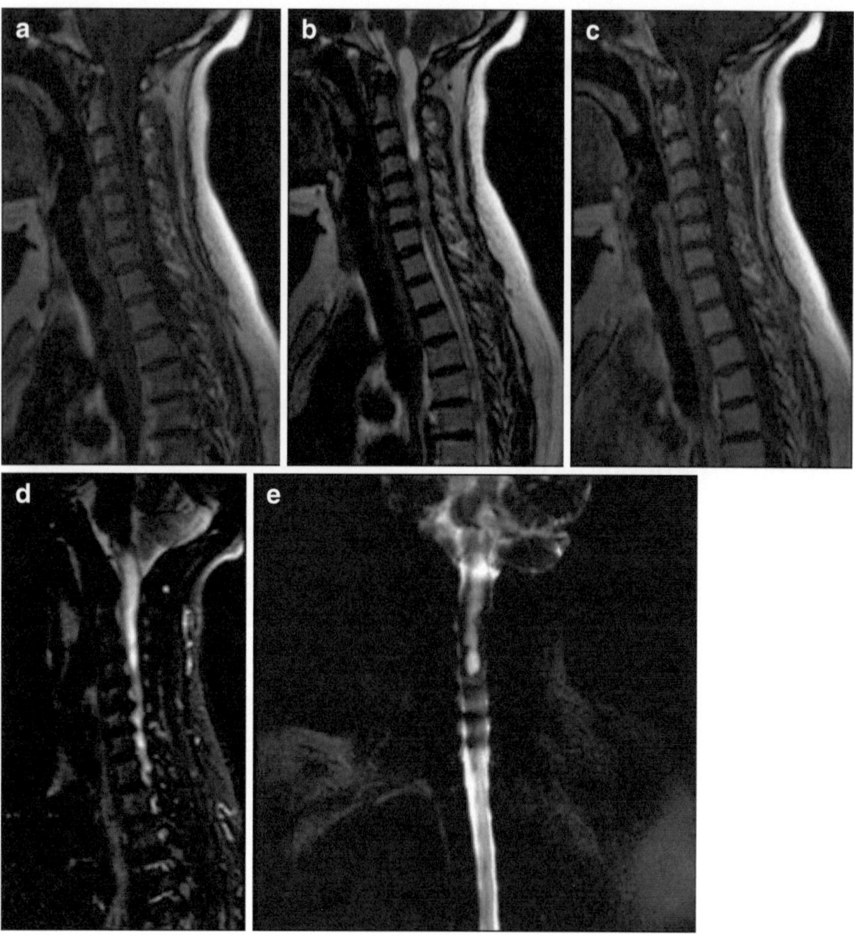

Fig. 7.1 Serial MRI images of the spine with the following: (**a**) Sagittal T1, (**b**) Sagittal T2, (**c**) Sagittal T1 C+, (**d**) Sagittal SPIR, (**e**) Myelography. (Figure courtesy of Dr. Samer Hoz)

Imaging Description

The cervical cord exhibits significant expansion due to cystic dilatation that extends to the medulla oblongata, representing a prominent syrinx. No solid or enhancing components can be identified.

Syringobulbia (SB)

SB is a pathological condition characterized by the presence of a syrinx extending into the medulla oblongata. However, some authors use the term SB to refer to a syrinx in any part of the brainstem, encompassing syringopontia and syringomesencephaly. This rare condition can have congenital or acquired origins and can occur either independently or secondary to neoplasms, cervical junction deformities (e.g., Chiari malformation), spinal cord traumas or lesions, meningitis, and other etiologies. It can also develop as an extension of syringomyelia above the foramen magnum, sharing a similar pathogenesis involving the accumulation of cerebrospinal fluid within the spinal cord with dissection through the ependymal lining of the central canal. On the other hand, SB can also occur due to the loss of neural tissue, which is subsequently replaced by fluid.

Radiological imaging of SB reveals thin, irregular, slit-like clefts or cavities within the brainstem without evidence of CSF flow void on T2W images and is best visualized on T1W images. The CSF-filled cavity may extend beyond the medulla and, in rare instances, involve the pons or midbrain. Communication between the SB lesion and the fourth ventricle, as well as syringomyelia, may or may not be apparent in MRI sequences, even in cases that are confirmed surgically. This variability could be attributed to the limitations of spatial resolution and partial volume averaging (PVA) artifacts in MRI. Slit-like SB cavities can often be as small as 1 mm, while slice thickness is typically greater than 3–5 mm [1–3].

Questions

1. **SB, the FALSE answer is:**
   A. SB occurs early in the disease progression of syringomyelia.
   B. The CSF-filled cavity can extend to the pons and midbrain.
   C. Vestibular symptoms and trigeminal sensory loss are the most common manifestations of syringobulbia.
   D. The bulbar root of the trigeminal nerve is also almost always involved.
   E. Cavities between the pyramid and the inferior olive interrupt the emerging fibers of the hypoglossal nerve.
   The answer is A.
   Extension of the syrinx into the medulla oblongata is a late progression of syringomyelia.

Case 61: Split Cord Malformation (SCM)

Case 61.1: SCM Type I (Diastematomyelia)

Case Scenario
A 16-year-old male presented with a history of bilateral leg weakness (Fig. 7.2).

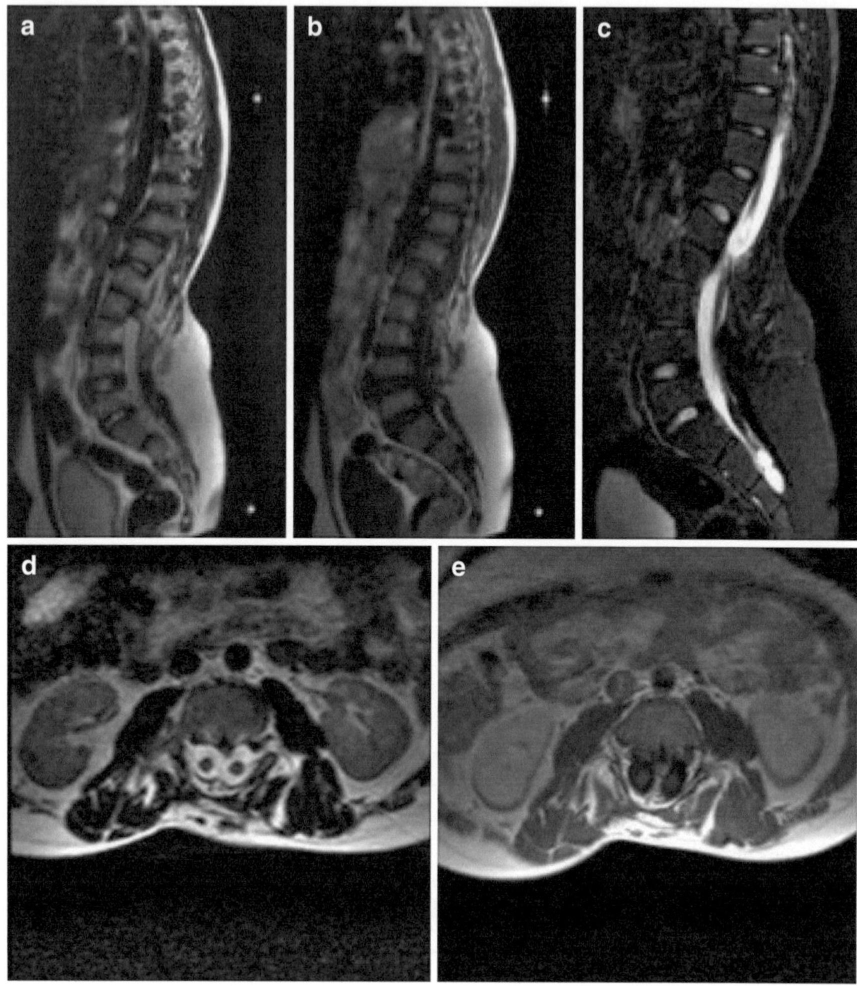

Fig. 7.2 Serial MRI images of the spine with the following: (**a**) Sagittal T2, (**b**) Sagittal T1, (**c**) Sagittal SPIR, (**d**) Axial T2, (**e**) Axial T1 C+. (Figure courtesy of Dr. Samer Hoz)

Image Description

The MRI findings reveal a low-lying spinal cord that terminates at the S1–S2 level, accompanied by small sacral Tarlov cysts. An axial T2-weighted MRI at the lumbar L1–L2 level demonstrates hemicords separated by a bony spur, characteristic of SCM type I. Spina bifida is evident at multiple levels, without an accompanying meningocele or myelomeningocele. Mild scoliosis is present.

Case 61.2: SCM Type II (Diastematomyelia)

Case Scenario

A 6-year-old female presented with a history of bilateral leg weakness, low back pain, and incontinence (Fig. 7.3).

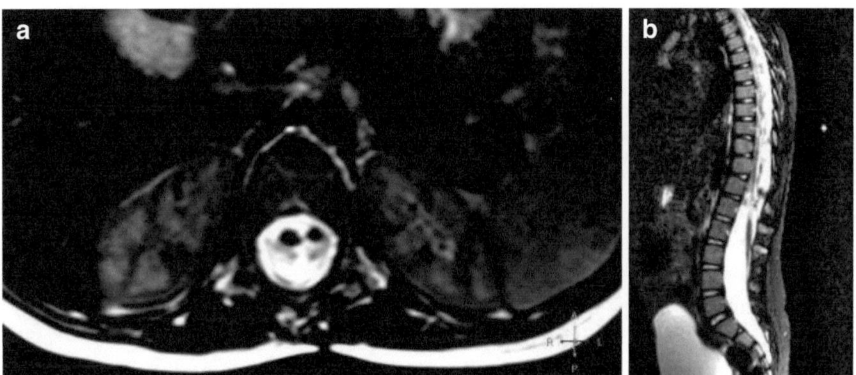

Fig. 7.3 Serial MRI images of the spine with the following: (**a**) Axial T2, (**b**) Sagittal T2. (Figure courtesy of Dr. Samer Hoz)

Image Description

The axial T2W MRI obtained at D9–D10 level demonstrates hemicords within a single dural sac with no spur/septum diplomyelia. Sagittal T2W MRI reveals conus medullaris terminating at L2–L3 with a distended thecal sac.

Split Cord Malformation (SCM)

SCM is rare, accounting for approximately 5% of all congenital spinal cord anomalies, and falls under the category of closed spinal dysraphism. SCM is characterized by a sagittal split of variable extent within the spinal cord, conus medullaris, or filum terminale. They are categorized based on a single versus duplicated dural sac and the presence of a dividing septum: In type I, each hemicord is contained in its own dural sac and is separated by a rigid osseous or osseocartilaginous septum. In type II, the two hemicords are contained within a single dural sac. The spinal cord may be partially or completely split in type II SCM, and a nonrigid fibrous septum may be present in the latter.

Patients with type I SCM experience symptoms similar to other types of spinal tethering lesions, including lower extremity weakness/loss of sensation, urological symptoms, and orthopedic deterioration. In contrast, type II patients are less symptomatic, and some may even be asymptomatic. Additionally, SCM type I may be associated with vertebral abnormalities such as hemivertebrae, butterfly vertebrae, spina bifida, and fusion of vertebrae and is often accompanied by scoliosis. On the other hand, type II may be associated with spina bifida, but other vertebral anomalies are far less common.

Cutaneous manifestations of dysraphism are common in SCM patients and most commonly include hypertrichosis at the level of the malformation. Other cutaneous manifestations include capillary angiomas, subdermal lipomas (strongly associated with type II SCM), dermal nevi, crypts, and sinuses.

MRI is the preferred imaging modality for assessing SCM, allowing visualization of the two hemicords and identification of associated anomalies such as hydromyelia. Bony malformations and dysplasias are generally recognizable on plain radiographs and, if clinically indicated, can be further investigated by CT, which may also reveal the bony septum.

In prenatal cases, ultrasound may reveal an extra echogenic focus in the midline between the fetal spinal posterior elements, which is considered a reliable sign. The confirmed diagnosis can be made through MRI [4–7].

Questions

1. **SCM, the FALSE answer is:**
   A. Considered as a rare spinal cord congenital anomaly (approximately 5%).
   B. Belongs to the spina bifida occulta group.
   C. Characterized by a longitudinal split in the spinal cord.
   D. Is more common in the upper cord.
   E. Divided into two types according to the presence of a rigid dividing septum and single versus dual dural sac.
   The answer is D.
   SCMs are more common in the lower cord but sometimes occur at multiple levels, with 50% occurring between L1 and L3 and 25% between D7 and D12.

2. **SCM types, the FALSE answer is:**
   A. Type I SCM has a rigid midline septum.
   B. Type II SCM does not have a rigid midline septum.
   C. In type II SCM, a nonrigid fibrous septum may present.
   D. Most type II SCMs do not have a septum.
   E. Type I SCMs have a common dural sac.
   The answer is E.
   Type I SCM is characterized by a duplicated dural sac, while the presence of a common dural sac is characteristic of type II diastematomyelia.

3. **SCM clinical picture, the FALSE answer is:**
   A. Patients may exhibit symptoms similar to those of tethered cord syndrome.
   B. Hydromyelia is common in type I.
   C. The most commonly associated spine deformity is lordosis.
   D. Vertebral abnormalities such as spina bifida may be present.
   E. The most common cuteaneous manifestation is hypertrichosis at the level of the malformation.
   The answer is C.
   Scoliosis is the most commonly associated spine deformity.

4. **SCM radiology, the FALSE answer is:**
   A. MRI is the preferred technique for assessment.
   B. MRI may reveal hydromyelia and numerous associated anomalies.
   C. In type I, CT may demonstrate the bony septum.
   D. X-rays may reveal bony malformations and dysplasia.
   E. Antenatal ultrasound may reveal a hypoechogenic focus between the fetal spinal posterior elements.
   The answer is E.
   Antenatal ultrasound may reveal a hyperechogenic focus in the midline between the fetal spinal posterior elements.

Case 62: Lateral Thoracic Meningocele (LTM)

Case Scenario

A 30-year-old female presented with back pain and scoliosis (Fig. 7.4).

Image Description

MRI reveals the presence of bilateral LTM extending from D8 to D10 with upper dorsal kyphoscoliosis convex to the left. The meningocele exhibits lateral expansion

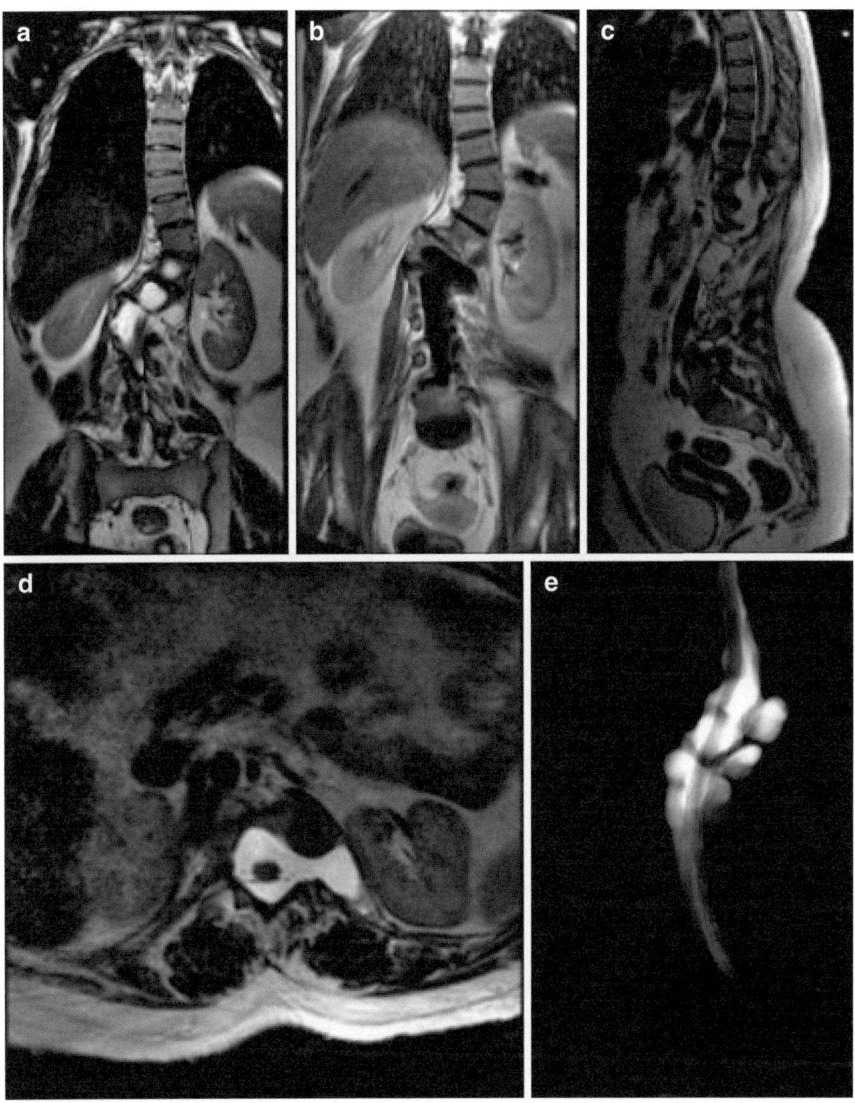

Fig. 7.4 (a) Coronal T2, (b) Coronal T1, (c) Sagittal T2, (d) Axial T2, (e) Myelography. (Figure courtesy of Dr. Samer Hoz)

through the left D11–D12 and D12–L1 foramina with a circumferential dural envelope, causing scalloping of the posterior aspect of the corresponding vertebrae. The spinal cord is displaced anteriorly to the right side. There is no lumbar spinal dysraphism, and the conus terminates normally at L1. There is no stigmata of NF1 or Marfan's syndrome.

Lateral Thoracic Meningocele (LTM)

LTM, like other meningoceles, involves the herniation of the meninges through a vertebral column defect or foramen. While it is commonly associated with neurofibromatosis (NF) and collagen disorders like Marfan's syndrome, isolated occurrences are rare. In cases with scoliosis or kyphosis, the meningocele is typically located at the apex of the convexity.

LTM is often asymptomatic and is frequently discovered incidentally during middle age as a paraspinal mass. However, some individuals may experience pain and dyspnea, with generalized backache or nerve root pain. Diffuse back pain is believed to result from the disruption of intervertebral joints due to spinal deformity.

On plain radiography, associated features such as kyphoscoliosis, posterior vertebral scalloping, and widened intervertebral foramina may be visible. CT may reveal a well-circumscribed, low-attenuation paravertebral mass, occasionally displaying peripheral rim enhancement. MRI typically shows a lesion originating from the intervertebral foramen, following CSF signal characteristics [8–10].

Questions

1. **LTM radiographic features, the FALSE answer is:**
   A. CT scan demonstrates a well-circumscribed, low-attenuation paravertebral mass.
   B. CT scan may reveal peripheral rim enhancement.
   C. MRI reveals a lesion that follows the CSF signal.
   D. Plain radiograph is not helpful in detecting vertebral anomalies.
   E. The mass is often visible at the apex of the scoliotic curve.
   The answer is D.

   A plain radiograph has the capability to detect rib or vertebral anomalies in patients with LTM.

Case 63: Tethered Cord Syndrome (TCS)

Case Scenario

A 15-year-old male presented with a history of urinary incontinence (Fig. 7.5).

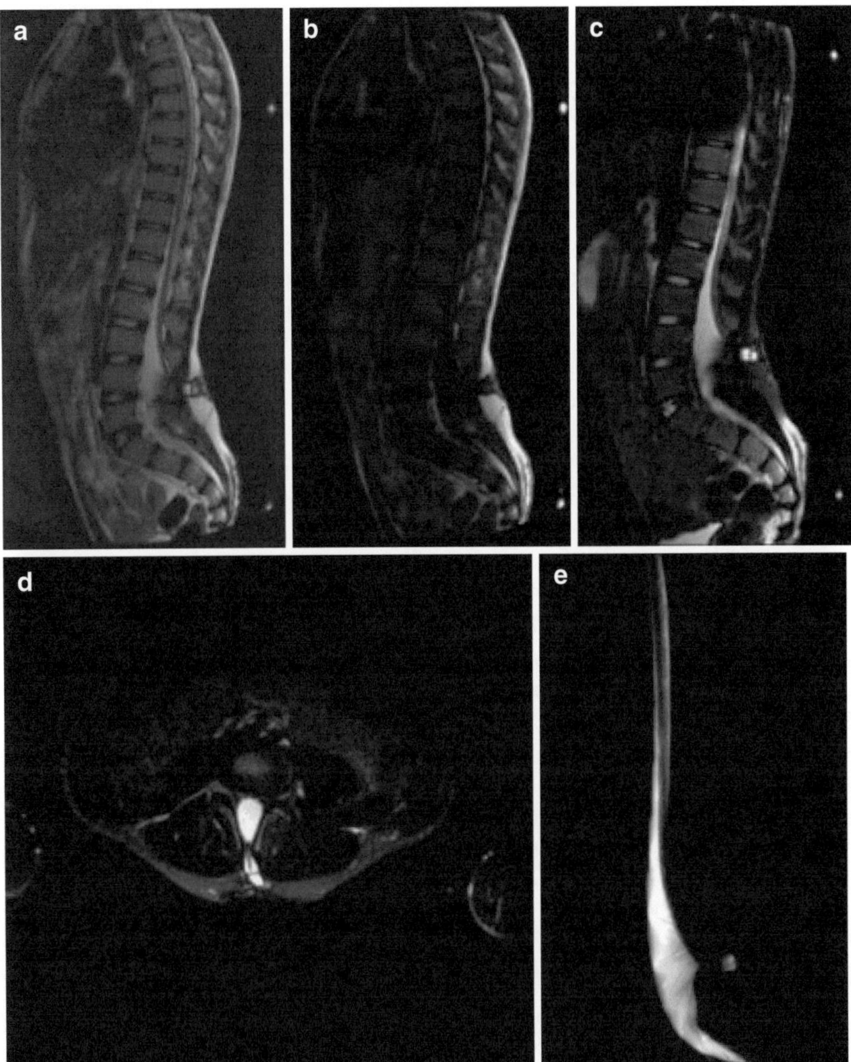

Fig. 7.5 (**a**) Sagittal T2, (**b**) Sagittal T1, (**c**) Sagittal SPIR, (**d**) Axial SPIR, (**e**) Myelography. (Figure courtesy of Dr. Samer Hoz)

Image Description

MRI reveals spina bifida at the L4–L5 level, showing an open posterior neural arch with a dorsal dermal sinus tract containing dysplastic neurogenic and fibrous tissues. At this level, there is a superficially located subcutaneous loculus with CSF-like intensities, suggestive of a small meningocele. The low-lying spinal cord is tethered posteriorly to the site of the defect.

TCS

TCS is a neurological disorder caused by the stretching of the spinal cord due to its caudal attachment to inelastic structures, which limits the vertical movement of the spinal cord. The term TCS was introduced in 1976 by the Canadian neurosurgeon Harold J. Hoffman and his colleagues as a clinical diagnosis based on neurological deterioration affecting the lower spinal cord.

Tethering of the spinal cord can occur either congenitally (primary TCS) or in association with other intraspinal pathologies or postoperative scarring (secondary TCS). The most common causes of TCS are closed spinal dysraphisms such as spinal lipomas, tight filum terminale, and diastematomyelia. In cases of open spinal dysraphism, such as myelomeningocele, TCS may develop due to adhesions rather than the dysraphism itself. Other potential etiologies include dermal sinus tract and spinal cord tumors.

TCS typically manifests in childhood, but there are instances where it goes undiagnosed until adulthood or is acquired later in life. Presenting symptoms include weakness, low back pain or sciatica, urinary and fecal incontinence, gait abnormality (spastic or wide-based gait), scoliosis, orthopedic deformities, and cutaneous stigmata of spinal dysraphism (e.g., hairy patch, dimple, subcutaneous lipoma).

MRI of the spine is the preferred radiographic modality for diagnosis and demonstrates the position of the conus medullaris, which commonly terminates below the level of the L2 vertebral body. However, a normal conus position does not rule out TCS. MRI also aids in assessing the thickness of the filum terminale, identifying traction lesions, and detecting associated underlying pathology. While supine MRI is the standard, prone imaging may be useful in cases with high clinical suspicion, even if supine MRI shows no abnormalities.

In the pediatric population, ultrasound serves as a useful screening tool. It is indicated when there is a lack of ossification of the posterior arch of the spine in normal infants and when a bony defect is present in patients with spina bifida [11–14].

Questions

1. **TCS, the FALSE answer is:**
   A. TCS is a stretch-induced neurological disorder.
   B. TCS is a clinical diagnosis based on neurologic deterioration involving the lower cord.
   C. It mostly presents in childhood.
   D. It only presents as a congenital disorder.
   E. Closed spinal dysraphisms are the most common cause.
   The answer is D.
   TCS could be congenital (primary TCS) or acquired (secondary TCS).
2. **TCS radiology, the FALSE answer is:**
   A. May occur in association with spinal cord tumors.
   B. In TCS, the conus medullaris most commonly terminates below L2 level.
   C. Prone MRI can be used as a tool to assess spine mobility.
   D. Normal conus medullaris position excludes TCS.
   E. Ultrasound can be used for screening.

The answer is D.

Although conus medullaris mostly terminates in a low position (below L2 level), a normal conus position does not exclude TCS.

Case 64: Caudal Regression Syndrome (CRS)

Case Scenario

An 8-year-old male presented with sensorimotor paresis in the lower limbs, hypoplastic gluteal muscles, shallow intergluteal cleft with mild foot deformities and gait abnormalities (Fig. 7.6).

Imaging Description

Agenesis of the sacrum is observed below the level of S2, with the thecal sac ending at the S1 level. The conus medullaris terminates at the mid-portion of the L1 level, displaying a truncated appearance. The cauda equina is normal, with no signs of hydronephrosis, and the rectum is distended.

Caudal Regression Syndrome (CRS)

CRS is characterized by structural malformations in the caudal region, ranging from isolated partial agenesis of the coccyx to lumbosacral agenesis. It may occur due to disease conditions during early pregnancy (<4th week of gestation), such as hyperglycemia, infection, toxic exposure, and ischemic conditions. Severe cases are typically identified in utero or at birth, while milder cases may remain undetected until adulthood. There is no gender predominance.

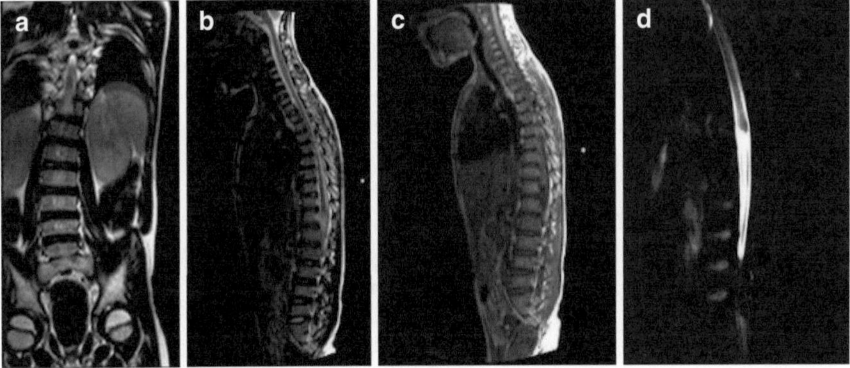

Fig. 7.6 Serial MRI images of the spine showing: (**a**) Coronal T2, (**b**) Sagittal T2, (**c**) Sagittal T1, (**d**) Myelography. (Figure courtesy of Dr. Samer Hoz)

While the majority of cases are sporadic, familial occurrences are observed in some instances. CRS is associated with VACTREL and Currarino triad syndromic complexes. Associations with congenital genitourinary anomalies (such as renal agenesis or hydronephrosis and various forms of duplication of the Mullerian ducts), anorectal abnormalities, diastematomyelia, terminal hydromyelia, myelomeningocele, sacral agenesis, spinal canal lipoma, and tethered cord (even when the cord is positioned high) have also been reported.

CRS was first defined in 1852 and has since been classified based on clinical presentation into five types of varying severity:

- Type I: Total or partial sacral agenesis.
- Type II: Total sacral agenesis with partial lumbar agenesis, maintaining continuity of the ilia and lower vertebrae.
- Type III: Complete lumbosacral agenesis with the caudal endplate resting on the iliac amphiarthrosis.
- Type IV: Complete absence of separation of the caudal soft tissues.
- Type V: Single midline femur and tibia, colloquially referred to as sirenomelia or "mermaid syndrome." Recent epidemiological investigations suggest that Type V may be better categorized as a specific disease entity.

The etiology for all classes remains inadequately understood and is under research.

The following radiographic features can be identified:

Dysgenesis or hypogenesis of the lumbosacral vertebral body.

- Atresia or dysgenesis usually occurs below L1 and is commonly confined to the sacrum.
- The spinal cord appears truncated and blunt, terminating beyond the expected level, taking on a wedged- or cigar-shaped conus medullaris.
- Significant canal narrowing occurs rostrally to the last intact vertebra.

Prenatal ultrasound findings include:

- A blunt and sharp ending cord, with the conus often extending well above the expected level (sometimes even higher than L1).
- A hypoplastic sacrum or absent sacral vertebrae (shield sign).
- Hypoplastic lower extremities with separated limbs (compare with sirenomelia).
- Fetal extremities may assume a "crossed leg tailor" or "Buddha" position. In early scans (first trimester), the crown-rump length may be smaller than anticipated for gestational age, serving as an indirect indicator.

MRI demonstrates comparable features to ultrasound but with finer detail, aiding in canal stenosis assessment. A characteristic wedge-shaped cord terminus may be apparent. Patients with CRS can be classified into two main groups based on imaging findings:

- Group 1: The conus medullaris is blunt and ends above the normal level. These patients may exhibit major sacral malformations, sometimes accompanied by a dilated central canal or a CSF-filled cyst at the inferior end of the conus.
- Group 2: The conus medullaris is elongated and tethered by a thickened filum terminale or intraspinal lipoma, terminating below the normal level. Neurologic abnormalities tend to be more severe in this group [15–19].

Questions

1. **CRS classification, the FALSE answer is:**
   A. CRS is traditionally classified into five subtypes.
   B. Type I CRS represents total or partial sacral agenesis.
   C. Type III CRS has complete lumbosacral agenesis.
   D. Type IV CRS is a complete lack of separation of the caudal soft tissues, referred to as sirenomelia or "mermaid syndrome".
   E. Type II CRS involves total sacral agenesis with partial lumbar agenesis, maintaining continuity of the ilia and lower vertebrae.
   The answer is D.
   Type V CRS is characterized by a single midline femur and tibia and is colloquially referred to as sirenomelia or "mermaid syndrome".
2. **CRS imaging, the FALSE answer is:**
   A. The level of atresia is usually below L1 and often limited to the sacrum.
   B. A characteristic wedge-shaped cord terminus may be seen on MRI.
   C. Prenatal ultrasound findings include a crown-rump length that is smaller than expected for gestational age.
   D. In Group 1 CRS, the conus medullaris is blunt and ends below the normal level.
   E. In Group 2 CRS, the conus medullaris is elongated and tethered by a thickened filum terminale.
   The answer is D.
   In Group 1 CRS, the conus medullaris is blunt and ends above the normal level.

References

1. Sherman JL, Barkovich AJ, Citrin CM. The MR appearance of syringomyelia: new observations. Am J Roentgenol. 1987;148(2):381–91.
2. Greenlee JD, Menezes AH, Bertoglio BA, Donovan KA. Syringobulbia in a pediatric population. Neurosurgery. 2005;57(6):1147–53.
3. Vandevelde M, Higgins R, Oevermann A. Veterinary neuropathology: essentials of theory and practice. Wiley; 2012.
4. Rossi A, Cama A, Piatelli G, Ravegnani M, Biancheri R, Tortori-Donati P. Spinal dysraphism: MR imaging rationale. J Neuroradiol. 2004;31(1):3–24.
5. Zaleska-Dorobisz U, Bladowska J, Biel A, Pałka LW, Hołownia D. MRI diagnosis of diastematomyelia in a 78-year-old woman: case report and literature review. Pol J Radiol. 2010;75(2):82.

6. Kobets AJ, Oliver J, Cohen A, Jallo GI, Groves ML. Split cord malformation and tethered cord syndrome: case series with long-term follow-up and literature review. Childs Nerv Syst. 2021;37:1301–6.

7. Prasad VS, Sengar RL, Sahu BP, Immaneni D. Diastematomyelia in adults modern imaging and operative treatment. Clin Imaging. 1995;19(4):270–4.

8. Maiuri F, Corriero G, Giampaglia F, Simonetti L. Lateral thoracic meningocele. Surg Neurol. 1986;26(4):409–12.

9. Ueda K, Honda O, Satoh Y, Kawai M, Gyobu T, Kanazawa T, Hidaka S, Yanagawa M, Sumikawa H, Tomiyama N. Computed tomography (CT) findings in 88 neurofibromatosis 1 (NF1) patients: prevalence rates and correlations of thoracic findings. Eur J Radiol. 2015;84(6):1191–5.

10. Hansell DM, Bankier AA, MacMahon H, McLoud TC, Muller NL, Remy J. Fleischner Society: glossary of terms for thoracic imaging. Radiology. 2008;246(3):697–722.

11. Yamada S, Won DJ, Yamada SM. Pathophysiology of tethered cord syndrome: correlation with symptomatology. Neurosurg Focus. 2004;16(2):1–5.

12. Michelson DJ, Ashwal S. Tethered cord syndrome in childhood: diagnostic features and relationship to congenital anomalies. Neurol Res. 2004;26(7):745–53.

13. Wang J, Zhou Q, Fu Z, Xiao X, Lu Y, Zhang G, Zhang H. MRI evaluation of fetal tethered-cord syndrome: correlation with ultrasound findings and clinical follow-up after birth. Clin Radiol. 2021;76(4):314–e1.

14. O'Connor KP, Smitherman AD, Milton CK, Palejwala AH, Lu VM, Johnston SE, Homburg H, Zhao D, Martin MD. Surgical treatment of tethered cord syndrome in adults: a systematic review and meta-analysis. World Neurosurg. 2020;137:e221–41.

15. Pang D. Sacral agenesis and caudal spinal cord malformations. Neurosurgery. 1993;32(5):755–8.

16. Semba K. Etiology of caudal regression syndrome. Human Genet Embryol. 2013;3(02):107.

17. Sharma P, Kumar S, Jaiswal A. Clinico-radiologic findings in Group II caudal regression syndrome. J Clin Imaging Sci. 2013;3:26.

18. Dayasiri K, Thadchanamoorthy V, Thudugala K, Ranaweera A, Parthipan N. Clinical and radiological characterization of an infant with caudal regression syndrome type III. Case Rep Neurol Med. 2020;2020:8827281.

19. Kang S, Park H, Hong J. Clinical and radiologic characteristics of caudal regression syndrome in a 3-year-old boy: lessons from overlooked plain radiographs. Pediatric Gastroenterol Hepatol Nutr. 2021;24(2):238.

Congenital Spinal Lesions

8

Sajjad G. Al-Badri, Ameer M. Aynona, Fatimah O. Ahmed,
Saleh A. Saleh, Khadija J. Ismael, Muslim M. Badr,
and Asmaa H. AL-Sharee

Case 65: Lipomyelomeningocele

Case Scenario

A male patient presented with motor delay and swelling involving the sacral area
(Fig. 8.1).

Imaging Description

MRI reveals classical findings of a tethered cord with an associated posterior spinal
defect and lipomyelocele. The tethered cord extends to the L4–5 level, with neural
elements extruding through the spina bifida. However, the defect appears to be covered with skin. T1 images display a bright fat signal within the canal, communicating with the prominent subcutaneous fat through the sacral defect. The wide-open
sacral bony defect is clearly visible, with neural elements protruding outside and a
subcutaneous lipoma extending into the canal (lipomyocele). Intact skin covering is

---

S. G. Al-Badri · S. A. Saleh
College of Medicine, University of Baghdad, Baghdad, Iraq

A. M. Aynona
College of Medicine, University of Babylon, Babylon, Iraq

F. O. Ahmed
College of Medicine, University of Mustansiriyah, Baghdad, Iraq

K. J. Ismael
College of Medicine, Diyala University, Diyala, Iraq

M. M. Badr · A. H. AL-Sharee (✉)
Department of Neuroradiology, Neurosurgery Teaching Hospital, Baghdad, Iraq

S. Hoz et al. (eds.), *Neuroradiology Board's Favorites*,
https://doi.org/10.1007/978-3-031-64261-6_8

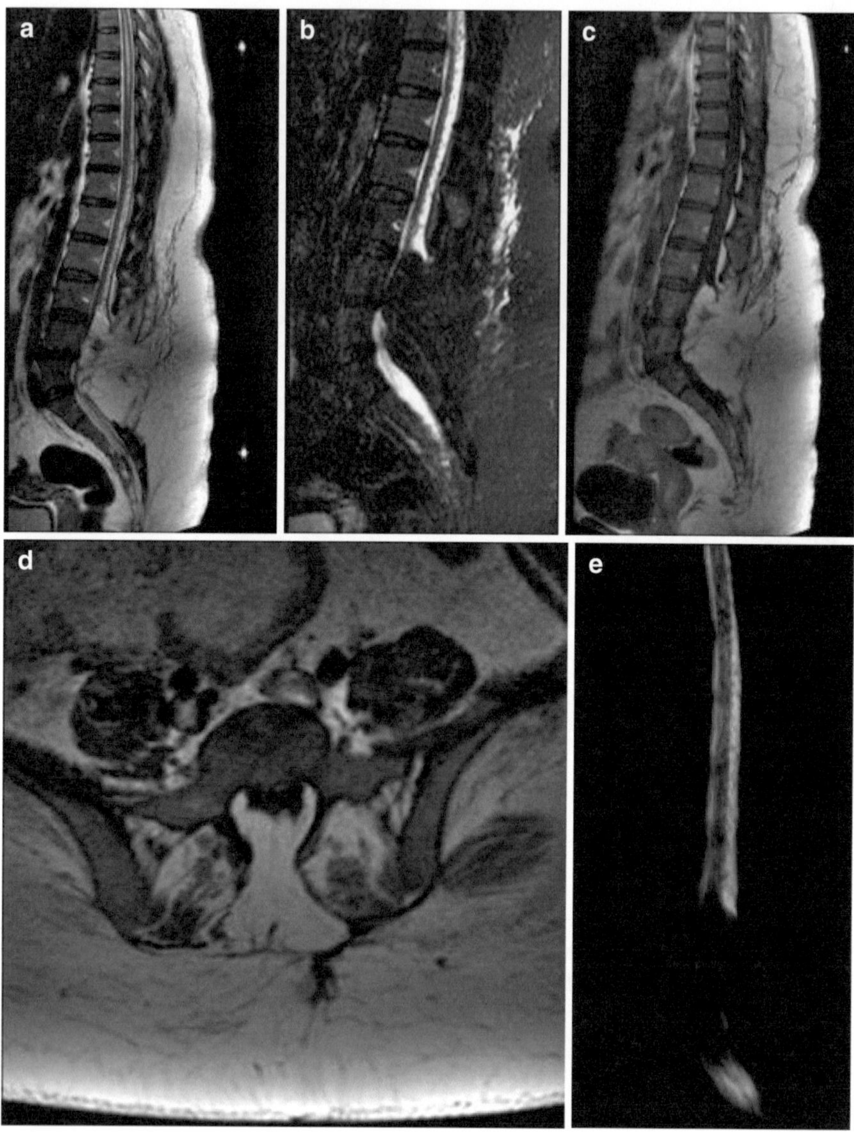

Fig. 8.1 Serial MRI images of the spine with the following: (**a**) Sagittal T2, (**b**) Sagittal STIR, (**c**) Sagittal T1 C+, (**d**) Axial T2, (**e**) Myelography. (Figure courtesy of Dr. Samer Hoz)

also evident. On T2 images, small flow voids are seen with intradural extramedullary involvement of the mid-dorsal spinal cord, consistent with a spinal AVM.

Lipomyelomeningocele

Lipomyelomeningoceles are characterized by the presence of a subcutaneous fatty mass, typically situated just above the intergluteal cleft, and are considered to be a form of closed spinal dysraphism. However, in some cases, these anomalies may occur at various locations along the spinal canal.

On plain radiographs, lipomyelomeningoceles appear similar to other spinal dysraphisms, displaying non-specific splaying and non-fusion of the posterior elements of the spine. Ultrasound, on the other hand, can visualize vertebral arch defects or defects associated with an elongated spinal cord, revealing a placode attached to an intrathecal echogenic fatty mass that protrudes into the dorsal subcutaneous tissue.

CT scans demonstrate spina bifida, focal enlargement of the spinal canal at the placode level, and a hypodense fat-attenuated mass in continuity with the subcutaneous fat planes. MRI reveals a spinal defect with lipomatous tissue covered by skin, where the neural placode-lipoma interface lies outside the spinal canal due to the enlargement of the subarachnoid space, often accompanied by a low-lying cord.

In the differential diagnosis of lipomyelomeningoceles, consideration should be given to lipomyelocele (where the placode-lipoma interface lies within the spinal canal), lipoma of the filum terminale, and spinal intradural lipoma [1, 2].

Questions

1. **Lipomyelomeningoceles, the FALSE answer is:**
   A. CT shows spina bifida and focal enlargement of the spinal canal at the placode level.
   B. CT shows a hypodense fat-attenuated mass in continuity with the subcutaneous fat planes.
   C. Lipomyelomeningoceles may occur at various locations along the spinal canal.
   D. Ultrasound is helpful in visualizing vertebral arch defects or defects associated with an elongated spinal cord.
   E. MRI reveals a neural placode-lipoma interface that lies inside the spinal canal due to the enlargement of subarachnoid space.
   The answer is E.

   MRI reveals a neural placode-lipoma interface that lies outside the spinal canal due to the enlargement of the subarachnoid space.

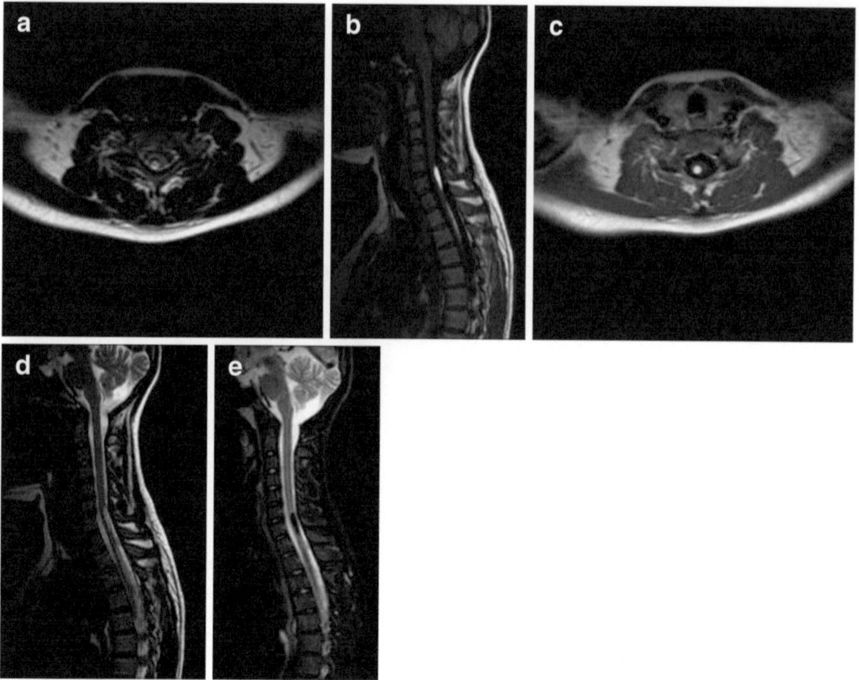

Fig. 8.2 Serial MRI images of the spine with the following: (**a**) Axial T2, (**b**) Sagittal T1, (**c**) Axial T1, (**d**) Sagittal T2, (**e**) Sagittal SPIR. (Figure courtesy of Dr. Samer Hoz)

Case 66: Spinal Dermoid Cyst

Case Scenario

A 26-year-old female presented with a history of pain and sensory disturbances (Fig. 8.2).

Imaging Description

An intramedullary hyper-intense mass lesion is observed in both T1 and T2 imaging, with suppression in SPIR, at the dorsal aspect of the spinal cord at C6, C7, D1, D2, and D3 levels. There is no enhancement noted after the administration of intravenous contrast. Histopathological examination subsequently confirmed the diagnosis of an intramedullary dermoid cyst.

Spinal Dermoid Cyst

Dermoid cysts present as unilocular or multilocular cystic tumors that are lined by squamous epithelium and contain skin appendages such as hair follicles, sweat glands, and sebaceous glands. These cysts have a congenital origin, with 40% being intramedullary and 60% extramedullary. Although spinal dermoid cysts are relatively uncommon, comprising nearly 20% of intradural tumors observed in patients younger than 1 year of age, they are predominantly diagnosed in individuals younger than 20 years old, affecting both males and females equally. There is a potential association with occult spinal dysraphism.

Radiographically, spinal dermoid cysts are typically detected through imaging studies and are often located in the lumbosacral region (60% of cases) and the cauda equina (20% of cases). Manifestations in the cervical or thoracic spine are rare. The imaging characteristics of these cysts can vary, but they commonly appear as a mass with CSF density/intensity and components of fat density/intensity.

CT scans reveal a well-defined mass isodense to CSF, often with hypodense components. Fat calcification may be present, and minimal enhancement can be observed. MRI signal intensity can be either homogeneous or heterogeneous. T1 signals may be hypo- or hyper-intense, with hypo-intensity indicating water content and hyper-intensity indicating fatty secretions. STIR sequences suppress the fat content, while T1 with contrast (Gadolinium) shows either no enhancement or mild rim enhancement [3, 4].

Differential Diagnosis

1. Spinal arachnoid cyst: exhibits CSF intensity in all sequences, with no restrictions on DWI.
2. Spinal epidermoid cyst: lacks fatty elements and demonstrates diffusion restriction on DWI.
3. Spinal neurenteric cyst: is typically located ventral to the spinal cord, displaying intense CSF in all sequences. More commonly found in the thoracic and cervical regions.
4. Spinal lipoma: is homogeneously hyper-intense on both T1 and T2 sequences. More prevalent in the thoracic and cervical regions.
5. Spinal teratoma: shows a heterogeneous appearance on both T1 and T2 sequences due to the presence of fat, soft tissue, fluid, and/or calcium.

Questions

1. **Spinal dermoid cysts, the FALSE answer is:**
   A. MRI signal intensity can be homogeneous or heterogeneous.
   B. CT scan typically shows a well-defined mass that is hypodense to CSF.
   C. In T1 C+ (Gd), there is either no enhancement or mild rim enhancement.

    D. Spinal dermoid cysts are most commonly situated in the lumbosacral region (60%) and the cauda equina (20%).

    E. T1 signals can exhibit hypo- or hyper-intensity, where hypo-intensity typically indicates water content and hyper-intensity indicates the presence of fatty secretions.

    The answer is B.

    CT scan typically shows a well-defined mass that is isodense to CSF.

2. **Spinal dermoid cysts, the FALSE answer is:**

    A. These cysts originate congenitally, with 40% occurring intramedullary and 60% extramedullary.

    B. No association with occult spinal dysraphism has been found.

    C. It is rare for them to manifest in the cervical or thoracic spine.

    D. They are uncommon overall but account for nearly 20% of intradural tumors seen in patients younger than 1 year of age.

    E. They generally present in patients younger than 20 years.

    The answer is B.

    Spinal dermoid cysts may be associated with occult spinal dysraphism.

Case 67: Multiple Spinal Arachnoid Cysts

Case 67.1

Case Scenario

A 14-year-old female presented with a progressive onset of symptoms, including spastic and flaccid paralysis, pain, weakness, numbness, and bladder incontinence (Fig. 8.3).

Image Description

MRI reveals a dilated neural canal with multiple lobulated CSF-filled extradural cysts within it. These cysts extend from the cervical spine to the lumbo-sacral spine, involving the entire spinal region. The spinal cord and cauda equina are anteriorly deviated and compressed, with some lesions demonstrating early extension through the neural foramina.

Case 67.2

Case Scenario

A 3-year-old female presented with progressive limb weakness (Fig. 8.4).

Imaging Description

The MRI reveals a well-defined cystic lesion with CSF intensity anterior to the cervical cord at the craniocervical junction. This lesion is associated with distortion and thinning of the cord, and there is posterior vertebral body scalloping. A

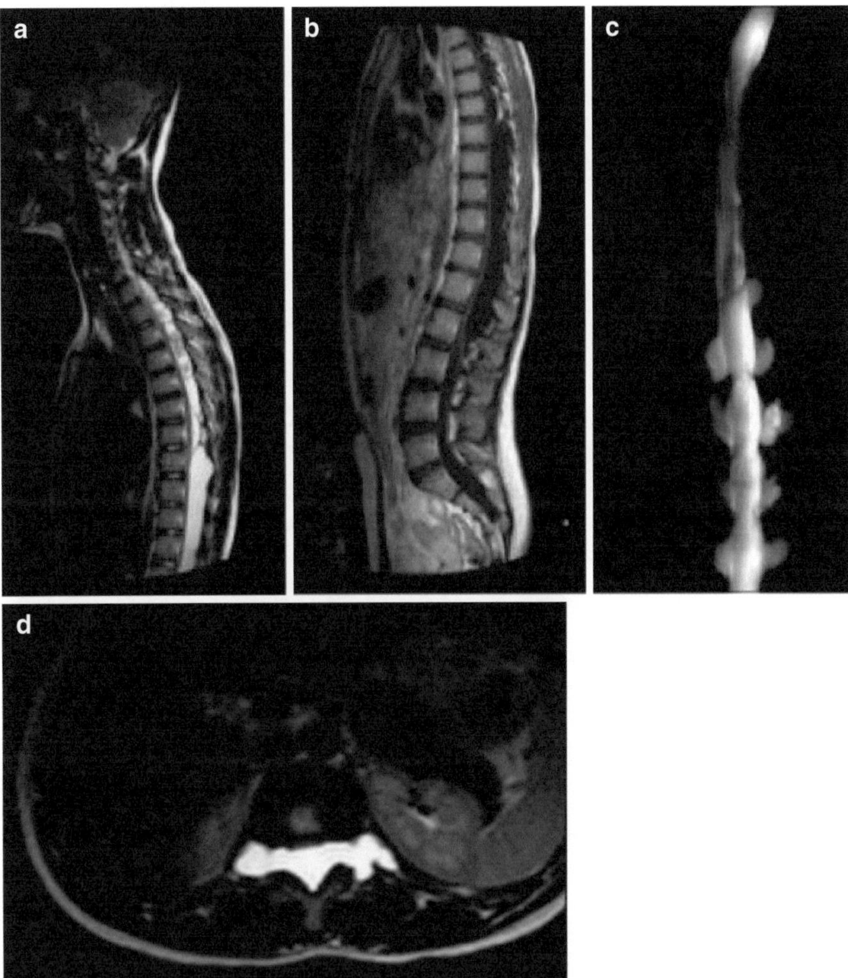

Fig. 8.3 Serial MRI images of the spine with the following: (**a**) Sagittal T2, (**b**) Sagittal T1, (**c**) Myelography, (**d**) Axial T2. (Figure courtesy of Dr. Samer Hoz)

prominent CSF flow signal artifact is observed within the abnormality, consistent with a spinal arachnoid cyst.

Spinal Arachnoid Cysts

Spinal arachnoid cysts are rare and may be congenital or acquired, with the potential to develop at any age. They can occur in both the intra- and extradural regions. In cases of intradural cysts, the majority are situated dorsally to the spinal cord, with approximately 80% located in the thoracic region, 15% in the cervical region, and 5% in the lumbar region. Secondary cysts can emerge in various locations and are associated with conditions such as spina bifida, diastematomyelia, and syrinx.

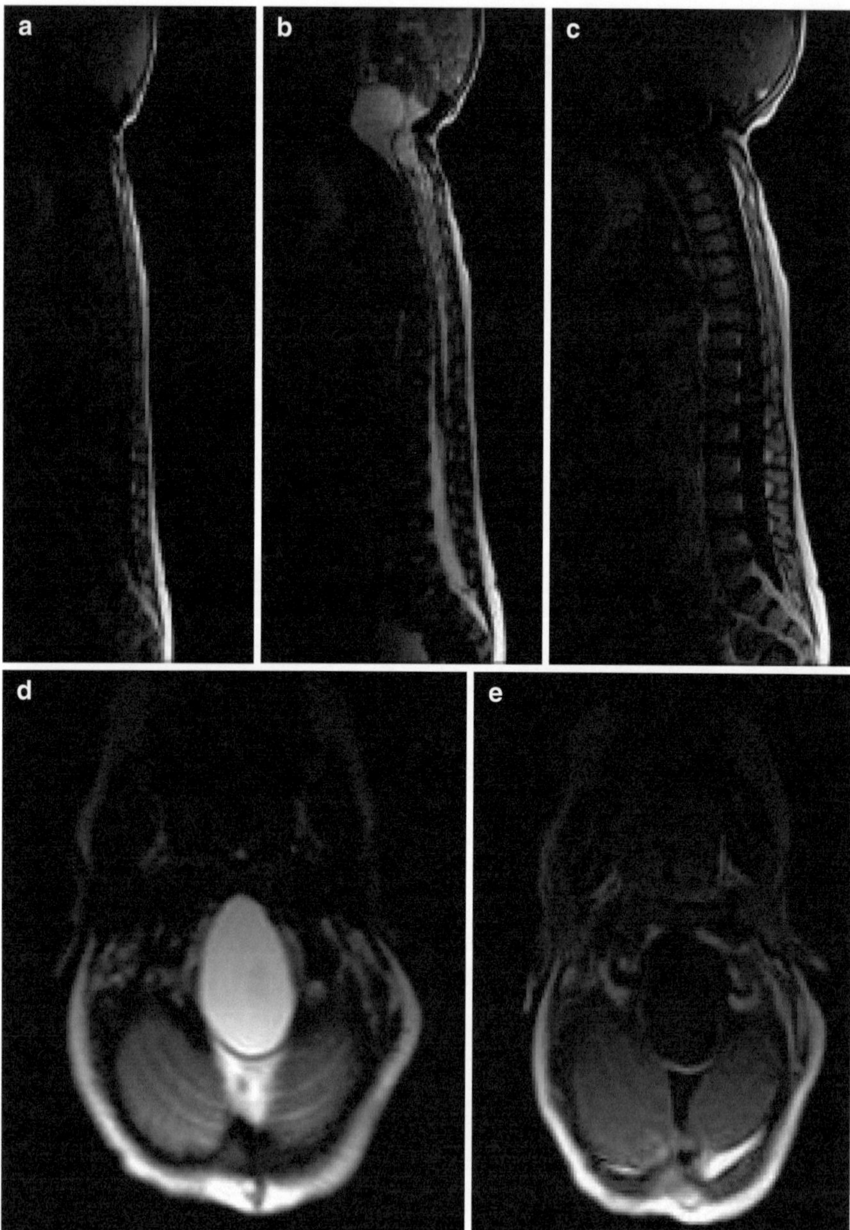

Fig. 8.4 Serial MRI images of the spine with the following: (**a**) Sagittal T1, (**b**) Sagittal T2, (**c**) Sagittal T1 C+, (**d**) Axial T2, (**e**) Axial T1 C+. (Figure courtesy of Dr. Samer Hoz)

With regards to imaging, CT myelography can reveal displaced and compressed cords, with arachnoid cysts appearing opacified with contrast. Early scanning is advisable to detect the cyst before it reaches isodensity with CSF. Delayed scanning may be necessary if appropriate CSF mixing is not achieved. On MRI, arachnoid cysts may not be visible unless the spinal cord is displaced. They follow CSF intensity on T1, may appear hyper-intense to CSF on T2, and typically show no enhancement with gadolinium contrast [5, 6].

Differential Diagnosis

1. Spinal epidermoid cysts: appear bright on DWI.
2. Spinal dermoid cysts: imaging features can vary but typically resemble fat.
3. Spinal endodermal cysts: are frequently found in the lower cervical and ventral part of the upper dorsal spine and are difficult to differentiate from arachnoid cysts.

Questions

1. **Multiple spinal arachnoid cysts, the FALSE answer is:**
   A. They follow CSF intensity on T1W imaging.
   B. CT myelography demonstrates a displaced and compressed cord, with arachnoid cysts appearing opacified with contrast.
   C. They may appear brighter than CSF on T2W imaging.
   D. They enhance with gadolinium contrast.
   E. Early scanning is recommended to detect the cyst before it becomes isodense with CSF.
   The answer is D.
   The cyst shows no enhancement with gadolinium contrast.
2. **Multiple spinal arachnoid cysts, the FALSE answer is:**
   A. They can occur in both intradural and extradural forms.
   B. Intradural cysts are most commonly located ventral to the spinal cord.
   C. They are associated with conditions like spina bifida, diastematomyelia, and syrinx.
   D. Eighty percent of them occur in the thoracic region, 15% in the cervical region, and 5% in the lumbar region.
   E. They can be congenital or acquired and can develop at any age.
   The answer is B.
   Intradural cysts are most commonly located dorsally to the spinal cord.

Case 68: Lipoma of the Filum Terminale

Case Scenario

A 39-year-old female presented with complaints of low back pain (Fig. 8.5).

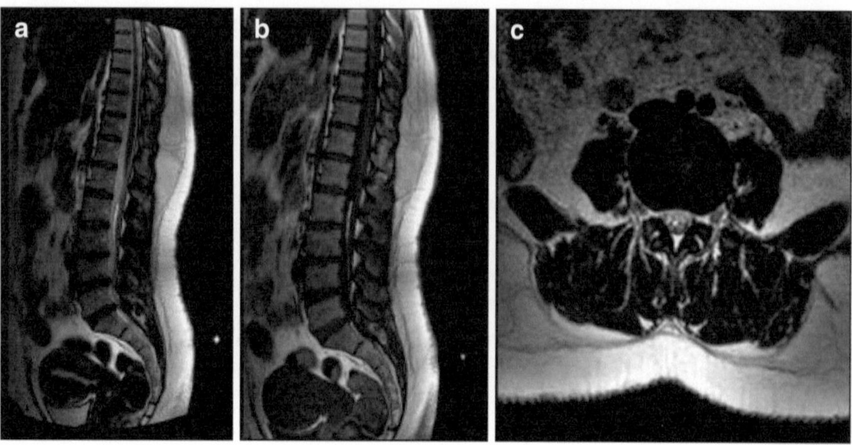

Fig. 8.5 Serial MRI images of the spine with the following: (**a**) Sagittal T2, (**b**) Sagittal T1, (**c**) Axial T2. (Figure courtesy of Dr. Samer Hoz)

Image Description

The lumbar canal and conus appear normal, with the tip of the conus located at the L1 level. Immediately below the conus, in the midline, there is a longitudinal streak of tissue with high T1 and high T2 signals.

Lipoma of the Filum Terminale

This condition, also known as fatty filum terminale or filar lipoma, is a relatively common incidental finding during lumbar spine imaging (approximately 5% of relevant imaging tests reveal fat in the filum terminale). In the majority of cases, it is considered a benign finding with no clinical significance. However, in some instances, individuals may present with symptoms suggestive of tethered cord syndrome, where a thickened filum and a low-lying conus are commonly observed.

In cases where the lipoma reaches a substantial size, it can be visualized on a CT scan as a region of fat density (−90 to −30 HU) below the conus level. However, the detection of small lipomas may be challenging and depends on factors such as the patient's size, the quality of the CT scanner, and the amount of quantum mottle.

On MRI, the lesion typically appears in a linear shape extending over some distance. The signal follows that of fat across all sequences and may exhibit a chemical shift artifact in T2*/gradient-weighted sequences. Both T1 and T2 sequences show hyper-intensity, while fat-saturated sequences demonstrate signal loss. No enhancement is usually observed in T1 with contrast (Gd) [7, 8].

Differential Diagnosis

1. Paraganglioma of the filum terminale appears iso- to hyper-intense on T2 and hypo-intense on T1, with intense enhancement following the administration of contrast.
2. Myxopapillary ependymoma is characterized by being iso- to hyper-intense on T2 and hypo-intense on T1, with enhancement observed after the administration of contrast.

Questions

1. **Lipoma of the filum terminale, the FALSE answer is:**
   A. In most cases, it is an incidental finding without clinical concern.
   B. On MRI, it usually appears as a linear structure extending over some distance.
   C. T1WI and T2WI show hypo-intense signals.
   D. In T1 with contrast (Gd), there is no enhancement.
   E. It may demonstrate a chemical shift artifact in T2*/gradient-weighted sequences.
   The answer is C.
   T1WI and T2WI show hyper-intense signals.

References

1. Sarris CE, Tomei KL, Carmel PW, Gandhi CD. Lipomyelomeningocele: pathology, treatment, and outcomes: a review. Neurosurg Focus. 2012;33(4):E3.
2. Wagner KM, Raskin JS, Hansen D, Reddy GD, Jea A, Lam S. Surgical management of lipomyelomeningocele in children: challenges and considerations. Surg Neurol Int. 2017;8:63.
3. De-La-Paz Y, Cherian I, Valencia-Bayona E, Alaswad M, Muñoz-Cobos A, Carrillo-Ruiz JD, Beltrán JQ. Lumbar dermoid cysts: 3 illustrative cases and a total review of the literature of the last two decades. Neurocirugia. 2021;32(6):300–4.
4. Falavigna A, Righesso O, Teles AR. Concomitant dermoid cysts of conus medullaris and cauda equina. Arq Neuropsiquiatr. 2009;67:293–6.
5. Wang MY, Levi AD, Green BA. Intradural spinal arachnoid cysts in adults. Surg Neurol. 2003;60(1):49–55.
6. Fam MD, Woodroffe RW, Helland L, Noeller J, Dahdaleh NS, Menezes AH, Hitchon PW. Spinal arachnoid cysts in adults: diagnosis and management. A single-center experience. J Neurosurg Spine. 2018;29(6):711–9.
7. Usami K, Lallemant P, Roujeau T, James S, Beccaria K, Levy R, Di Rocco F, Sainte-Rose C, Zerah M. Spinal lipoma of the filum terminale: review of 174 consecutive patients. Childs Nerv Syst. 2016;32:1265–72.
8. Al-Omari MM, Eloqayli HM, Qudseih HM, Al-shinag MK. Isolated lipoma of filum terminale in adults: MRI findings and clinical correlation. J Med Imaging Radiat Oncol. 2011;55(3):28.

Spinal Oncology: Intradural Tumors

Maliya Delawan, Usama AlDallal, Sajjad G. Al-Badri,
Fatimah K. Al-Kishawi, Saja A. Albanaa, Ali M. Neamah,
and Asmaa H. AL-Sharee

Case 69: Spinal Ependymomas

Case Scenario

A 12-year-old male with a 4-month history of lumbar pain radiating down to the lower extremities presented with progressive worsening of the pain associated with numbness in the thighs and legs (Fig. 9.1).

Imaging Description

MRI reveals a well-demarcated central intramedullary tumor at L1–L2, which is iso-intense on T1, heterogeneously hyper-intense on T2, and has intense contrast enhancement. There is a small polar cyst component at the center of the tumor. These features suggest an intramedullary soft tissue neoplasm. Histopathological examination confirmed the diagnosis of spinal ependymoma.

M. Delawan
College of Medicine, Gulf Medical University, Ajman, United Arab Emirates

U. AlDallal
School of Medicine, Royal College of Surgeons in Ireland—Medical University of Bahrain, Al Sayh Muharraq Governorate, Bahrain

S. G. Al-Badri · S. A. Albanaa · A. M. Neamah
College of Medicine, University of Baghdad, Baghdad, Iraq

F. K. Al-Kishawi
Al-Kindy College of Medicine, University of Baghdad, Baghdad, Iraq

A. H. AL-Sharee (✉)
Department of Neuroradiology, Neurosurgery Teaching Hospital, Baghdad, Iraq

S. Hoz et al. (eds.), *Neuroradiology Board's Favorites*,
https://doi.org/10.1007/978-3-031-64261-6_9

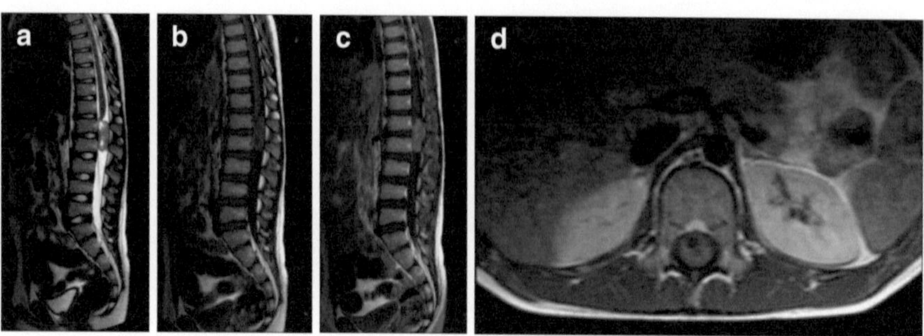

Fig. 9.1 Serial MRI images of the spine with the following: (**a**) Sagittal T2, (**b**) Sagittal T1, (**c**) Sagittal T1 C+, (**d**) Axial T1 C+. (Figure courtesy of Dr. Samer Hoz)

Spinal Ependymomas

Spinal ependymomas can be seen in both adults and children and are overall the most common tumor affecting the spinal cord. They contribute to 60% of all glial tumors involving the spinal cord, making them the most common intramedullary neoplasm in adults. In children, spinal ependymomas are the second most common intramedullary spinal neoplasm after astrocytoma, comprising 30% of all such neoplasms. The peak incidence is in the fourth decade of life, affecting more males than females. Spinal ependymomas are also found to be associated with NF2.

Spinal ependymomas are predominantly WHO grade 2, and grade 3 tumors are exceedingly rare. They most commonly occur in the cervical cord (44%). Twenty-six percent of spinal ependymomas involve the thoracic cord only, and 23% extend from the cervical cord into upper thoracic segments.

On plain radiographs, spinal ependymomas can manifest with features including scoliosis, non-specific widening of the spinal canal, posterior vertebral body scalloping, erosion into the pedicles, as well as thinning of the lamina. On CT, the lesion usually appears iso- to slightly hyper-attenuating with intense enhancement with contrast. CT may also show spinal canal widening and, for large lesions, vertebral scalloping and enlargement of the neural foramina.

MRI is the modality of choice and can demonstrate a symmetrically expanded spinal cord, tumoral cysts in 22% of the cases, non-tumoral cysts in 62% of the cases, and syringohydromyelia in 9–50% of the cases. The lesion is usually well-circumscribed and spans the length of four vertebral bodies on average. Unlike intracranial ependymomas, they are not typically associated with calcification. On T1, most ependymomas appear iso- to hypo-intense but can have mixed signals in the presence of a cyst, necrosis or hemorrhage. On T2, they appear hyper-intense, and edema around the tumor can be seen in 60% of the cases. In 20–33% of cases, the "cap sign" can be seen, which is a hypo-intense rim of hemosiderin surrounding the tumor due to hemorrhage. However, the "cap sign" can also be seen in hemangioblastomas and paragangliomas and is not specific to ependymomas. Almost all spinal ependymomas enhance strongly on T1 C+ (Gd) [1, 2].

Differential Diagnosis

- Astrocytoma: In contrast to ependymomas, astrocytomas tend to be ill-defined on imaging and are eccentrically located in the spinal canal. Astrocytomas are associated with patchy irregular enhancement and hemorrhage and bone changes are uncommon. They also affect longer segments of the spinal cord and involve the entire cord diameter.
- Spinal cavernous malformations: Do not enhance and are seen with a complete surrounding hemosiderin ring, as well as diffuse midline glioma, H3 K27-altered.

Questions

1. **Spinal ependymomas, the FALSE answer is:**
   A. Spinal ependymomas typically cause symmetric cord expansion.
   B. Spinal ependymoma is the most common tumor affecting the spinal cord.
   C. Spinal ependymomas appear hypo-intense on T2 MRI.
   D. Spinal ependymomas are usually well-circumscribed.
   E. Spinal ependymoma most commonly occurs in the cervical cord.
   The answer is C.
   Spinal ependymomas appear hyper-intense on T2 MRI.
2. **Spinal ependymomas, the FALSE answer is:**
   A. Spinal ependymoma is the most common intramedullary neoplasm in adults.
   B. Spinal ependymomas typically enhance strongly on T1 C+ (Gd).
   C. Spinal ependymoma exhibits intense enhancement with contrast on CT.
   D. On CT, the lesion usually appears iso- to slightly hyper-attenuating.
   E. Spinal ependymoma is typically associated with calcification.
   The answer is E.
   Unlike intracranial ependymomas, they are not typically associated with calcification.

Case 70: Meningioma (WHO Grade 1)

Case 70.1

Case Scenario

A 50-year-old female with progressive right side upper limb weakness and paresthesia (Fig. 9.2).

Imaging Description

MRI reveals a homogeneously enhancing intradural, extramedullary mass with a broad dural base and dural tail in the vertebral canal. It is situated anteriorly on the right at the level of C4, resulting in significant cord compression with flattening of the cord and obliteration of the CSF space. The cord is displaced laterally to the left.

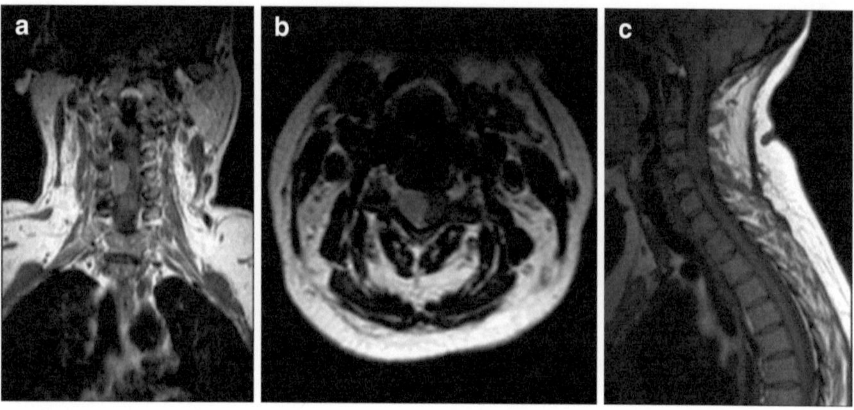

Fig. 9.2 Serial MRI images of the spine with the following: (**a**) Coronal T1 C+, (**b**) Axial T1 C+, (**c**) Sagittal T1. (Figure courtesy of Dr. Samer Hoz)

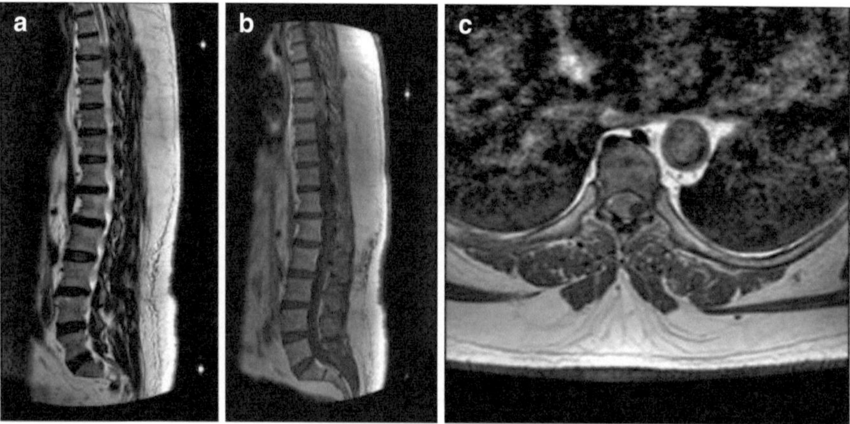

Fig. 9.3 Serial MRI images of the spine with the following: (**a**) Sagittal T2, (**b**) Sagittal T1 C+, (**c**) Axial T1 C+. (Figure courtesy of Dr. Samer Hoz)

The histopathological examination confirmed the diagnosis of meningioma (WHO grade 1).

Case 70.2: Spinal Meningioma

Case Scenario
A 54-year-old male presented with severe back pain and radiculopathy (Fig. 9.3).

Imaging Description

Within the spinal canal at D7–D8, there is a slightly T1 hyper-intense, T2 iso-intense, and intensely enhancing extramedullary intradural mass. The lesion displaces the spinal cord towards the left, with complete effacement of the CSF signal at this level and significant cord compression. However, no abnormal cord signal is definitively identified. This mass appears to be solitary within the imaged spinal canal.

Spinal Meningioma

Meningioma is a slow-growing tumor that arises from arachnoidal cap cells of the meninges, and is the second most common intradural extramedullary spinal tumor after spinal ependymoma, representing 25–30% of all such tumors and about 12% of all meningiomas cases. The peak incidence is between 60 and 80 years old. While spinal meningiomas are mostly seen in females (75–90%) among adults, they do not show sex predilection in children. Exposure to high doses of ionizing radiation is one of the well-known risk factors contributing to the development of spinal meningiomas. Previous trauma is another risk factor, and spinal meningiomas are often associated with NF2.

The majority (90%) of spinal meningiomas are located in the intradural extramedullary space, while approximately 5% are purely extradural, and the remainder 5% are mixed and have both intradural and extradural components, resulting in a dumbbell appearance. The most common location of spinal meningioma is the thoracic spine (80%), followed by the cervical spine (15%), and lastly the lumbosacral spine (5%). Meningiomas are often located lateral to the spinal cord (60–70%), but those in the cervical spine are more likely to be located anteriorly. Most meningioma cases are solitary lesions (98%), and multiple meningiomas are often associated with NF2.

Plain radiographs usually appear normal, and bone erosion or calcification are rare. On CT scans, meningiomas typically appear as an isodense or moderately hyperdense mass. Hyperostosis and calcification may be present but are rare. CT myelography may reveal an arachnoid isolation sign, indicating that the intradural tumor is separated from the spinal cord with contrast in the subarachnoid space.

On MRI, meningiomas are usually well-circumscribed and exhibit a broad-based dural attachment. The dural tail sign is present in 60–70% of cases, and meningiomas that arise lateral or ventrolateral to the spinal cord may display a ginkgo leaf sign. The signal characteristics of spinal meningiomas are similar to those of typical intracranial meningiomas, appearing iso-intense to slightly hypo-intense and possibly heterogeneous with T1W images and appearing iso-intense to slightly hyper-intense on T2W images. Moderate homogeneous enhancement may be observed on T1W images with gadolinium contrast (Gd). In some cases, densely calcified meningiomas may be hypo-intense on T1 and T2 with little contrast enhancement [3].

Differential Diagnosis

Nerve sheath tumors (spinal schwannoma, spinal neurofibroma): they are usually located anteriorly, compared to meningioma, which is typically located posterolaterally. Nerve sheath tumors are also often multiple and have central regions displaying low intensity on post-contrast T1 and T2W images. Unlike meningiomas, they are not typically associated with a broad dural base or exit foraminal widening.

Questions

1. **Spinal meningioma, the FALSE answer is:**
   A. Spinal meningiomas are most commonly seen in females among adults.
   B. Ionizing radiation is a risk factor for developing spinal meningioma.
   C. Spinal meningiomas are associated with NF2.
   D. About 15% of the cases are located at the cervical spine and are anterior to the spinal cord.
   E. Calcification is a very common finding in spinal meningiomas.
   The answer is E.
   Calcification may be seen in only 5% of spinal meningioma cases.
2. **Spinal meningioma, the FALSE answer is:**
   A. On MRI, spinal meningiomas are usually well-circumscribed.
   B. The dural tail sign is present in 60–70% of cases, and meningiomas that arise lateral or ventrolateral to the spinal cord may display a ginkgo leaf sign.
   C. On CT scans, meningiomas typically appear as an isodense or moderately hyperdense mass.
   D. CT myelography may reveal an arachnoid isolation sign.
   E. Spinal meningioma appears iso-intense to slightly hypo-intense on T2 MRI.
   The answer is E.
   Spinal meningioma appears iso-intense to slightly hyper-intense on T2 MRI.

Case 71: Spinal Neurofibroma

Case Scenario

A 30-year-old female presented with neck pain and radiculopathy (Fig. 9.4).

Imaging Description

There is a large solid mass with an intra- and extraspinal component, centered at the RT C2/C3 foramen, appearing heterogeneous on T1/T2. It is mainly T2-hyperintense and markedly T1-hypointense. There is heterogeneous enhancement of the mass and marked mass effect/compression of the cervical cord with leftward deviation. The lesion extends within the spinal canal and occupies more than 50% of the surface.

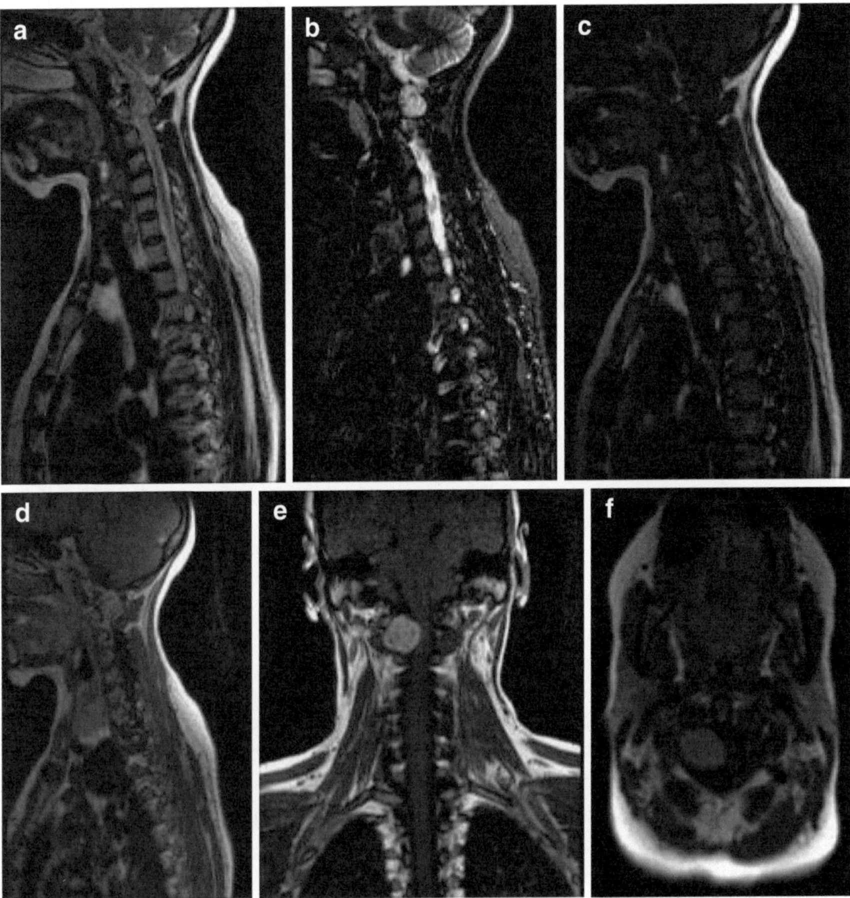

Fig. 9.4 Serial MRI images of the spine with the following: (**a**) Sagittal T2, (**b**) Sagittal SPIR, (**c**) Sagittal T1, (**d**) Sagittal T1 C+, (**e**) Coronal T1 C+, (**f**) Axial T1 C+. (Figure courtesy of Dr. Samer Hoz)

Spinal Neurofibroma

Neurofibromas are benign peripheral nerve sheath tumors that may be single or multiple. However, the majority occur in association with NF1, especially for the plexiform variant or if multiple neurofibromas are present. They are generally divided into five subtypes depending on morphology: localized/nodular intraneural neurofibroma, localized/nodular cutaneous neurofibroma, diffuse cutaneous neurofibroma, plexiform neurofibroma, and massive diffuse soft tissue neurofibroma. Localized intraneural neurofibromas contribute to 90% of neurofibromas, making them the most common form. Their peak incidence is between 20 and 30 years of age but appears earlier in childhood when occurring in the setting of NF1, with no sex predilection. Most of them are solitary and do not occur in association with NF1.

On CT, they appear as a well-defined hypodense lesion with minimal or no contrast enhancement. Neural foramina widening, vertebral scalloping, thinning of the pedicle, and extension into the extradural space, resulting in a dumbbell shape, can often be seen. However, the dumbbell appearance is not specific to neurofibromas and can be seen with other tumors such as schwannomas and meningiomas. On MRI, neurofibromas are hypo-intense on T1 and hyper-intense on T2 with heterogeneous enhancement on T1 C+ (Gd). The target sign can often be seen on T2, which appears as a hyper-intense rim with a central area of hypo-intensity representing collagenous stroma. However, this sign is not specific and can also be seen in schwannomas and malignant peripheral nerve sheath tumors [4, 5].

Differential Diagnosis

Spinal neurofibromas most commonly involve the cervical cord and are difficult to distinguish from schwannomas. Spinal meningioma is also another important differential diagnosis for neurofibromas.

Questions

1. **Spinal neurofibroma, the FALSE answer is:**
   A. Spinal neurofibromas can extend into the extradural space.
   B. Most neurofibromas occur in association with NF1.
   C. Spinal neurofibromas most commonly involve the cervical cord.
   D. The dumbbell appearance on MRI is not specific to neurofibromas.
   E. On MRI, neurofibromas are hyper-intense on T1 and hypo-intense on T2.
   The answer is E.
   On MRI, neurofibromas are hypo-intense on T1 and hyper-intense on T2.
2. **Spinal neurofibroma, the FALSE answer is:**
   A. Spinal neurofibromas may be single or multiple.
   B. Spinal neurofibromas appear as a well-defined hyperdense lesion on CT scan.
   C. The target sign can often be seen on T2.
   D. Localized intraneural neurofibromas are the most common form.
   E. It exhibits heterogeneous enhancement on T1 C+ (Gd).
   The answer is B.
   Spinal neurofibromas appear as a well-defined, hypodense lesion on CT scan.

Case 72: Intradural Lipoma

Case Scenario

A 50-year-old female presented with low back pain radiating to the right thigh (Fig. 9.5).

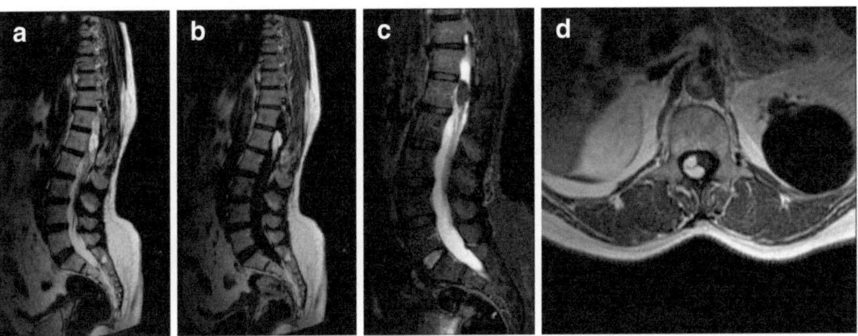

Fig. 9.5 Serial MRI images of the spine with the following: (**a**) Sagittal T2, (**b**) Sagittal T1, (**c**) Sagittal SPIR, (**d**) Axial T1 C+. (Figure courtesy of Dr. Samer Hoz)

Imaging Description

The MRI findings describe a well-defined fusiform-shaped lesion with high T1 and T2 signal intensity and low signal in the T2 fat-suppressed images. At the L1 levels, the mass can be seen positioned towards the right side in the extramedullary intradural space and slightly pushing the adjacent cord segment to the left side. These MRI features are consistent with an intradural lipoma.

Intradural Lipoma

Intradural lipomas are usually in the subpial juxtamedullary region, but intramedullary lesions have also been reported in the literature. Peak incidence is in the second and third decades of life, with no sex predilection. Non-dysraphic intradural lipomas are not usually associated with vertebral and dermal abnormalities.

The most common site for intradural lipomas is the thoracic region in adults and the cervical cord in children, although they can occur anywhere in the spine. They are usually positioned in the dorsal midline of the spinal cord and compress the adjacent segment ventrally. On plain film, canal widening and thinning of the pedicles may be seen. CT can reveal a homogeneously hypoattenuating lesion. On MRI, intradural lipomas appear as well-circumscribed masses that are hyper-intense on T1 and T2 but hypo-intense on fat-suppressed sequences, with no enhancement on T1 C+ (Gd) [6, 7].

Differential Diagnosis

- Lipomyelomeningocele, a type of occult spinal dysraphism that presents with a palpable mass and possible cutaneous stigmata.
- Lipoma of the terminal filum as well as spinal dermoid cyst, which demonstrates mixed-signal intensity and may present with a dermal sinus.

Questions

1. **Intradural lipoma, the FALSE answer is:**
   A. Intradural lipomas are hypo-intense on fat-suppressed MRI sequences.
   B. Intradural lipomas are hyper-intense on T1 and T2.
   C. Intradural lipomas are most commonly subpial intramedullary lesions.
   D. Intradural lipomas are typically well-circumscribed.
   E. Non-dysraphic intradural lipomas are not usually associated with vertebral or dermal abnormalities.
      The answer is C.
      Intradural lipomas are most commonly subpial juxtamedullary lesions.
2. **Intradural lipoma, the FALSE answer is:**
   A. The most common site for intradural lipomas is the thoracic region in adults.
   B. The cervical cord is the most commonly affected site in children.
   C. They are usually positioned in the ventral midline of the spinal cord and compress the adjacent segment dorsally.
   D. CT can reveal a homogeneously hypoattenuating lesion.
   E. On plain film, canal widening and thinning of the pedicles may be seen.
      The answer is C.
      They are usually positioned in the dorsal midline of the spinal cord and compress the adjacent segment ventrally.

Case 73: Myxopapillary Ependymoma

Case Scenario

A 21-year-old male presented with low back pain, sacral pain, leg weakness, and sphincter dysfunction (Fig. 9.6).

Imaging Description

An ovoid intradural mass below the conus demonstrates heterogeneous intensity extending from the lumbar cord to D11–D12 and causes scalloping of the vertebral bodies. In this particular case, the patient refused administration of contrast. The most likely diagnosis is a myxopapillary ependymoma with a differential diagnosis of a schwannoma or a paraganglioma.

Myxopapillary Ependymoma

Myxopapillary ependymomas are a variant subtype of Ependymoma that are primarily found in the filum terminale and/or conus medullaris regions. These tumors make up approximately 13% of all spinal ependymomas and are the most frequent

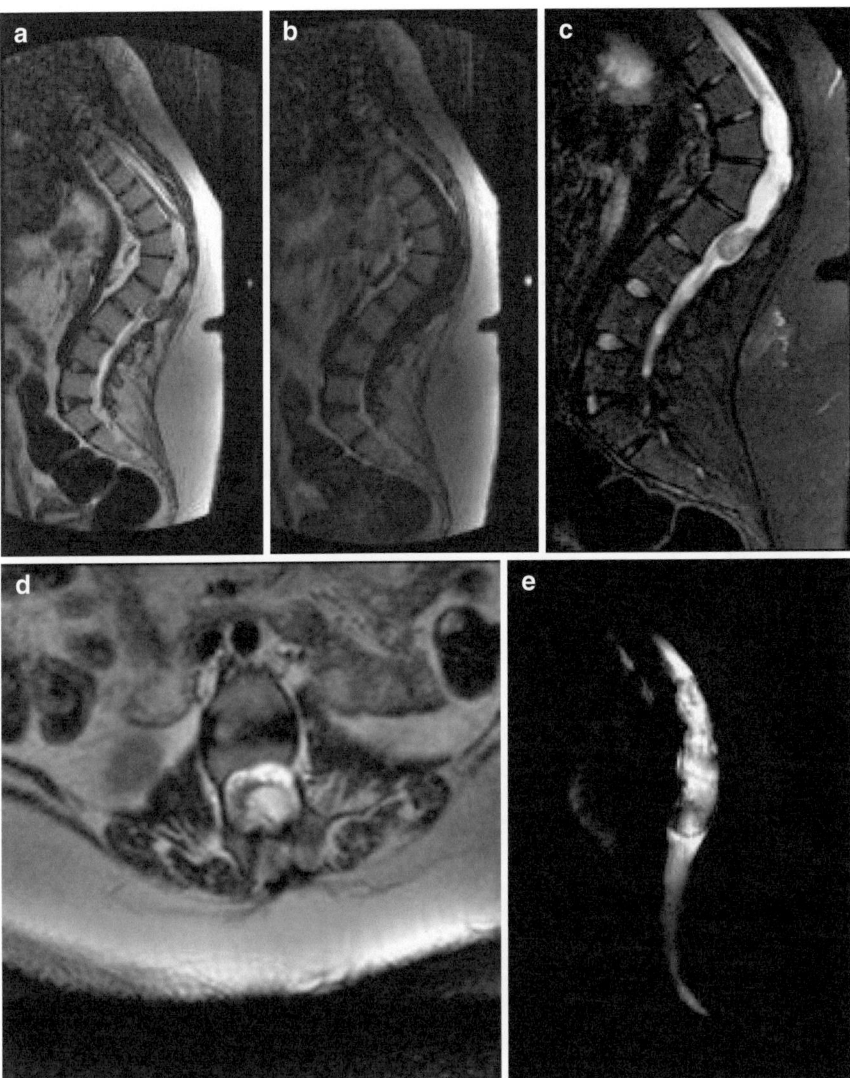

Fig. 9.6 Serial MRI images of the spine with the following: (**a**) Sagittal T2, (**b**) Sagittal T1, (**c**) Sagittal SPIR, (**d**) Axial T2, (**e**) Myelography. (Figure courtesy of Dr. Samer Hoz)

type of tumors found in the cauda equina area. They tend to manifest clinically at an earlier age compared to other spinal ependymomas, with an average presentation age of 35. They are slightly more prevalent in males and may occasionally present with subarachnoid hemorrhage.

The majority of cases are located in the lumbosacral spine, particularly in the filum terminale and/or conus medullaris. Although uncommon, they can also extend from the lumbar to the thoracic spine, as seen in some cases, or even arise in the cervicothoracic spine or fourth ventricle, although such occurrences are rare. Previously, myxopapillary ependymomas were classified as grade 1 tumors. Formerly classified as grade 1 tumors, the latest WHO classification of CNS tumors (5th edition, 2021) upgraded them to grade 2 due to their comparable local recurrence rates to other spinal cord ependymomas.

When myxopapillary ependymomas attain a larger size, plain radiographs or CT scans may reveal spinal canal widening, scalloping of vertebral bodies, and extension beyond the neural canal. On MRI, they present as well-defined intradural tumors, often appearing as large, sausage-shaped masses spanning multiple vertebral levels, though smaller, oval-shaped tumors are also observed. Smaller tumors displace nerve roots of the cauda equina, while larger ones may compress or encase them. The T1 signal is usually iso-intense, but a notable mucinous component can lead to T1 hyper-intensity. Hemorrhage and calcification may cause areas of hyper- or hypo-intensity on T1. In T2-weighted imaging, high or low intensity may be observed at the tumor margins due to hemorrhage, known as the cap sign (myxopapillary ependymomas being the subtype most predisposed to hemorrhage), and calcification may also result in areas of low T2 signal. Enhancement is consistently seen on T1 with contrast (Gd), typically displaying a homogeneous pattern. However, the presence of hemorrhage can introduce variability in the enhancement pattern [8].

Differential Diagnosis

The differential diagnosis depends on whether the tumor is small or large. The differential diagnosis for small conus and filum myxopapillary ependymoma includes schwannoma and paraganglioma. For large myxopapillary ependymoma that damages the sacrum, the differential diagnosis includes aneurysmal bone cyst, chondroma, and GCT that involves the spine.

Questions

1. **Myxopapillary ependymoma, the FALSE answer is:**
   A. Myxopapillary ependymomas are the most frequent type of tumors found in the cauda equina area.
   B. Myxopapillary ependymomas are known to have an earlier clinical manifestation compared to other spinal ependymomas.
   C. The presence of a prominent mucinous component can lead to T1 hypo-intensity.
   D. Enhancement is almost always seen on T1 C+ (Gd).
   E. The differential diagnosis of large myxopapillary ependymoma includes aneurysmal bone cyst, chondroma, and GCT that involves the spine.
   The answer is C.
   The presence of a prominent mucinous component can lead to T1 hyper-intensity.

2. **Myxopapillary ependymoma, the FALSE answer is:**
   A. Myxopapillary ependymoma sometimes presents as subarachnoid hemorrhage.
   B. The majority of cases are located in the thoracic spine.
   C. Myxopapillary ependymomas appear as large, sausage-shaped masses that span more than one vertebral level on MRI.
   D. Enhancement seen on T1 C+ (Gd) is typically homogeneous.
   E. CT scans may reveal spinal canal widening, scalloping of vertebral bodies, and extension beyond the neural canal.

   The answer is B.

   The majority of cases are located in the lumbosacral spine, particularly in the filum terminale and/or conus medullaris.

Case 74: Spinal Schwannomas

Case Scenario

A 69-year-old woman presented with neck pain, weakness, and radicular sensory changes (Fig. 9.7).

Imaging Description

MRI reveals a well-circumscribed intradural extramedullary ovoid soft tissue mass of the spinal canal at the C5–C6, lateralized to the left with mild extension along the C5 nerve root. It displays an isosignal to the spinal cord on T1 and a moderately high signal on T2 sequences. The spinal cord is compressed and displaced to the right, with no sign of myelopathy.

Spinal Schwannomas

Spinal schwannoma are benign nerve sheath tumors that arise from Schwann cells of the spinal nerve roots. They represent the most common type of nerve sheath tumor in the spine, constituting 15–50% of intradural extramedullary spinal tumors. The peak incidence of these tumors is from the fifth through seventh decades of life, with no specific sex predilection.

Most spinal schwannomas are sporadic and solitary, accounting for 95% of cases. However, they can be associated with NF2, and in individuals with NF2, nearly all spinal nerve root tumors are either schwannomas or mixed tumors. The presence of multiple schwannomas in a young adult without the NF2 mutation may suggest schwannomatosis.

Radiologically, distinguishing schwannomas from neurofibromas can be challenging. Spinal schwannomas are typically located in the intradural extramedullary

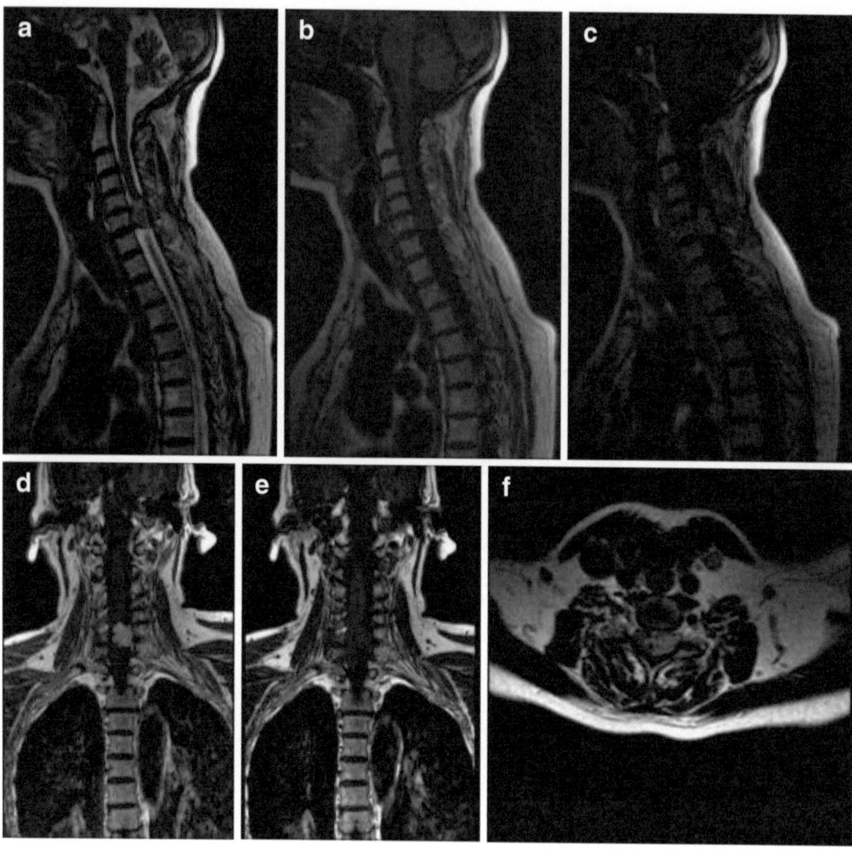

Fig. 9.7 Serial MRI images of the spine with the following: (**a**) Sagittal T2, (**b**) Sagittal T1, (**c**) Sagittal T1 C+, (**d**) Coronal T1 C+, (**e**) Coronal T1, (**f**) Axial T1 C+. (Figure courtesy of Dr. Samer Hoz)

space and are most commonly found in the cervical and lumbar regions, with less frequency in the thoracic spine. They often appear as well-defined, solid, rounded lesions, often accompanied by adjacent bony remodeling. As they grow larger, they may adopt either a sausage-shaped configuration aligning with the spinal cord's long axis or a dumbbell shape protruding from the neural foramen.

MRI can aid in differentiating between schwannomas and neurofibromas, even in cases of overlap. Schwannomas are frequently associated with features such as hemorrhage, intrinsic vascular changes (thrombosis, sinusoidal dilatation), cyst formation, and fatty degeneration, which are rare in neurofibromas. On T1, the majority of schwannomas are iso-intense (75%), with about 25% being hypo-intense. On T2, over 95% exhibit hyper-intensity, often with mixed signals. Virtually all schwannomas enhance on T1 with contrast, except for a specific subtype known as melanotic schwannomas, which are T1 hyper-intense and T2 hypo-intense [9].

Differential Diagnoses

Differential diagnoses of spinal schwannomas include the following: neurofibroma, meningioma, paraganglioma, myxopapillary ependymoma, and intradural extramedullary metastases.

Questions

1. **Spinal schwannomas, the FALSE answer is:**
   A. Most spinal schwannomas are sporadic and solitary in nature.
   B. The most common locations are the cervical and lumbar regions.
   C. Spinal schwannomas occur more frequently in females.
   D. Melanotic schwannomas are hyper-intense on T1 and hypo-intense on T2.
   E. Radiographic features of schwannomas are often difficult to distinguish from neurofibromas.
   The answer is C.
   There is no specific sex predilection.
2. **Spinal schwannomas, the FALSE answer is:**
   A. The vast majority of spinal schwannomas are intradural extramedullary in location.
   B. Schwannomas appear as well-defined, solid, rounded lesions, often with associated adjacent bony remodeling.
   C. Spinal schwannomas are the most common nerve sheath tumor of the spine.
   D. Differential diagnoses of spinal schwannomas include neurofibroma and meningioma.
   E. Schwannomas are not commonly associated with hemorrhage.
   The answer is E.
   Schwannomas are frequently associated with hemorrhage.

Case 75: Spinal Astrocytoma

Case Scenario

A 46-year-old man presented with a history of persistent pain, sensory changes, tetraparesis, and muscle weakness (Fig. 9.8).

Imaging Description

The MRI sequences reveal an eccentric intramedullary mass lesion situated on the left side of the dorso-lateral aspect of the spinal cord, spanning from C2 to C5. It appears hypo-intense on T1 and hyper-intense on T2, with peripheral patchy enhancement in the post-contrast sequence. No hemorrhagic component is observed,

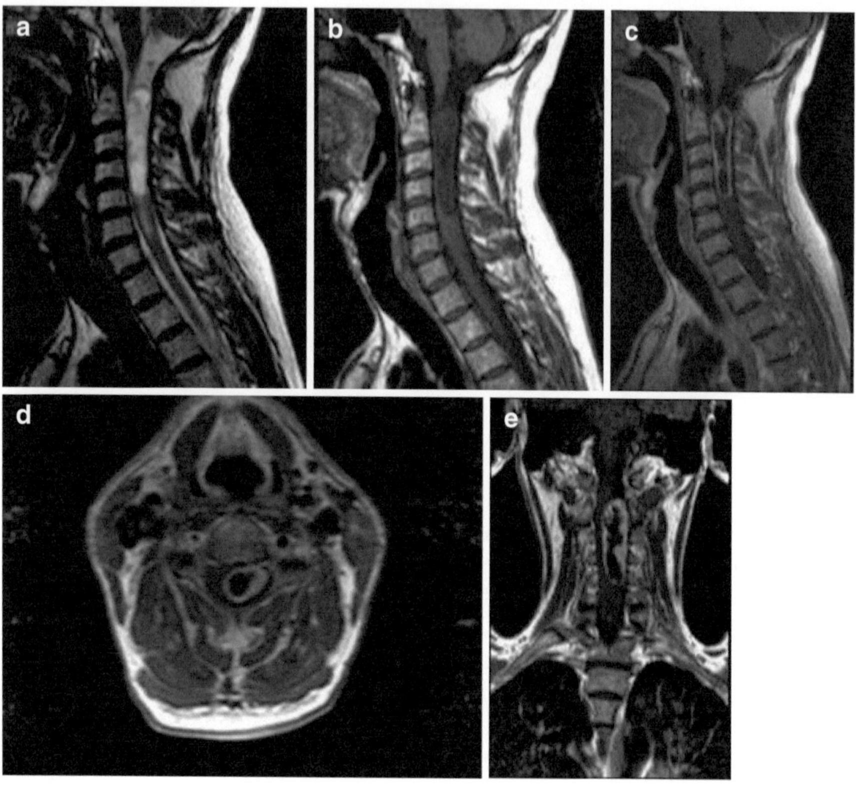

Fig. 9.8 Serial MRI images of the spine with the following: (**a**) Sagittal T2, (**b**) Sagittal T1, (**c**) Sagittal T1 C+, (**d**) Axial T1 C+, (**e**) Coronal T1 C+. (Figure courtesy of Dr. Samer Hoz)

and there is no evidence of posterior vertebral body scalloping or thinning of the pedicles.

Spinal Astrocytoma

Spinal astrocytomas account for 40% of intramedullary tumors, ranking as the second most common spinal cord tumor in adults. In children, they are the most common spinal cord tumor and represent 60% of pediatric intramedullary tumors. They occur most frequently in the third decade of life, with a predilection for males over females (M:F = 3:2) and an increased incidence in individuals with NF1.

On average, spinal astrocytomas span the length of 4–7 vertebral body segments and most frequently involve the thoracic region (67%), followed by cervical region (49%). Involvement of the entire length of the spinal cord can occur and is more common in children than in adults. They are typically intramedullary and cause diffuse expansion of the spinal cord with osseous changes. Plain radiograph and CT can be normal or show osseous remodeling as the only clue to the tumor, such as vertebral scalloping and pedicle or laminar thinning. Cord expansion can appear subtle on CT and may be better visualized with myelography, which may also show a block in the flow of contrast across the lesion.

On MRI, they are typically located in an eccentric position since they arise from the parenchyma, in contrast to ependymomas, which arise from the spinal canal. They can also appear largely extramedullary and are typically poorly circumscribed. In 40% of the cases, peritumoral edema can be seen. Twenty percent of the cases present with intratumoral cysts, and 15% of individuals have peritumoral cysts, but hemorrhage is uncommon. They appear iso to hypo-intense on T1, hyper-intense on T2, and most frequently enhancing on T1 C+ (Gd) [10, 11].

Differential Diagnosis

Spinal ependymoma: Imaging features that can help distinguish between the two are that ependymomas, which are more common in adults, are typically centrally located in the spinal canal, well-circumscribed, and commonly associated with hemorrhage, which can manifest as hemosiderin capping. In comparison, scoliosis, osseous remodeling, and cysts are present more frequently with ependymomas than with spinal astrocytomas.

Questions

1. **Spinal astrocytoma, the FALSE answer is:**
   A. Spinal astrocytomas typically cause diffuse expansion of the spinal cord.
   B. Spinal astrocytoma is the most common spinal cord tumor in children.
   C. Spinal astrocytomas are typically poorly circumscribed.
   D. On average, spinal astrocytomas span the length of 4–7 vertebral body segments.
   E. Spinal astrocytomas are typically located in a midline position.
   The answer is E.
   Spinal astrocytomas are typically located in an eccentric position.

2. **Spinal astrocytoma, the FALSE answer is:**
   A. In 40% of the cases, peritumoral edema can be seen.
   B. Spinal astrocytoma appears iso- to hypo-intense on T1 MRI.
   C. Spinal astrocytomas account for 40% of intramedullary tumors, ranking as the second most common spinal cord tumor in adults.
   D. Spinal astrocytoma appears hypo-intense on T2.
   E. They most frequently involve the thoracic region (67%), followed by cervical region (49%).
   The answer is D.
   Spinal astrocytoma appears hyper-intense on T2.

Case 76: Spinal Cord Metastasis

Case Scenario

A 78-year-old woman with breast cancer presented with upper limb weakness and neck pain (Fig. 9.9).

Imaging Description

The cord demonstrates extensive edema extending from C6 to C7, with a small, homogeneously enhancing intramedullary lesion. There is no evidence of hemorrhage within the cord, extrinsic cord compression or convincing sizable bony metastases.

Spinal Cord Metastasis

Intramedullary spinal metastasis occurs only in about 1% of cancer patients and contributes to 8.5% of CNS metastasis, far less common than leptomeningeal spread. In such patients, one-third have accompanying cerebral metastasis and 25% have leptomeningeal spread. Multiple mechanisms of metastasis have been proposed, from hematogenous dissemination to extension via Virchow-Robin spaces or the leptomeninges. The mean age is 55 years. 50% of the cases of intramedullary spinal metastasis are due to primary lung cancer, with breast cancer, melanoma,

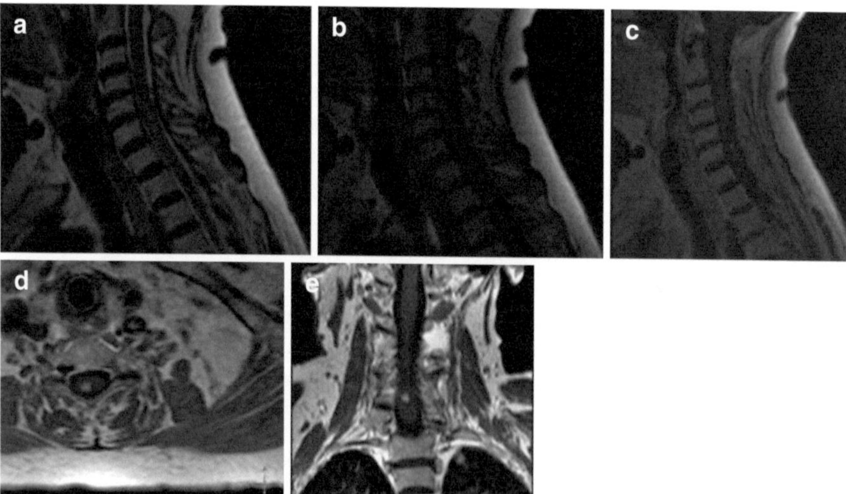

Fig. 9.9 Serial MRI images of the spine with the following: (**a**) Sagittal T2, (**b**) Sagittal T1, (**c**) Sagittal T1 C+, (**d**) Axial T1 C+, (**e**) Coronal T1 C+. (Figure courtesy of Dr. Samer Hoz)

colorectal cancer, lymphoma, leukemia, renal cell cancer, and prostate cancer being some of the many other causes for metastasis.

The cervical cord is the most common site of involvement, followed by the thoracic, then the lumbar spine. Intramedullary metastatic lesions are typically solitary and span the length of 2–3 vertebral body segments. Plain radiographs and CT are usually normal, although rarely hypervascular lesions may appear enhancing. CT myelography is also typically normal but can show nodularity of the spinal cord and focal expansion. On MRI, intramedullary metastatic lesions are well-defined and cause cord expansion. They are hypo-intense on T1 and hyper-intense on T2, often with surrounding edema. In more than 80% of cases, they demonstrate avid enhancement on T1 C+ (Gd). The rim sign, which is seen as an enhancing region surrounding the tumor, is helpful in making the diagnosis. There can also be an ill-defined region of enhancement extending above and/or the tumor, representing the flame sign [12, 13].

Differential Diagnosis

The differential diagnosis for intramedullary spinal metastasis includes intramedullary spinal tumors such as ependymoma, astrocytoma, and hemangioblastoma. Inflammatory lesions such as transverse myelitis are also on the differential but usually involve longer segments of the spinal cord, demonstrate variable enhancement with contrast, and symptoms are rapidly progressive. MS causes multiple lesions and causes milder cord expansion and edema. Other differentials include syrinx, which does not enhance with contrast, intrathecal infections, and spinal cord tuberculoma.

Questions

1. **Spinal cord metastasis, the FALSE answer is:**
    A. The cervical cord is the most common site of intramedullary spinal metastasis.
    B. Intramedullary spinal metastatic lesions are typically multiple.
    C. Intramedullary spinal metastatic lesions are well-defined on MRI.
    D. Intramedullary spinal metastatic lesions are hypo-intense on T1 and hyper-intense on T2.
    E. Intramedullary spinal metastasis is most commonly due to primary lung cancer.
    The answer is B.
    Intramedullary metastatic lesions are typically solitary.
2. **Imaging for spinal cord metastasis, the FALSE answer is:**
    A. Plain radiographs and CT scans typically demonstrate hypervascular lesions.
    B. Although CT myelography is typically normal, it might reveal localized expansion and nodularity of the spinal cord.
    C. The diagnosis is aided by the presence of the rim sign.

D. The flame sign, a poorly defined zone of enhancement extending above the tumor, can be seen.

E. Inflammatory lesions like transverse myelitis are on the differential which typically affect longer portions of the spinal cord.

The answer is A.

Plain radiographs and CT scans are typically unremarkable, although rarely, hypervascular lesions may demonstrate enhancement, which could be visible on plain radiographs or CT.

References

1. Celano E, Salehani A, Malcolm JG, Reinertsen E, Hadjipanayis CG. Spinal cord ependymoma: a review of the literature and case series of ten patients. J Neurooncol. 2016;128(3):377–86.
2. Kobayashi K, Ando K, Kato F, Kanemura T, Imagama S, Sato K, Kamiya M, Ito K, Tsushima M, Matsumoto A, Morozumi M. MRI characteristics of spinal ependymoma in WHO grade II: a review of 59 cases. Spine. 2018;43(9):E525–3.
3. Oyemolade TA, Adeolu AA, Malomo AO, Shokunbi MT, Salami AA. Spinal meningioma: clinical profile and outcome of surgical management. Pan Afr Med J. 2022;43:44.
4. Grover DS, Kundra DR, Grover DH, Gupta DV, Gupta DR. Imaging diagnosis of Plexiform neurofibroma—unravelling the confounding features: a report of two cases. Radiol Case Rep. 2021;16(9):2824–33.
5. Nguyen R, Dombi E, Akshintala S, Baldwin A, Widemann BC. Characterization of spinal findings in children and adults with neurofibromatosis type 1 enrolled in a natural history study using magnetic resonance imaging. J Neurooncol. 2015;121:209–15.
6. Massimi L, Feitosa Chaves TM, Legninda Sop FY, Frassanito P, Tamburrini G, Caldarelli M. Acute presentations of intradural lipomas: case reports and a review of the literature. BMC Neurol. 2019;19(1):1–8.
7. Ikeda N, Odate S, Shikata J, Yamamura S, Kawaguchi S. Surgical strategies and outcomes for intradural lipomas over the past 20 years. J Clin Neurosci. 2019;60:107–11.
8. Pesce A, Palmieri M, Armocida D, Frati A, Miscusi M, Raco A. Spinal myxopapillary ependymoma: the Sapienza University experience and comprehensive literature review concerning the clinical course of 1602 patients. World Neurosurg. 2019;129:245–53.
9. Harimaya K, Matsumoto Y, Kawaguchi K, Okada S, Saiwai H, Matsushita A, Iida K, Kumamaru H, Saito T, Nakashima Y. Clinical features of multiple spinal schwannomas without vestibular schwannomas. J Orthop Sci. 2022;27(3):563–8.
10. Fakhreddine MH, Mahajan A, Penas-Prado M, Weinberg J, McCutcheon IE, Puduvalli V, Brown PD. Treatment, prognostic factors, and outcomes in spinal cord astrocytomas. Neurooncology. 2013;15(4):406–12.
11. Butenschoen VM, Hubertus V, Janssen IK, Onken J, Wipplinger C, Mende KC, Eicker SO, Kehl V, Thomé C, Vajkoczy P, Schaller K. Surgical treatment and neurological outcome of infiltrating intramedullary astrocytoma WHO II–IV: a multicenter retrospective case series. J Neurooncol. 2021;151:181–91.
12. Lv J, Liu B, Quan X, Li C, Dong L, Liu M. Intramedullary spinal cord metastasis in malignancies: an institutional analysis and review. Onco Targets Ther. 2019;12:4741.
13. Mizuta H, Namikawa K, Nakama K, Yamazaki N. Intramedullary spinal cord metastasis of malignant melanoma: two cases with rim signs in contrast-enhanced magnetic resonance imaging: a case report. Mol Clin Oncol. 2021;14(3):47.

Maliya Delawan, Ameer M. Aynona, Oday Atallah,
Sajjad N. Majeed, Ahmed Muthana, Minaam Farooq,
and Asmaa H. AL-Sharee

Case 77: Giant Cell Tumors (GCTs) of the Bone

Case Scenario

A 40-year-old woman presented with a history of severe back pain (Fig. 10.1).

Imaging Description

MRI reveals a mild expansile lytic lesion of the transverse process of D4 extending
to the left pedicle with a large soft tissue component. It appears hypo-intense on
both T1 and T2, with homogeneous enhancement of the solid component. There is
an extension of soft tissue into the epidural space and left extraforaminal zone

M. Delawan
College of Medicine, Gulf Medical University, Ajman, United Arab Emirates

A. M. Aynona
College of Medicine, University of Babylon, Babylon, Iraq

O. Atallah
Hannover Medical School, Hannover, Germany

S. N. Majeed
College of Medicine, Al-Mustansiriyah University, Baghdad, Iraq

A. Muthana
College of Medicine, University of Baghdad, Baghdad, Iraq

M. Farooq
Mayo Hospital, King Edward Medical University, Lahore, Pakistan

A. H. AL-Sharee (✉)
Department of Neuroradiology, Neurosurgery Teaching Hospital, Baghdad, Iraq

© The Author(s), under exclusive license to Springer Nature
Switzerland AG 2024
S. Hoz et al. (eds.), *Neuroradiology Board's Favorites*,
https://doi.org/10.1007/978-3-031-64261-6_10

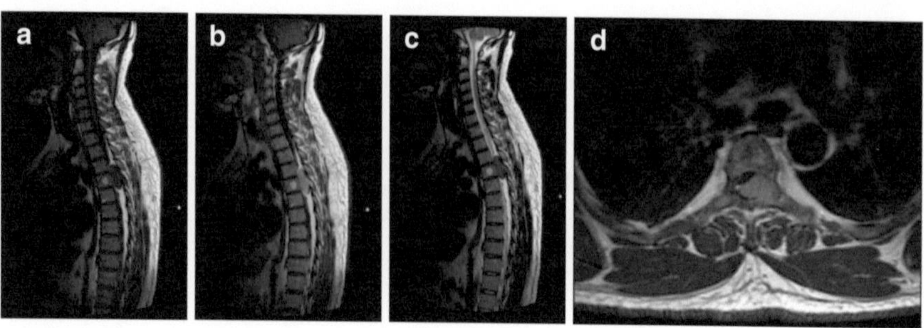

Fig. 10.1 Serial MRI images of the spine with the following: (**a**) Sagittal T1, (**b**) Sagittal T1 C+, (**c**) Sagittal T2, (**d**) Axial T1 C+. (Figure courtesy of Dr. Samer Hoz)

through an enlarged D4 foramina. Histopathological examination confirmed the diagnosis of GCT.

Giant Cell Tumors (GCTs)

GCTs are locally aggressive benign neoplasms that affect the bone. The peak incidence is between the ages of 20 and 30. Although there seems to be a mild female predilection, malignant transformation is more common in men (M:F = 3:1). Multicentric GCTs are associated with various syndromes, including pheochromocytoma-paraganglioma and GCT syndrome, Paget disease, Gorlin-Goltz syndrome, and Jaffe-Campanacci syndrome. Complications include pathologic fracture, nerve root compression, metastasis, and recurrence of the tumor.

Only 3–6% of GCTs occur in vertebral bodies, with the thoracic spine being the most common site, followed by cervical and lumbar regions. Plain radiographs may reveal an eccentrically located osteolytic or radiolucent lesion. The zone of transition is relatively narrow but can be broader in more aggressive tumors. Surrounding sclerosis is only seen in 15–20% of the cases. CT is better at delineating pathological fractures, periosteal reaction, soft tissue extension, an absence of matrix mineralization, cortical thinning, and expansile remodeling with breakthrough of the cortex.

On MRI, the solid component demonstrates a low to intermediate signal on T1 with a low signal rim. GCTs are also heterogenous on T2 and T1 C+ (Gd) shows enhancement of solid components. Other MRI features include hemorrhagic regions, bone marrow edema around the lesion, fluid-fluid levels, as well as soft tissue component and extension. In bone scintigraphy, GCTs usually show increased uptake on delayed images. The doughnut sign represents central photopenia in the lesion with increased uptake in the periphery. Adjacent bone also exhibits hyperemia [1, 2].

Differential Diagnosis

The differential diagnosis for lytic bone lesions such as GCT includes chondroblastoma, aneurysmal bone cyst, chondromyxoid fibroma, brown tumor, chondrosarcoma, and osteosarcoma.

Questions

1. **GCTs, the FALSE answer is:**
   A. Surrounding sclerosis is seen in most cases of GCTs.
   B. GCTs are heterogenous in T1 and T2.
   C. The thoracic spine is the most common site for GCTs involving the spine.
   D. On bone scintigraphy, GCTs usually show increased uptake on delayed images.
   E. Malignant transformation of GCT is more common in males than females.
   The answer is A.
   Surrounding sclerosis is only seen in 15–20% of the cases of GCT.
2. **Regarding GCTs imaging, the FALSE answer is:**
   A. Plain radiograph may demonstrate the "soap bubble" sign due to pseudotrabeculation.
   B. A CT scan can more precisely identify cortical abnormalities.
   C. MRI often demonstrates a well-defined lesion surrounded by a band of low signal intensity.
   D. Aneurysmal bone cysts typically affect the vertebral body, while GCTs mostly affect the vertebral arch.
   E. The presence of a high signal on T2W images would favor a diagnosis of chordoma or chondrosarcoma rather than GCT.
   The answer is D.
   Aneurysmal bone cysts usually affect the vertebral arch, while GCTs typically affect the vertebral body.

Case 78: Secondary Bone Lymphoma (SBL)

Case Scenario

A 17-year-old boy undergoing chemotherapy for Hodgkin lymphoma affecting the thorax presented with severe back pain, weakness, and sensory alterations (Fig. 10.2).

Imaging Description

MRI reveals a large plaque of mildly enhancing contiguous soft tissue in the epidural space surrounding the spinal cord. There is involvement of bilateral para-vertebral and pre-vertebral soft tissue, more pronounced on the right, without vertebral body

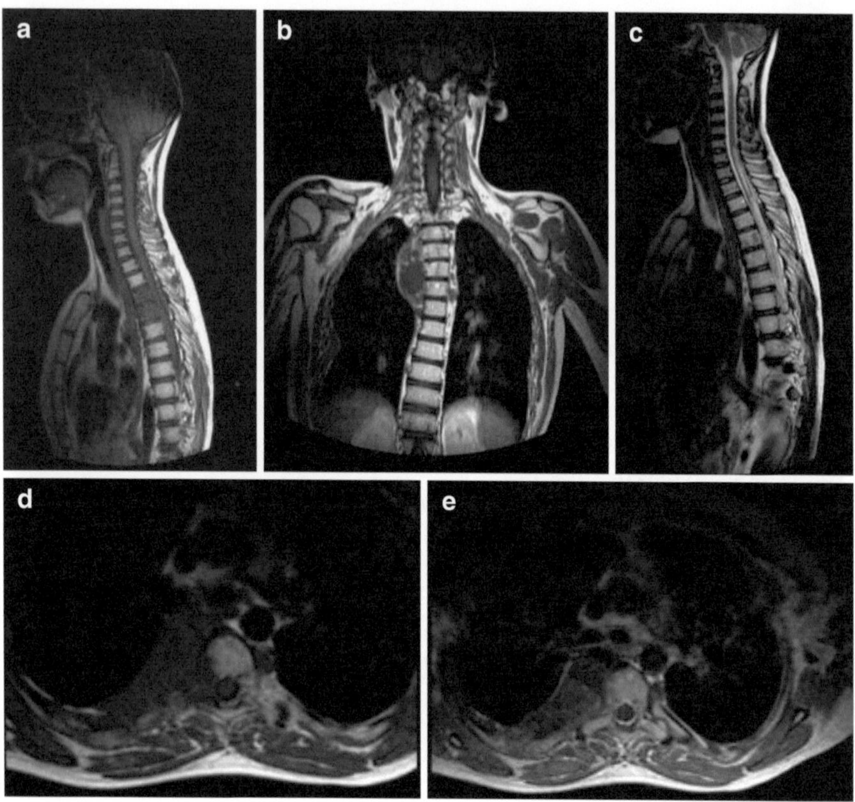

Fig. 10.2 Serial MRI images of the spine with the following: (**a**) Sagittal T1, (**b**) Coronal T1, (**c**) Sagittal T2, (**d**) Axial T1, (**e**) Axial T1 C+. (Figure courtesy of Dr. Samer Hoz)

or rib destruction. Abnormal signal intensity involves the bone marrow of adjacent vertebral bodies D4, D5, and D6 and posterior element on the right side, appearing hypo-intense on T1 and hyper-intense on T2. Abnormal signal intensity along the spinal cord is also observed, and all the lesions enhance after contrast study. These MRI findings suggest lymphoma deposits with compression myelopathy.

Secondary Bone Lymphoma (SBL)

SBL occurs in approximately 15% of cases of disseminated lymphoma. It occurs far more frequently than primary bone lymphoma and is more common in pediatric populations. SBL is defined as bone lymphoma with nodal involvement occurring within 6 months or primary soft tissue lymphoma with secondary bone involvement at least 6 months after diagnosis.

SBL occurs more commonly in the axial skeleton when compared to the appendicular skeleton. Plain radiograph or CT demonstrate lytic lesions that cause widespread bone destruction with cortical breaching. Adjacent soft tissue masses can often be seen [3–5].

Questions

1. **SBL, the FALSE answer is:**
   A. SBL occurs more commonly in the axial skeleton when compared to the appendicular skeleton.
   B. SBL is more common than primary bone lymphoma.
   C. SBL occurs more commonly in adults than children.
   D. SBL typically causes widespread bone destruction.
   E. Adjacent soft tissue masses can often be seen with SBL.
   The answer is C.
   SBL occurs more commonly in children than adults.
2. **SBL, the FALSE answer is:**
   A. Plain radiographs demonstrate lytic lesions.
   B. Lymphoma is hypo-intense on T1W sequences and hyper-intense on T2.
   C. Lymphoma demonstrates no enhancement after paramagnetic contrast administration.
   D. MRI may demonstrate the "wrap around" sign.
   E. The "floating aorta" sign occurs when the soft-tissue mass surrounding the vertebral body is sufficiently prominent.
   The answer is C.
   Following the administration of paramagnetic contrast, lymphoma typically exhibits enhancement.

Case 79: Aneurysmal Bone Cysts (ABC)

Case Scenario

A 12-year-old female presented with a history of back pain and swelling (Fig. 10.3).

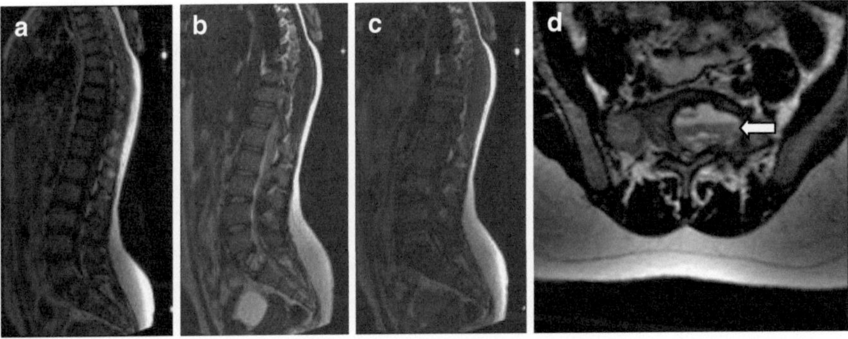

Fig. 10.3 Serial MRI images of the spine with the following: (**a**) Sagittal T1, (**b**) Sagittal T2, (**c**) Sagittal T1 C+, (**d**) Axial T2. (Figure courtesy of Dr. Samer Hoz)

Imaging Description

The MRI shows a large lobulated right extradural paravertebral mass lesion that involves the vertebral body of S1 and extends partially into the spinal canal. It exhibits a fluid-fluid level in T2WI (arrow). The appearance is consistent with an aneurysmal bone cyst.

Aneurysmal Bone Cysts (ABCs)

ABCs are benign expansile osteoclastic giant cell-rich bony neoplasms, consisting of many channels and cystic spaces that are filled with blood. It can occur at any age but is mostly seen in children and adolescents, with ~20% above the age of 20 years. Both genders are equally affected. ABCs can lead to many complications, including pathological fracture, nerve compression syndrome, spinal canal stenosis, and sub-articular zone stenosis with nerve root compression. Only ~20–30% of ABCs occur in the spine.

Plain radiographs may reveal a sharply defined, expansile, solitary lucent bone lesion with thin-walled cavities. CT shows lucent bone lesions with a mean density higher than fat. Specific features, such as cortical breach or soft tissue extension, may be seen. ABCs can also demonstrate fluid-fluid levels, which are more difficult to appreciate than on MRI and require viewing with a narrow window width.

MRI reveals the characteristic fluid-fluid levels remarkably clearly. It may also identify the presence of a solid component and concerning features of other tumor entities that have an ABC-like appearance. It is crucial to note that the presence of fluid-fluid levels, although characteristic of ABC, is by no means pathognomonic and can be seen in other lesions as well, both benign and malignant (e.g., GCT, chondroblastoma, simple bone cysts, and telangiectatic osteosarcomas). ABCs exhibit variable signal intensity on T1, are hyper-intense on T2, and may show enhancement of septations on T1 post-contrast. The cysts may display various signals, with a surrounding rim of low T1 and T2 signals. Focal areas of high T1 and T2 signals can also be seen, likely representing blood of varying ages. In bone scintigraphy, the doughnut sign indicates central photogenic uptake, with increased uptake in the periphery. On DSA, ABCs appear poorly vascularized [6–8].

Differential Diagnosis

The differential diagnosis for lytic bone lesions such as ABC includes chondroblastoma, fibrous dysplasia, GCT, osteosarcoma (especially telangiectatic osteosarcoma).

Questions

1. **ABC, the FALSE answer is:**
   A. ABC consists of many channels and cystic spaces that are filled with blood.
   B. MRI scan reveals the characteristic fluid-fluid levels remarkably clearly.
   C. On bone scintigraphy, the doughnut sign represents a central photogenic region in the lesion with increased uptake in the periphery.
   D. It can occur at any age but is mostly seen in children and adolescents, with ~20% occurring above the age of 20 years.
   E. The presence of fluid-fluid levels is considered pathognomonic of ABC.
      The answer is E.
      The presence of fluid-fluid levels, although characteristic of ABC, is by no means pathognomonic as it can be found in simple bone cysts, telangiectatic osteosarcoma, giant cell tumors, osteoblastomas, sarcomas, chondroblastoma, and fibroxanthomas.

2. **ABC imaging, the FALSE answer is:**
   A. On plain radiographs, there may be characteristic ballooning of the posterior elements with a thin rim.
   B. CT scan shows multiloculated lytic lesions with numerous interior septations.
   C. MRI shows multiseptate lesions with a thin, well-defined rim of low signal intensity in the periphery.
   D. The cysts exhibit various signals, likely representing blood of varying ages.
   E. CT scan is the most sensitive technique.
      The answer is E.
      MRI is the most sensitive technique with remarkably clearly visible distinctive fluid-fluid levels. It may also recognize the presence of a solid component and alarming characteristics that point to another tumor entity.

Case 80: Vertebral Metastases from Breast Cancer with Pathological Fracture

Case Scenario

A 55-year-old female newly diagnosed with breast cancer presented with back pain (Fig. 10.4).

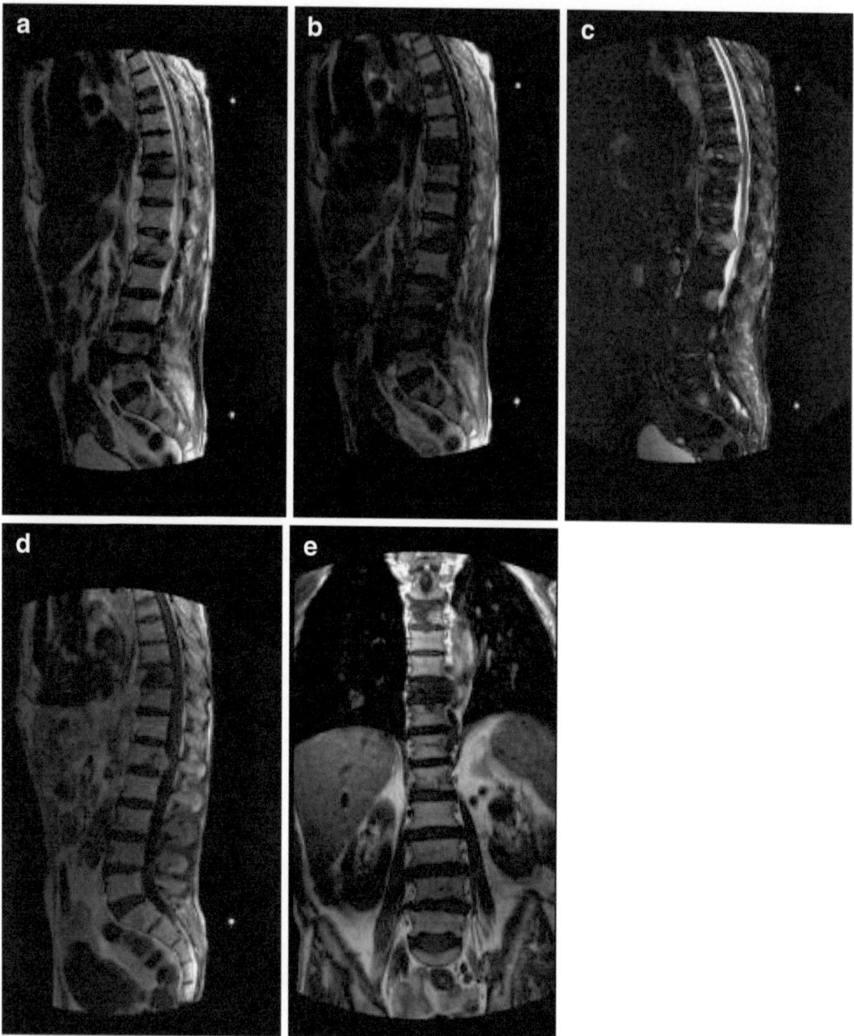

Fig. 10.4 Serial MRI images of the spine with the following: (**a**) Sagittal T1, (**b**) Sagittal T2, (**c**) Sagittal SPIR, (**d**) Sagittal T1 C+, (**e**) Coronal T1. (Figure courtesy of Dr. Samer Hoz)

Imaging Description

MRI reveals multiple enhancing vertebral lesions, consistent with breast cancer metastases. D10 and L1 vertebral bodies are massively infiltrated with the involvement of posterior elements. With compression fracture of the D10 and L1 vertebral body is seen, with posterior bulging of the posterior cortex of the vertebral body, subsequent compression of the anterior thecal sac, and abutment of the spinal nerve root.

Vertebral Metastases

Vertebral metastasis is the secondary involvement of the vertebral spine by a hematogenous route of metastasis and occurs in approximately 10% of newly diagnosed cancers. Although it is more prevalent in individuals aged over 50, it should be considered in the differential diagnosis of spinal bone lesions in any patient older than 40. Breast cancer, lung cancer, prostate cancer, lymphoma, renal cell carcinoma, gastrointestinal tract malignancies, melanoma, pancreatic cancer, thyroid carcinoma, and carcinoid are considered the most common primary malignancies that involve vertebra.

Metastastic lesions are classified as osteolytic, osteoblastic, and mixed lesions. The primary malignancies with predominantly osteoblastic metastases (sclerotic extradural bone lesions) include prostate carcinoma, osteosarcoma, and medullary thyroid carcinoma, while primary malignancies with predominantly osteolytic metastases that may rarely become osteoblastic (mixed sclerotic and lytic extradural bone lesions) include breast cancer, lymphoma, and urothelial carcinoma. Primary malignancies with osteolytic metastases include lung cancer, gastrointestinal tract cancers, renal cell carcinoma, malignant melanoma, and multiple myeloma.

Metastatic lesions exhibit variable radiographic features, ranging from benign to aggressive primary bone tumors. Plain radiographs are insensitive to small lytic lesions, rendering it difficult to determine their origin. CT scans reveal lytic and sclerotic lesions, with lytic lesions showing soft tissue attenuation and irregular margins. MRI is sensitive for metastatic disease and assesses cord compression. Osteoblastic metastases are hypo-intense on T1 and T2, while mixed sclerotic and lytic extradural bone lesions are hypo-intense on T1 and hyper- or iso-intense on T2 [9–11].

Differential Diagnosis

The differential diagnosis for osteoblastic metastases includes bone islands (enostoses), spondylosclerosis hemispherica, primary bone tumors (e.g., osteoblastoma and osteoid osteoma), and therapy effects from radiation, chemotherapy, and vertebroplasty. Other differentials for lytic extradural bone lesions include primary bone tumors, aneurysmal bone cyst, neurilemmoma/schwannoma, infective spondylitis, and atypical hemangioma.

Questions

1. **Vertebral metastases, the FALSE answer is:**
   A. Metastasic lesions can be osteolytic, osteoblastic, or mixed.
   B. Breast cancer is one of the most common primary malignancies that involve the vertebra.
   C. The absent pedicle sign is a useful sign to look for on AP films when assessing for metastasis.

D. Primary malignancies with predominantly osteolytic metastases that may rarely become osteoblastic (mixed sclerotic and lytic extradural bone lesions) include breast cancer.
E. Mixed sclerotic and lytic extradural bone lesions are hyper-intense on T1 and iso- and/or hyper-intense on T2.

The answer is E.

Mixed sclerotic and lytic extradural bone lesions are hypo-intense on T1 and iso- and/or hyper-intense on T2.

2. **Vertebral metastases, the FALSE answer is:**
   A. On radiography, before a lesion can be clearly detected, at least 50% of the cortical bone mass must be destroyed.
   B. Although CT is less sensitive for cortical invasion, it is sensitive enough to detect medullary bone invasion.
   C. MRI is effective in detecting intramedullary metastases in bones with extensive marrow cavities, such as vertebral bodies.
   D. T2-weighted images are not deemed beneficial because of the diverse appearance of metastases.
   E. Spine DWI can be used to distinguish between compression fractures caused by osteoporosis and pathologic fractures.

   The answer is B.

   CT scan is less sensitive for detecting medullary bone invasion; however, it is sensitive enough to identify mild cortical invasion.

References

1. Shi LS, Li YQ, Wu WJ, Zhang ZK, Gao F, Latif M. Imaging appearance of giant cell tumour of the spine above the sacrum. Br J Radiol. 2015;88(1051):20140566.
2. Yuan B, Zhang L, Yang S, Ouyang H, Han S, Jiang L, Wei F, Yuan H, Liu X, Liu Z. Imaging features of aggressive giant cell tumors of the mobile spine: retrospective analysis of 101 patients from single center. Global Spine J. 2022;12(7):1449–61.
3. Wu H, Bui MM, Leston DG, Shao H, Sokol L, Sotomayor EM, Zhang L. Clinical characteristics and prognostic factors of bone lymphomas: focus on the clinical significance of multifocal bone involvement by primary bone large B-cell lymphomas. BMC Cancer. 2014;14(1):1–9.
4. Xu D, Wang B, Chen L, Zhang H, Wang X, Chen J. The incidence and mortality trends of bone lymphoma in the United States: an analysis of the Surveillance, Epidemiology, and End Results database. J Bone Oncol. 2020;24:100306.
5. Moulopoulos LA, Dimopoulos MA, Vourtsi A, Gouliamos A, Vlahos L. Bone lesions with soft-tissue mass: magnetic resonance imaging diagnosis of lymphomatous involvement of the bone marrow versus multiple myeloma and bone metastases. Leuk Lymphoma. 1999;34(1–2):179–84.
6. Abrar WA, Sarmast A, Singh AR, Khursheed N, Ali Z. Aneurysmal bone cysts of spine: an enigmatic entity. Neurol India. 2020;68(4):843.
7. Muratori F, Mondanelli N, Rizzo AR, Beltrami G, Giannotti S, Capanna R, Campanacci DA. Aneurysmal bone cyst: a review of management. Surg Technol Int. 2019;35:325–35.
8. Zileli M, Isik HS, Ogut FE, Is M, Cagli S, Calli C. Aneurysmal bone cysts of the spine. Eur Spine J. 2013;22:593–601.

9. Cho JH, Ha JK, Hwang CJ, Lee DH, Lee CS. Patterns of treatment for metastatic patho-logical fractures of the spine: the efficacy of each treatment modality. Clin Orthop Surg. 2015;7(4):476–82.
10. Schwarzenbach O, Boos N, Aebi M. Spinal metastases and metastasis-induced pathological fractures of the spine. Unfallchirurg. 1990;93(10):457–66.
11. Taoka T, Mayr NA, Lee HJ, Yuh WT, Simonson TM, Rezai K, Berbaum KS. Factors influ-encing visualization of vertebral metastases on MR imaging versus bone scintigraphy. Am J Roentgenol. 2001;176(6):1525–30.

Spinal Trauma

11

Rokaya H. Abdalridha, Sajjad G. Al-Badri, Sama Albairmani,
Zainab K. A. Al-araji, Fatimah O. Ahmed, Ahmed Muthana,
and Asmaa H. AL-Sharee

Case 81: Traumatic Spinal Cord Injury (TSCI)

Case Scenario

A 62-year-old male presented to the emergency department following a road traffic accident, complaining of severe mid-back pain, bilateral lower limb weakness, and urinary incontinence (Fig. 11.1).

Image Description

The sagittal T2 STIR sequence reveals a notable hyper-intense signal at the D12–L1 level, suggesting an edematous and expanded cord. While there is a slight anterior narrowing of the spinal canal at this level, the canal itself remains spacious, and there is no compression of the cord. Significant edema is observed in the posterior

R. H. Abdalridha
College of Medicine, Babylon University, Babylon, Iraq

S. G. Al-Badri · A. Muthana
College of Medicine, University of Baghdad, Baghdad, Iraq

S. Albairmani
College of Medicine, University of Al-Iraqia, Baghdad, Iraq

Z. K. A. Al-araji
College of Medicine, Al-Nahrain University, Baghdad, Iraq

F. O. Ahmed
College of Medicine, University of Mustansiriyah, Baghdad, Iraq

A. H. AL-Sharee (✉)
Department of Neuroradiology, Neurosurgery Teaching Hospital, Baghdad, Iraq

© The Author(s), under exclusive license to Springer Nature
Switzerland AG 2024
S. Hoz et al. (eds.), *Neuroradiology Board's Favorites*,
https://doi.org/10.1007/978-3-031-64261-6_11

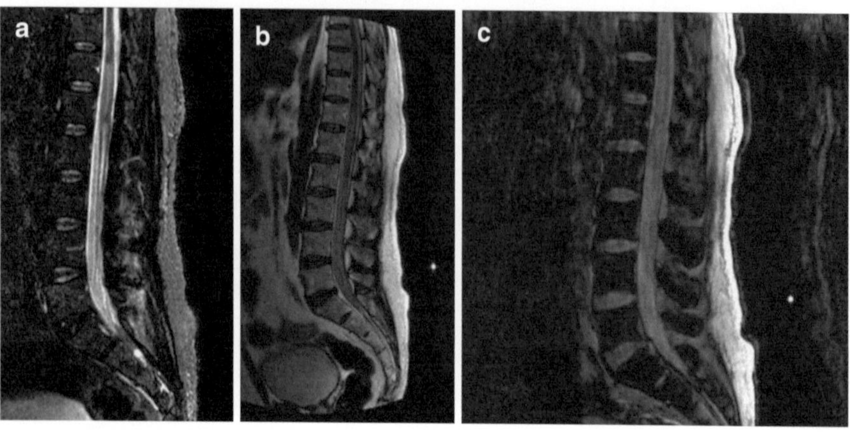

Fig. 11.1 Serial MRI images of the spine with the following: (**a**) Sagittal T2 STIR, (**b**) Sagittal T2 FFE, (**c**) Sagittal T2 GRE. (Figure courtesy of Dr. Samer Hoz)

paraspinal muscles and interspinous ligaments along the lumbar vertebrae. No evidence of epidural or cord hemorrhage is present in both the T2 FFE and T2 GRE sequences, and the intervertebral discs appear normal.

Traumatic Spinal Cord Injury (TSCI)

TSCI is a devastating neurological condition with significant socio-economic implications for the affected individuals, caregivers, and the healthcare system. TSCI can result from various incidents, including sports injuries, falls, assaults, or road accidents. In some cases, even mild trauma can lead to TSCI if the spine is already compromised due to conditions like ankylosing spondylitis or pre-existing spinal stenosis.

The clinical presentation of TSCI varies based on factors such as the extent and completeness of the injury, the severity of the primary injury, and the progression of secondary damage. This diverse range of clinical syndromes may manifest with neurological symptoms, including altered sensation, limb weakness, autonomic dysfunction, and sphincter disturbances. Moreover, patients often experience pain arising from damage to the musculoskeletal system linked to the injury.

In terms of pathophysiology, the most common mechanism of injury is in the context of fracture-dislocation injuries, where bone fragments from burst fractures cause persistent compression of the spinal cord. Hyperextension injuries, in contrast to fracture-dislocation injuries, lead to an impact on the spinal cord with transient compression and are less common. The third mechanism is distraction injury, which occurs when two adjacent vertebrae are pulled apart, causing strain and tearing of the spinal cord in its axial plane. Finally, transection injuries result from splintered bones and severe dislocations.

In the evaluation of TSCI, plain radiographs have limited sensitivity in detecting spinal cord trauma and associated bony injuries. For a comprehensive assessment of bony injuries, a CT scan is considered the most effective imaging modality. In addition to standard axial and sagittal T1 and T2 MRI sequences, it is crucial to include additional MRI sequences, such as the T2* sequence (high sensitivity to bleeding) and the STIR sequence (high sensitivity to ligamentous injury). These sequences provide valuable information about various findings associated with TSCI. On MRI, spinal cord edema at the level of trauma is characterized by focal cord expansion without signal anomalies, best visualized on a sagittal T1 MRI. The T2* sequence can reveal blooming, which is indicative of both bleeding and contusion, aiding in the diagnosis and understanding of the injury [1–3].

Questions

1. **TSCI, the FALSE answer is:**
   A. Can be complete or incomplete.
   B. Hyperextension injuries cause transient compression of the spinal cord.
   C. The most frequent mechanism is persistent compression in the setting of fracture-dislocation injuries.
   D. Clinical presentation varies depending on the extent and completeness of the injury.
   E. TSCI is always caused by severe trauma.
   The answer is E.
   In some cases, even mild trauma can lead to TSCI if the spine is already compromised due to conditions like ankylosing spondylitis or pre-existing spinal stenosis.

2. **TSCI radiological features, the FALSE answer is:**
   A. For associated bony injuries, a CT scan is the preferred technique.
   B. Spinal cord edema is characterized by hyper-intensity on T2-weighted MRI images.
   C. In spinal cord contusion, there is a T2 hyper-intense thick rim surrounding a small central T1 hypo-intense focus.
   D. In intramedullary hemorrhage, MRI demonstrates a T2 hyper-intense thin rim surrounding a large central T1 hypo-intense focus.
   E. STIR is performed as it is more sensitive for hemorrhage.
   The answer is E.
   In TSCI assessment, additional sequences include STIR sequence, which is more sensitive for ligamentous injury, while T2* sequence is more sensitive for hemorrhage.

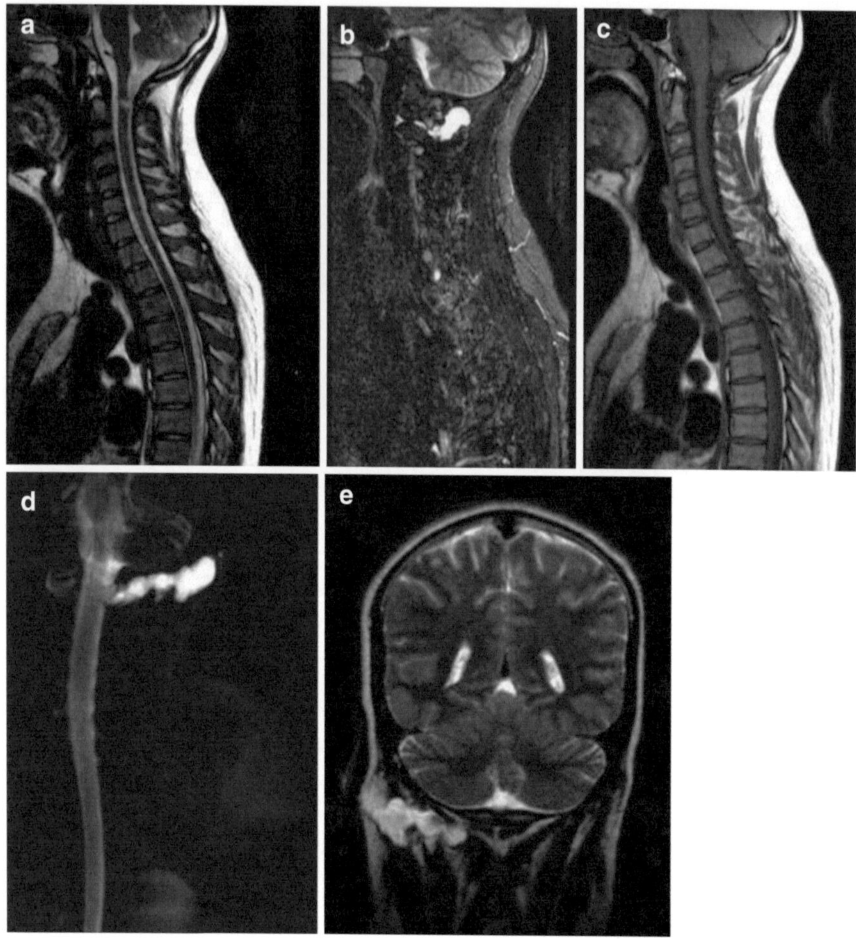

Fig. 11.2 Serial MRI images of the spine with the following: (**a**) Sagittal T2-MRI, (**b**) Sagittal STIR, (**c**) Sagittal T1-MRI, (**d**) Myelography, (**e**) Coronal T2-MRI. (Figure courtesy of Dr. Samer Hoz)

Case 82: Spinal Cord Transection

Case Scenario

A 30-year-old male presented with complete tetraplegia and loss of sphincter control following a stab wound by knife to the posterior upper neck (Fig. 11.2).

Imaging Description

On T2W MRI, a distinct and prominent horizontal line of high signal intensity is observed in the upper cervical spinal cord at the C1–C2-disc space level. This finding indicates a spinal cord transection. Additionally, there is an adjacent fluid intensity collection located within the right paraspinal muscle posterior to the lamina of C2. The presence of a connection with the CSF canal suggests meningeal injury and CSF leak.

Spinal Cord Transection

Spinal cord transection refers to the complete severing of white matter tracts, segmental gray matter, and nerve roots, occurring anywhere between the cervicomedullary junction and the tip of the conus medullaris. This traumatic event disrupts the circulation of both blood and CSF within the cord. Clinically, spinal cord transection is typically associated with severe trauma and presents with distinct motor and sensory neurological symptoms that correspond to the level of cord injury. At the level of injury, lower motor neuron paralysis is evident, while below the injury level, upper motor neuron paralysis (or spasticity) is observed.

MRI is the preferred imaging modality for the acute assessment of recent traumatic spinal injuries, often used with or without confirmation by trauma CT. On sagittal T2-weighted images, edema of the spinal cord exhibits high signal intensity. In contrast, acute hemorrhage appears hypo-intense with a thin rim of high signal intensity. Additionally, transection of the spinal cord presents as a discontinuity on both T1-weighted and T2-weighted images. STIR images are effective in detecting edema and ligamentous injuries, especially in the interspinous and supraspinous ligaments. While fat-suppressed T2-weighted images can also detect edema, STIR images provide superior fat suppression [4, 5].

Questions

1. **Spinal cord transection, the FALSE answer is:**
    A. Complete transaction accounts for a minority of spinal cord injuries.
    B. At the level of the injury, lower motor neuron paralysis is observed.
    C. Acute hemorrhage appears hyper-intense with a thin rim on T2W.
    D. STIR images are effective for detecting edema and ligamentous injuries.
    E. Fat-suppressed T2W provides superior fat suppression for detecting edema.
        The answer is C.
        Acute hemorrhage appears hypo-intense with a thin rim of high signal intensity on T2W.

References

1. Ellingson BM, Salamon N, Holly LT. Imaging techniques in spinal cord injury. World Neurosurg. 2014;82(6):1351–8.
2. Talekar K, Poplawski M, Hegde R, Cox M, Flanders A. Imaging of spinal cord injury: acute cervical spinal cord injury, cervical spondylotic myelopathy, and cord herniation. Semin Ultrasound CT MRI. 2016;37(5):431–47.
3. Talbott JF, Huie JR, Ferguson AR, Bresnahan JC, Beattie MS, Dhall SS. MR imaging for assessing injury severity and prognosis in acute traumatic spinal cord injury. Radiol Clin North Am. 2019;57(2):319–39.
4. Lammertse D, Dungan D, Dreisbach J, Falci S, Flanders A, Marino R, Schwartz E. Neuroimaging in traumatic spinal cord injury: an evidence-based review for clinical practice and research: report of the National Institute on Disability and Rehabilitation Research Spinal Cord Injury Measures Meeting. J Spinal Cord Med. 2007;30(3):205–14.
5. Alkadeem RM, El-Shafey MH, Eldein AE, Nagy HA. Magnetic resonance diffusion tensor imaging of acute spinal cord injury in spinal trauma. Egypt J Radiol Nucl Med. 2021;52(1):1–3.

Spinal Vascular Pathologies

12

Mostafa H. Algabri, Zainab Q. Saadi,
Sadeem A. Albulaihed, Wafa M. Almuallim,
Tabarek F. Mohammed, Fatimah O. Ahmed,
and Samer S. Hoz

Case 83: Spinal Dural Arteriovenous Fistula (SDAVF)

Case Scenario

A 21-year-old female with an insidious onset of symptoms, exhibiting progressive thoracic back pain, bilateral lower limb weakness with a notable emphasis on the right side, sensory loss, and urinary retention (Fig. 12.1).

Imaging Description

In T1-weighted imaging, hypointensity is evident within the lower half of the thoracic cord surface, along with prominent flow voids. T2-weighted imaging reveals extensive central cord edema, manifesting as diffuse multilevel intramedullary hyperintensity. Multiple flow voids are observed on the dorsal surface of the cord, and T1-weighted imaging with contrast highlights serpentine enhancing veins on the cord surface.

M. H. Algabri · Z. Q. Saadi · T. F. Mohammed
College of Medicine, University of Baghdad, Baghdad, Iraq

S. A. Albulaihed
College of Medicine, Alfaisal University, Riyadh, Saudi Arabia

W. M. Almuallim
College of Medicine, Al-Rayan College, Al-Madinah Al-Monawarah, Saudi Arabia

F. O. Ahmed
College of Medicine, University of Mustansiriyah, Baghdad, Iraq

S. S. Hoz (✉)
University of Pittsburgh Medical Center (UPMC), Pittsburgh, PA, USA

S. Hoz et al. (eds.), *Neuroradiology Board's Favorites*,
https://doi.org/10.1007/978-3-031-64261-6_12

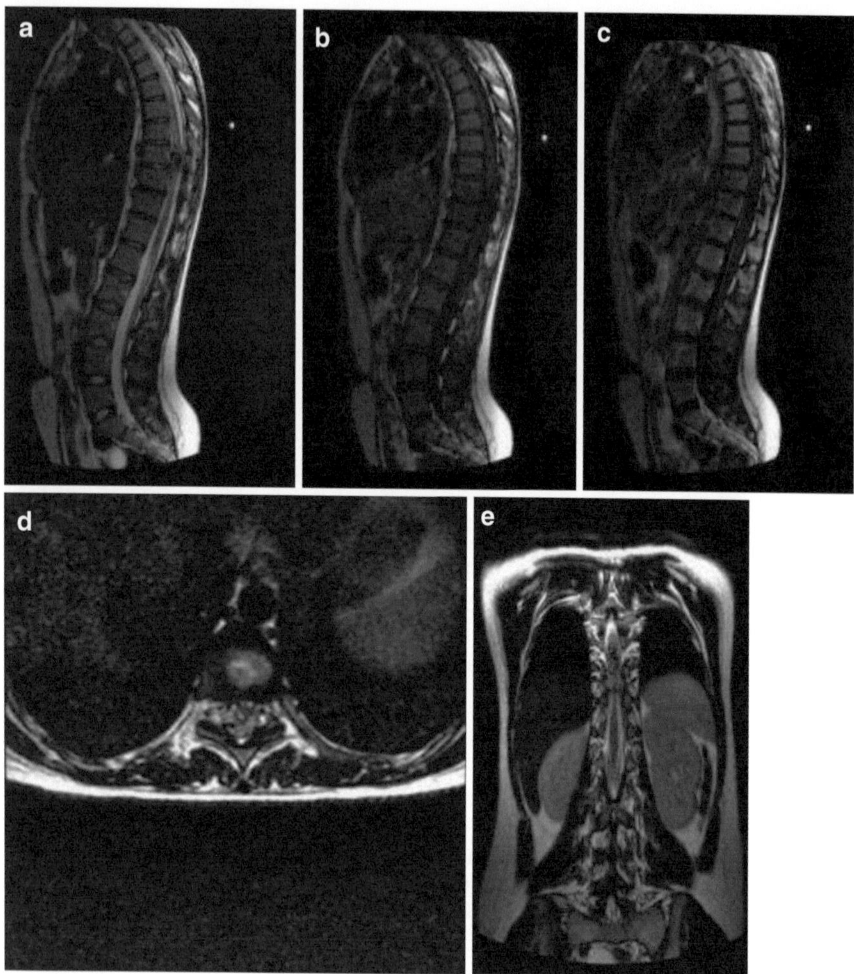

Fig. 12.1 Serial MRI images of the spine with the following: (**a**) Sagittal T2, (**b**) Sagittal T1, (**c**) Sagittal T1, (**d**) Axial T1 C+, (**e**) Coronal T2. (Figure courtesy of Dr. Samer Hoz)

Spinal Dural Arteriovenous Fistula (SDAVF)

SDAVF represents the most common type of spinal vascular malformation, constituting approximately 70% of all such lesions. It predominantly affects males and exhibits an incidence peak in the fifth and sixth decades of life. Around 60% of SDAVFs are spontaneous, with the remaining 40% resulting from trauma. Typical clinical manifestations include gradually worsening pain, weakness, or sensory alterations in the lower extremities. Sphincter dysfunction may also be present. The onset of symptoms is gradual, and the condition progresses over several years. Diagnosis is often delayed, with a significant gap between the appearance of symptoms and a formal diagnosis. While hemorrhage is infrequent, it can occur sporadically, and in some cases, it may be the underlying cause of unexplained subarachnoid hemorrhage.

SDAVFs are abnormal connections between a radicular or radiculomeningeal artery and a radicular or pial vein within the dura of an adjacent nerve root sleeve. In 85% of cases, a single transdural arterial feeder is involved, but there are instances with multiple arterial feeders originating from single or multiple levels, and they may be either unilateral or bilateral. The direct arterial inflow into the valveless coronal venous plexus raises the pressure, leading to dilation of the plexus and reduced venous drainage of the cord. This, in turn, causes venous congestion and intramedullary edema, known as congestive or venous hypertensive myelopathy. It is possible that cord ischemia and infarction may occur as a result.

MRI typically reveals cord enlargement in the lower thoracic region and conus medullaris, with signal changes affecting multiple spinal segments. Interestingly, the segmental level of cord enlargement and signal change does not necessarily correspond to the location of the fistula. Regarding signal characteristics, T1W images may display intramedullary hypointensity and flow voids on the cord surface. On T2W images, the most sensitive finding is diffuse multilevel intramedullary hyperintensity (edema), which is observed in up to 90% of cases involving the conus medullaris due to orthostasis. However, SDAVFs of the upper cervical spine (C1–C2) might behave differently, often draining intracranially and presenting more commonly with subarachnoid hemorrhage.

T2 hypointensity at the cord's periphery is thought to indicate pial capillaries containing deoxyhemoglobin due to venous hypertension. A highly specific finding in SDAVFs is the presence of prominent serpiginous intradural extramedullary flow voids, which typically span more than three segments along the dorsal and sometimes ventral aspects of the cord. These vessels may be large enough to give the cord surface a scalloped appearance. However, with significant cord swelling, the veins can be compressed by the mass effect, making them undetectable on imaging. With T1W contrast-enhanced (Gd) images, patchy intramedullary enhancement is often observed, resulting from the breakdown of the blood-brain barrier due to chronic infarction or a capillary leak phenomenon secondary to venous hypertension. Additionally, serpentine-enhancing veins may be visualized on the cord surface. Contrast-enhanced MRA is helpful in determining the segmental level of the fistula, aiding in guiding selective catheter angiography.

If MRI is not feasible, alternative imaging modalities include CT angiography, which may successfully localize the fistula in up to 75% of cases, and CT myelography, which may demonstrate tortuous filling defects due to dilated veins. DSA is the most dependable test for confirming the diagnosis and providing treatment options, but it is time-consuming and hazardous due to the possibility of dissecting a vessel, which could lead to cord ischemia [1, 2].

Differential Diagnosis

- Intramedullary neoplasm.

- Spinal AVM: often presents with hemorrhaging, has an abrupt onset of symptoms, affects both males and females equally, and typically manifests in the third decade of life.

Questions

1. **SDAVF, the FALSE answer is:**
   A. Is the most common form of spinal vascular malformation.
   B. Occurs more commonly in males than females.
   C. The incidence peak is in the second and third decades.
   D. Subarachnoid hemorrhage is a potential complication of SDAVF.
   E. Venous congestion and intramedullary edema are potential causes of cord ischemia in SDAVF.
   The answer is C.
   The incidence peak is in the fifth and sixth decades.
2. **SDAVF imaging, the FALSE answer is:**
   A. MRI typically shows cord enlargement in the lower thoracic region.
   B. T2W images may display intramedullary hypointensity and flow voids on the cord surface.
   C. T2 hypointensity at the cord's periphery is specific for SDAVFs.
   D. With significant cord swelling, the veins are easily detectable on imaging.
   E. Contrast-enhanced MRA is helpful in determining the segmental level of the fistula.
   The answer is C.
   T2 hypointensity at the cord's periphery is not specific for SDAVFs. It is thought to indicate pial capillaries containing deoxyhemoglobin due to venous hypertension, which can be seen in other conditions as well.

Case 84: Spinal AVM

Case 84.1

Case Scenario
A 49-year-old male reported difficulty walking, increased clumsiness, tingling sensations, and weakness in his legs. Upon examination, signs of spasticity, hyperreflexia, and lower extremity weakness were noted (Fig. 12.2).

Imaging Description
T2W MRI of the mid-dorsal spinal cord reveals intradural-extramedullary involvement. Multiple small areas of signal voids are observed throughout the spinal cord, consistent with flow voids, indicating the presence of a spinal AVM.

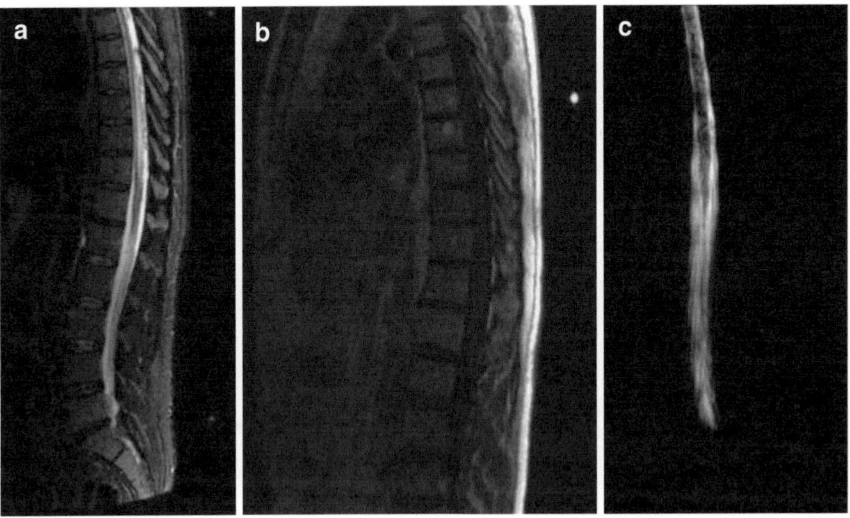

Fig. 12.2 Spinal MRI images with the following: (**a**) Sagittal T2, (**b**) Sagittal T1, (**c**) Myelography. (Figure courtesy of Dr. Samer Hoz)

Case 84.2

Case Scenario

A 9-year-old female presented with a gradual onset of difficulty in coordination, stumbling, and weakness in her limbs. Despite initial thoughts attributing these symptoms to developmental factors, further examination revealed signs of spasticity and impaired reflexes (Fig. 12.3).

Imaging Description

MRI demonstrates an edematous and swollen spinal cord, including the conus medullaris. On T1W images, it exhibits hypointense intramedullary signals with small central foci of hyperintensity. T2W images show diffuse multilevel intramedullary hyperintensity, with involvement of the conus medullaris starting at level D8. Surrounding this area, prominent serpiginous intradural extramedullary flow voids are observed, indicating dilated perimedullary vessels extending around the filum terminale. These vessels are large enough to create a tangle of vessels on the posterior surface of the vertebral body, causing erosion and a scalloped appearance at levels D12 and L2 (black). Additionally, there is patchy intramedullary enhancement observed at the level D12 nidus.

Spinal AVM

Spinal AVMs are characterized by arteriovenous shunting with a true nidus. They make up around 25% of the vascular malformations in the spine. Although different spinal AVM types present at different times, on average, 80% do so between the ages of 20 and 60. Spinal AVMs can be divided into intramedullary and

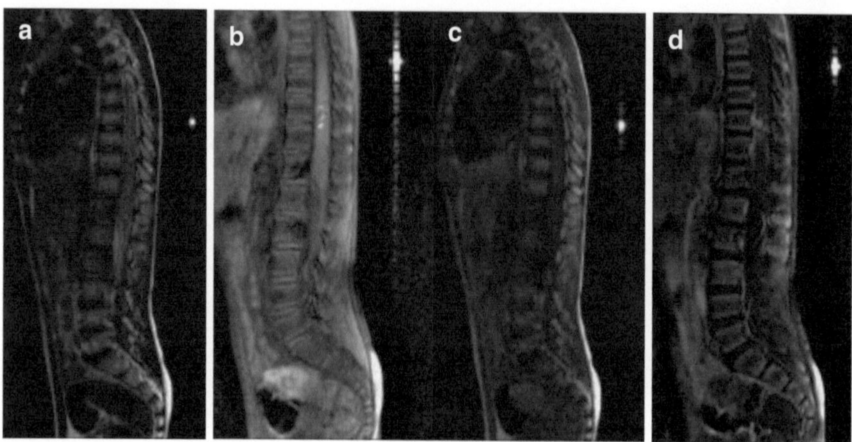

Fig. 12.3 Serial MRI images of the spine with the following: (**a**) Sagittal T2, (**b**) Sagittal T1 SPIR, (**c**) Sagittal T1, (**d**) Sagittal T1 C+. (Figure courtesy of Dr. Samer Hoz)

extramedullary types (80%). Spetzler et al. introduced a classification system for spinal AVMs (AVMs) with the following categories:

1. Intradural AVMs
   (a) Compact type
   (b) Diffuse type
2. Intramedullary AVMs
3. Extradural-intradural AVMs
4. Conus medullaris AVMs

The clinical presentation of the condition varies, encompassing a spectrum from progressive myelopathy, also known as Foix-Alajouanine syndrome, which may be characterized by a delayed diagnosis, to severe and sudden spinal subarachnoid hemorrhage, presenting as a catastrophic event.

MRI is the preferred initial imaging modality before angiography, although the latter remains the gold standard test for a definitive diagnosis. It is essential to perform angiography with proper technique because the site of arterial supply in spinal AVMs can vary widely, extending from the upper thoracic to the sacral regions, with little correlation to the clinical level or apparent nidus. On MRI, T1W images show signal voids caused by high-velocity flow and dilated perimedullary vessels, which can lead to cord indentation or scalloping. T2W images reveal signal voids resulting from high-velocity flow and may show either cytotoxic edema or myelomalacia, leading to an increase in the cord's signal intensity.

Spinal AVMs can give rise to several complications, including hemorrhage, which may occur within the cord parenchyma or in the subarachnoid space. Venous congestion or hypertension can also lead to myelopathy, and high-flow AVMs may cause adjacent spinal cord segments to experience arterial steal. The presence of a swollen and edematous spinal cord and conus medullaris on MRI, along with prominent serpiginous intradural extramedullary flow voids, supports the diagnosis of a spinal dural arteriovenous fistula or malformation. For definitive confirmation, selective spinal DSA remains the gold standard test [3, 4].

Differential Diagnosis

1. Intramedullary neoplasm.
2. CSF flow artifact: The observation of CSF flow artifact in the subarachnoid space behind the cord indicates a normal cord signal without any abnormality.
3. SDAVF: Hemorrhage is rare, and symptoms typically progress gradually. It is more commonly seen in males than females and often presents in the fifth or sixth decade of life.

Questions

1. **Spinal AVM, the FALSE answer is:**
   A. Characterized by arteriovenous shunting with a true nidus.
   B. They contribute to 25% of the vascular malformations in the spine.
   C. The majority of spinal AVMs present between the ages of 20 and 60.
   D. The intramedullary type is the most common type.
   E. May present as a severe and sudden spinal subarachnoid hemorrhage.
   The answer is D.
   Spinal AVMs are most commonly extramedullary (80%).
2. **Spinal AVM imaging, the FALSE answer is:**
   A. MRI is the preferred initial imaging modality before angiography.
   B. Angiography is the gold standard test for a definitive diagnosis.
   C. The site of arterial supply in spinal AVMs correlates with the clinical level.
   D. Spinal AVMs can give rise to several complications, including hemorrhage.
   E. T2W images show signal voids resulting from high-velocity flow.
   The answer is C.
   Spinal AVMs can vary widely, extending from the upper thoracic to the sacral regions, with little correlation to the clinical level or apparent nidus.

Case 85: Spinal Hemangioblastomas

Case Scenario

A 65-year-old male presented with upper limb weakness, pain, and tenderness at the site of a previous operation in the upper back of the neck. The patient has a documented history of spinal hemangioblastoma, which had been surgically treated in the past. The current symptoms suggest a recurrence of the hemangioblastoma at the same site (Fig. 12.4).

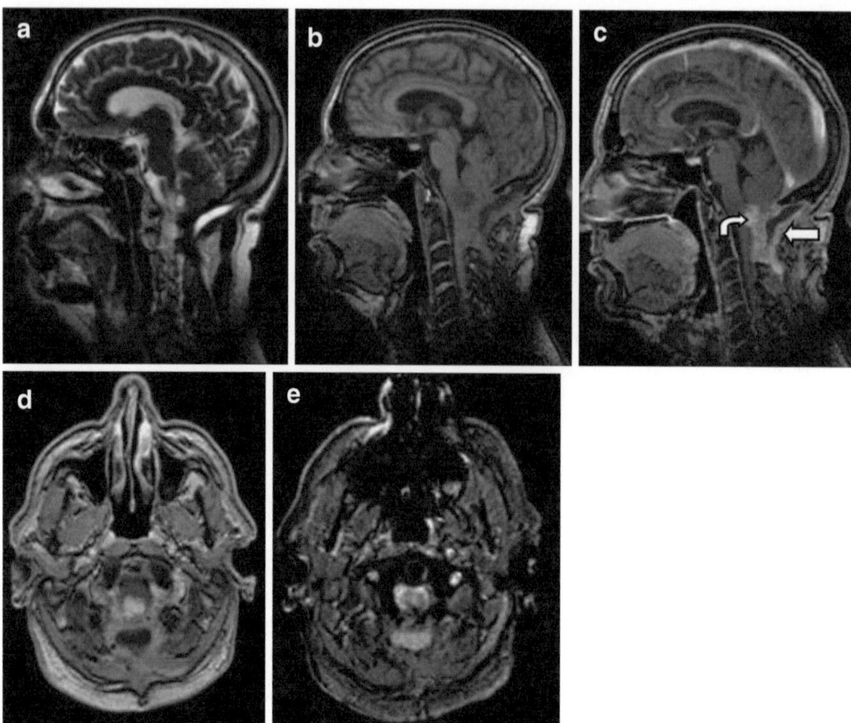

Fig. 12.4 Serial MRI image of the spine with the following: (**a**) Sagittal T2, (**b**) Sagittal T1, (**c**) Sagittal T1 C+, (**d**) Axial T1 C+, (**e**) Axial GER. (Figure courtesy of Dr. Samer Hoz)

Imaging Description

MRI reveals a thick, enhancing collection (46 × 22 × 30 mm) at the site of a previous brain tumor operation (arrow), accompanied by distortion in the surrounding soft tissue and muscle. At the C2 level, there is an enhancing, intramedullary exophytic nodule causing extensive cord edema, along with a small cystic lesion (curve arrow). The nodule exhibits evidence of blood products and vivid enhancement, featuring a signal void vessel in GRE. The measurements of the nodule are 19 × 13 × 14 mm, while the cyst measures 6 × 10 mm in the case of recurrent spinal hemangioblastoma.

Spinal Hemangioblastomas

Spinal hemangioblastomas, comprising 2–6% of all intramedullary tumors, rank as the third most common intramedullary spinal neoplasm. These tumors typically peak in presentation during the fourth decade, with about two-thirds being sporadic cases. It is unusual to encounter sporadic hemangioblastomas in children, and both males and females are equally affected.

Hemangioblastomas are benign vascular lesions categorized as WHO Grade 1 in the WHO classification of CNS tumors (5th Edition, 2021). In 80% of cases, they appear as solitary growths and do not progress into high-grade tumors. However, if multiple lesions are observed, Von Hippel-Lindau syndrome should be considered. This syndrome affects approximately one-third of hemangioblastoma patients (not limited to the spine) and often leads to earlier presentations with multiple tumors. Hemangioblastomas most commonly occur in the thoracic cord (50% of cases), followed by the cervical cord (40%). These tumors typically display an exophytic component and are eccentrically located in two-thirds of cases, with the intramedullary component often found along the dorsum of the cord.

On imaging, hemangioblastomas can manifest as a soft tissue nodule with a hypodense cyst-like component on non-contrast CT. Contrast administration leads to vivid enhancement of the solid component. MRI findings may reveal a distinct nodule or diffuse cord expansion, often accompanied by a tumor cyst or syrinx. The solid component shows variable T1 signal features, ranging from hypo- to isointensity, and hyperintensity in 25% of cases. On MRI-T2, it is typically iso- to hyperintense, and larger lesions may exhibit focal flow voids. Additional features such as hemosiderin capping, surrounding edema, and an accompanying syrinx may be present. On T1 with contrast (Gd), the tumor nodule shows significant enhancement. Angiogram reveals the characteristic features of a hemangioblastoma, including a highly enhancing nidus, dilated arteries, and prominent draining veins [5, 6].

Differential Diagnosis

1. Neoplasms of the spinal canal with an enhancing component:
   (a) Spinal meningioma
   (b) Myxopapillary ependymoma
   (c) Leptomeningeal spinal metastases
   (d) Neurogenic tumors (spinal schwannoma and spinal neurofibroma)
   (e) Spinal paraganglioma
2. Vascular malformations of the spinal cord with enlarged vessels:
   (a) Spinal AVM
   (b) Spinal dural arteriovenous fistula
   (c) Spinal cavernous malformations

Questions

1. **Spinal hemangioblastomas, the FALSE answer is:**
   A. They are malignant WHO grade 4 tumors that can metastasize.
   B. Peak in presentation during the fourth decade.
   C. If multiple, Von Hippel-Lindau syndrome should be considered.

    D. Sporadic hemangioblastomas are relatively rare in children.

    E. Most commonly occur in the thoracic cord.

The answer is A.

They are usually benign WHO grade 1 tumors.

2. **Spinal hemangioblastomas imaging, the FALSE answer is:**

    A. Appear as a soft tissue nodule with a hypodense cyst-like component on non-contrast CT.

    B. Contrast-enhanced MRI shows vivid enhancement of the solid component of hemangioblastomas.

    C. The solid component is consistently hyperintense on T1WI.

    D. May be accompanied by a tumor cyst or syrinx on MRI results.

    E. Angiogram reveals a highly enhancing nidus, dilated arteries, and prominent draining veins.

The answer is C.

The solid component exhibits variable T1 signal features, with hypo- to isointensity and hyperintensity (in 25% of cases) on T1WI.

References

1. Ronald AA, Yao B, Winkelman RD, Piraino D, Masaryk TJ, Krishnaney AA. Spinal dural arteriovenous fistula: diagnosis, outcomes, and prognostic factors. World Neurosurg. 2020;144:e306–15.
2. Shimizu K, Takeda M, Mitsuhara T, Tanaka S, Nagano Y, Yamahata H, Kurisu K, Yamaguchi S. Asymptomatic spinal dural arteriovenous fistula: case series and systematic review. J Neurosurg Spine. 2019;31(5):733–41.
3. Zozulya YP, Slin'ko EI, Al-Qashqish II. Spinal arteriovenous malformations: new classification and surgical treatment. Neurosurg Focus. 2006;20(5):1–7.
4. Zhan PL, Jahromi BS, Kruser TJ, Potts MB. Stereotactic radiosurgery and fractionated radiotherapy for spinal arteriovenous malformations—a systematic review of the literature. J Clin Neurosci. 2019;62:83–7.
5. Mossel P, van der Horst-Schrivers AN, Olderode-Berends MJ, Groen RJ, Hoving EW, Appelman AP, Links TP. Radiologic characteristics of spinal hemangioblastomas in von Hippel Lindau disease as guidance in clinical interventions. World Neurosurg. 2022;168:e67–75.
6. Bukhari SS, Bari ME, Ahmad Z, Din NU. Spinal cord hemangioblastomas with a focus on clinical presentation, diagnosis, and treatment at a tertiary care hospital of Karachi, Pakistan: a retrospective chart review. Surg Neurol Int. 2021;12:24.

Spinal Infection

Ali Q. Al-asady, Mahmood H. AlObaidy,
Alkawthar M. Abdulsada, Haneen A. Salih,
Hussein M. Hasan, Ahmed Muthana,
and Asmaa H. AL-Sharee

Case 86: Spinal Hydatid Disease

Case Scenario

A 28-year-old female presents with a chief complaint of severe back pain, bilateral radiculopathy, and paraparesis affecting both lower limbs (Fig. 13.1).

Imaging Description

The MRI findings reveal low signal intensity on T1WI and high signal intensity on T2WI with a multivesicular appearance and collapsed D9–D10. There is involvement of the posterior vertebral elements and intraspinal extension (extradural). Within the psoas muscles, numerous intramuscular multivesicular cysts are present, appearing as multiple daughter cysts within larger cysts. These intramuscular cysts exhibit a thin peripheral rim enhancement, predominantly on the right side. The imaging also shows notable fatty deterioration and atrophy of the right paraspinal

A. Q. Al-asady · H. M. Hasan · A. Muthana
College of Medicine, University of Baghdad, Baghdad, Iraq

M. H. AlObaidy
College of Medicine, Al-Nahrain University, Baghdad, Iraq

A. M. Abdulsada
Azerbaijan Medical University, Baku, Azerbaijan

H. A. Salih
Department of Molecular and Medical Biotechnology, College of Biotechnology, Al-Nahrain University, Baghdad, Iraq

A. H. AL-Sharee (✉)
Department of Neuroradiology, Neurosurgery Teaching Hospital, Baghdad, Iraq

S. Hoz et al. (eds.), *Neuroradiology Board's Favorites*,
https://doi.org/10.1007/978-3-031-64261-6_13

267

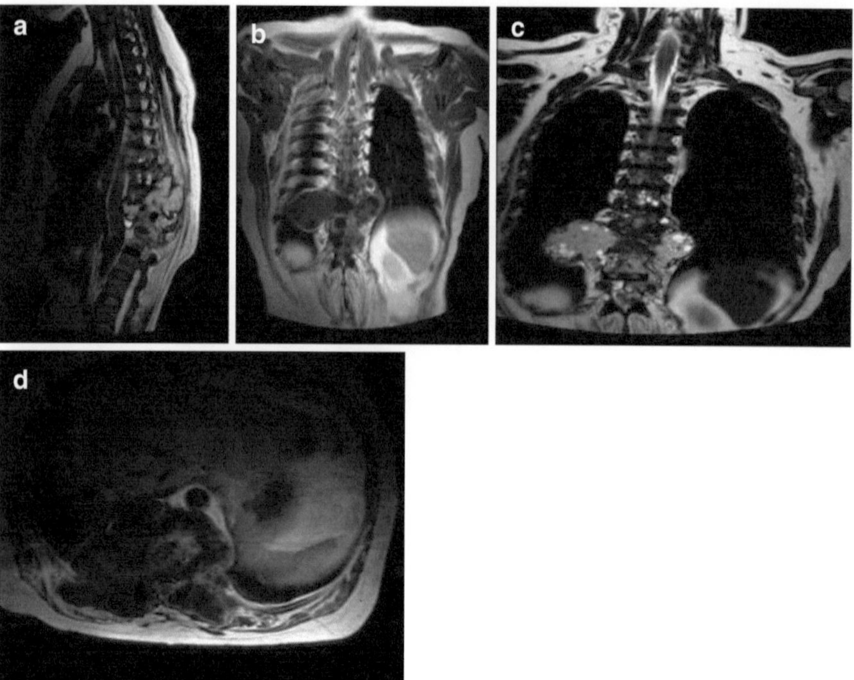

Fig. 13.1 Serial MRI images of the spine with the following: (**a**) Sagittal T2, (**b**) Coronal T1 C+, (**c**) Coronal T2, (**d**) Axial T1 C+. (Figure courtesy of Dr. Samer Hoz)

muscles. Moreover, a hyper-intense signal is observed at the spinal cord adjacent to the lesion, indicative of compression myelopathy. Confirmation of the diagnosis of spinal hydatid cysts was obtained through histopathological examination.

Spinal Hydatid Disease

Hydatid disease affecting the spinal cord or spine is a rare manifestation known as spinal hydatid disease. It is caused by the larval form of *Echinococcus granulosus* and can be classified anatomically according to the Dew/Braithwaite and Lees classification:

- Type 1: Intramedullary
- Type 2: Intradural
- Type 3: Extradural
- Type 4: Vertebral
- Type 5: Paravertebral

Approximately 90% of cases are Types 3 and 4; Types 1 and 2 are exceptionally rare. However, various types can coexist, as seen in instances involving Types 2, 4, and 5.

On CT scans, bony involvement resembles vertebral osteomyelitis but is less likely to exhibit signs of osteoporosis, sclerosis, or vertebral disc injury. The cysts have a density similar to CSF but rarely show calcification. If spinal hydatid disease results from direct invasion from extraspinal structures, a CT scan may reveal the underlying cause, but MRI is a more sensitive diagnostic tool. Unlike typical spherical cysts found elsewhere in the body without perilesional edema, spinal hydatid cysts typically exhibit a tube-like structure. In all pulse sequences, they show signals similar to CSF: T1-weighted images depict a hypo-intense cyst with an iso- or hypo-intense cystic wall, T1W post-contrast shows wall enhancement (although mild), and T2-weighted images depict a hyper-intense cyst with a hypo-intense cystic wall [1, 2].

Differential Diagnosis

- Pott disease
- Vertebral pyogenic infection (osteomyelitis)
- Spinal malignancy or metastasis
- Cystic lesions of the spinal cord (including a long list of differentials)

Questions

1. **Spinal hydatid disease, the FALSE answer is:**
   A. It can affect all parts of the spine.
   B. Vertebral pyogenic infections, like osteomyelitis, are a differential diagnosis.
   C. Osteoporosis is a common sign of spinal hydatid disease.
   D. Type 3 of the Dew/Braithwaite and Lees classification is extradural.
   E. Types 1 and 2 are extremely rare manifestations.
      The answer is C.
      Spinal hydatid disease does not commonly show signs of osteoporosis, sclerosis, or injury to the vertebral discs.
2. **Spinal hydatid disease imaging, the FALSE answer is:**
   A. T1WI shows a hypo-intense cyst.
   B. On CT scan, spinal hydatid disease may resemble vertebral osteomyelitis.
   C. On CT scan, the cysts exhibit a density similar to CSF.
   D. T2WI depicts hypo-intense cyst contents.
   E. In T2W FLAIR, the cyst wall appears hypo-intense.
      The answer is D.
      T2WI depicts hyper-intense cyst contents.

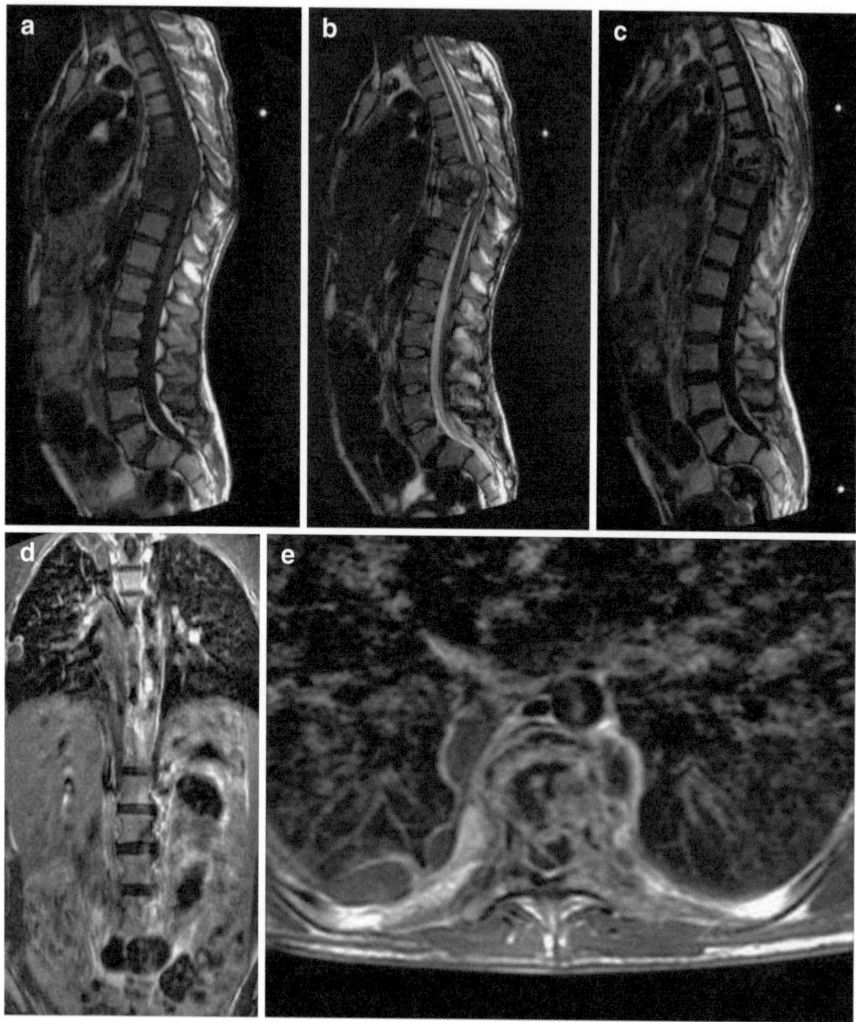

Fig. 13.2 Serial MRI images of the spine with the following: (**a**) Sagittal T1, (**b**) Sagittal T2, (**c**) Sagittal T1 C+, (**d**) Coronal T1 C+, (**e**) Axial T1 C+. (Figure courtesy of Dr. Samer Hoz)

Case 87.1: Tuberculous Spondylitis

Case Scenario

A 21-year-old individual with a history of prior incarceration presents complaining of back pain, paraparesis, kyphotic deformity, and weight loss (Fig. 13.2).

Imaging Description

MRI reveals a wedge collapse at vertebrae D8 and D9 and a malformation of the dorsal spine known as gibbus deformity. In the post-gadolinium assessment, vertebrae D8–D10 display patchy enhancement indicative of inflammatory alterations. A substantial multiloculated collection spanning vertebrae D7–D9 involves both prevertebral and paravertebral structures, with clearly marked enhancing margins. There is an epidural extension through the neuroforamina, suggesting additional collections. There is significant spinal cord compression and a multiloculated subpleural collection in the chest, particularly on the right side.

Considering the clinical history, demographics, and MRI results, the provisional diagnosis of tuberculous spondylitis was established, and this diagnosis was subsequently confirmed by acid-fast bacilli culture.

Case 87.2: TB Spine

Case Scenario

A 50-year-old patient presented with persistent back pain, along with a gradual onset of weakness and numbness in the lower limbs. Physical examination revealed localized tenderness along the spine without apparent deformities (Fig. 13.3).

Imaging Description

Bilateral paravertebral masses with intermediate/high signal on T2 at the L2/L3 level in the lumbar spine suggest fluid accumulation. Peripheral enhancement on post-contrast images indicates organized collections. The collections in the anterior spinal canal have an epidural component that is compressing the spinal cord. On T2 and STIR sequences, the L2 and L3 vertebral bodies exhibit abnormally strong signals, while T1 shows a hypo-intense signal. There is no loss of L2/L3 disc space. Additionally, the posterior part of the L1 vertebral body displays abnormal signal intensity. The right psoas muscle reveals a peripherally enhancing paravertebral collection (arrow).

Tuberculous Spondylitis

Vertebral body osteomyelitis and discitis caused by TB are referred to as tuberculous spondylitis, often known as Pott disease. More than 50% of cases of musculoskeletal TB involve the spine. The lumbar segments are most frequently affected, followed by the thoracic regions. Clinical manifestations include back discomfort, paraplegia or weakness of the lower limbs, kyphotic deformity, and constitutional symptoms, as observed in this case. The majority of cases of TB are seen in children

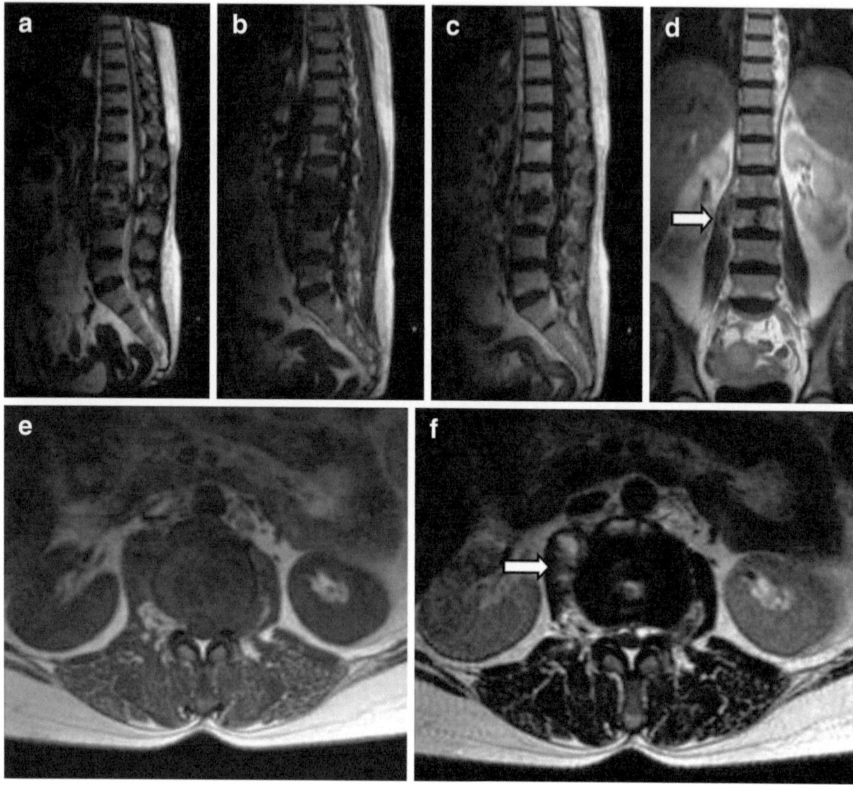

Fig. 13.3 Serial MRI images of the spine with the following: (**a**) Sagittal T2, (**b**) Sagittal T1, (**c**) Sagittal T1 C+, (**d**) Coronal T1 C+, (**e**) Axial T1, (**f**) Axial T1 C+. (Figure courtesy of Dr. Samer Hoz)

under the age of 20, and it is the most prevalent cause of vertebral body infection in many developing countries. The meninges around the spine can also be affected by TB, which can result in severe pachymeningitis that enhances dramatically [3, 4].

Differential Diagnosis

- Pyogenic infection: In contrast to pyogenic infection, tuberculous spondylitis relatively preserves disk-space height, results in massive paraspinal abscesses with a smooth enhancing wall, and typically involves multiple organs throughout the body.
- Brucellosis: May manifest as granulomatous osteomyelitis of the spine, which can be challenging to differentiate from TB. They are both acid-fast bacilli that might result in caseating granulomas.
- Fungal infection (e.g., cryptococcosis), sarcoidosis, dialysis-related amyloidosis, spondyloarthropathy, and metastasis.

Questions

1. **Tuberculous spondylitis, the FALSE answer is:**
   A. Can also involve the meninges of the spinal cord.
   B. Vertebral body osteomyelitis and discitis caused by TB are known as Pott disease.
   C. Symptoms include back discomfort, paraplegia, and weakness of the lower limbs.
   D. The majority of cases of TB are seen in children under the age of 20.
   E. The thoracic segments are more commonly affected than the lumbar regions.
   The answer is E.
   The lumbar segments are more commonly affected than the thoracic regions.
2. **Tuberculous spondylitis, the FALSE answer is:**
   A. TB is the most common cause of vertebral body infection in developing countries.
   B. In contrast to pyogenic infection, tuberculous spondylitis relatively preserves disk-space height.
   C. More than 50% of cases of musculoskeletal TB involve the spine.
   D. Brucellosis may manifest as granulomatous osteomyelitis of the spine.
   E. Acid-fast bacilli are positive in TB and negative in brucellosis.
   The answer is E.
   Both Brucella and TB are acid-fast bacilli that can result in caseating granulomas.

Case 88: Spinal Epidural Abscess

Case Scenario

A 28-year-old female presented with neck pain, fever, and radiculopathy affecting the upper limbs, along with an increased white cell count and elevated C-reactive protein (Fig. 13.4).

Imaging Description

MRI reveals the presence of a heterogeneous mass in the right anterior epidural space of the upper cervical canal. The mass exhibits areas of T2 brightening with peripheral fluid enhancement, suggestive of an abscess. Stronger-enhancing regions resembling a phlegmon are also observed. There is no evidence of swelling or enhancement in the vertebral bodies and discs.

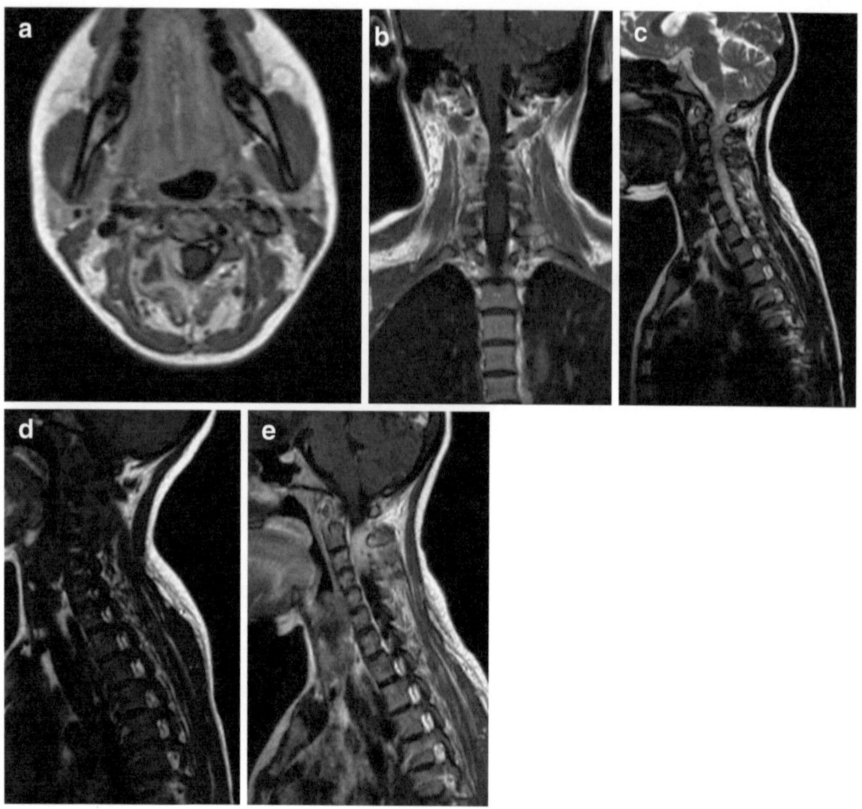

Fig. 13.4 Serial MRI images of the spine with the following: (**a**) Axial T1 C+, (**b**) Coronal T1 C+, (**c**) Sagittal T2, (**d**) Sagittal T1, (**e**) Sagittal T1 C+. (Figure courtesy of Dr. Samer Hoz)

Spinal Epidural Abscess

Spinal epidural abscess is considered a rare disorder, with an estimated 2–3 cases reported for every 10,000 hospital admissions. It exhibits a male predominance and typically peaks in incidence during the fifth to seventh decades of life. This age-related pattern may be associated with the higher prevalence of risk factors and predisposing conditions in older individuals. Risk factors include the following:

- Comorbid conditions such as diabetes, alcoholism, and HIV infection.
- Spinal deformities, surgery, or degenerative joint diseases.
- IV drug use.

The primary focus of spinal epidural abscesses is often in close proximity, indicating direct dissemination from nearby structures. CT scans, despite offering detailed bone anatomy, may be limited in visualizing small abscesses. However, they can reveal changes in facet joints secondary to septic arthritis and discitis-osteomyelitis, such as bone or joint destruction and soft tissue stranding.

The preferred imaging method for diagnosing spinal epidural infection is gadolinium-enhanced MRI. DWI is particularly useful to confirm the presence of an infection. Two main imaging patterns can be identified:

1. During the phlegmonous stage of infection, there is homogenous enhancement in the abnormal region, indicating granulomatous-thickened tissue with embedded micro-abscesses but no significant pus accumulation.
2. A liquid abscess surrounded by inflamed tissue shows varying degrees of peripheral enhancement.

Liquid abscesses exhibit a strong T2 signal, a low T1 signal, and no enhancement. These abscesses are typically surrounded by a rim of enhancement and the content frequently exhibits limited diffusion on DWI/ADC. Surgical drainage is necessary for these lesions [5, 6].

Differential Diagnosis

- Vertebral metastases: can extend extradurally and mimic the appearance of a phlegmonous epidural infection.
- Epidural hematoma.
- Lumbar disc disease: migrated or extruded disc.
- Spinal seroma following surgery.

Questions

1. **Spinal epidural abscess, the FALSE answer is:**
   A. Peak incidence is in the fifth to seventh decades of life.
   B. Risk factors include comorbid conditions such as diabetes.
   C. It typically manifests with gradually declining neurological function.
   D. The primary focus is typically in close proximity to the spinal epidural abscess, indicating direct dissemination from nearby structures.
   E. Liquid abscess surrounded by inflamed tissue shows varying degrees of peripheral enhancement.
   The answer is C.
   Due to compression, it manifests with rapidly declining neurological function.

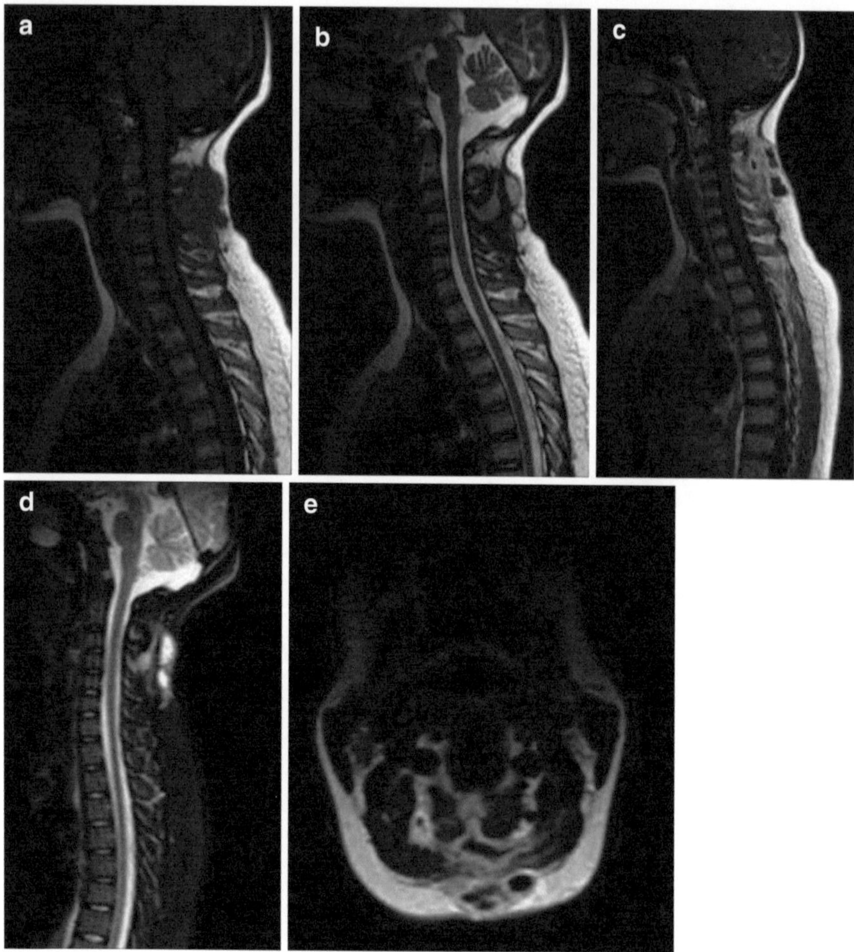

Fig. 13.5 Serial MRI images of the spine with the following: (**a**) Sagittal T1, (**b**) Sagittal T2, (**c**) Sagittal T1 C+, (**d**) Sagittal STIR, (**e**) Axial T1 C+. (Figure courtesy of Dr. Samer Hoz)

Case 89: Tuberculous Abscesses

Case Scenario

A 21-year-old female presented with neck pain and tenderness, along with a palpable mass in the upper neck and an increase in C-reactive protein (Fig. 13.5).

Imaging Description

MRI reveals a multioculated collection with thick peripheral rim enhancement in the paraspinal region and subcutaneous tissue at the C2, C3, C4, and C5 vertebrae. The diagnosis of tuberculous abscesses was confirmed through histopathological examination.

Tuberculous Abscesses

Tuberculous abscesses are an unusual manifestation of tuberculosis, characterized by the presence of pus containing identifiable organisms, a feature often absent in more common tuberculomas. Unlike tuberculomas, the granulomatous reaction typical of tuberculosis is lacking in the capsule that surrounds the necrotic, purulent core, giving it a resemblance to typical bacterial abscesses. Furthermore, these abscesses lack the characteristic pus and neutrophils found in pyogenic abscesses. Spectroscopy often reveals peaks for lactate and lipid.

Generally, soft tissue abscesses are localized collections of pus within the soft tissues of the body, enclosed by an abscess membrane or peripheral rim. They can occur in various age groups and encompass subcutaneous, intramuscular, and intermuscular abscesses, as well as those within the deep soft tissues in the fascial planes. Several factors increase the risk of developing a soft tissue abscess, including lymphedema, chronic venous insufficiency, immunosuppression, diabetes mellitus, obesity, cardiovascular disease, intravenous drug use, alcoholism, lacerations, surgical wounds, skin breaches, and infections elsewhere in the body. Diseases such as skin ulcers, osteomyelitis, cellulitis, pyomyositis, and infective endocarditis have also been associated with the development of soft tissue abscesses.

X-rays often provide limited benefits in identifying soft tissue abscesses, but they may reveal soft tissue gas or foreign objects that could suggest an infectious process or other underlying causes of soft tissue swelling.

Ultrasonography typically shows a fluid collection that is avascular in the center and irregular, surrounded by a hypervascular rim. Internal echoes within the central core may be observed and could be mobilized upon compression.

CT scan typically depicts an abnormal fluid collection with low attenuation and rim enhancement, which may appear irregular and thick compared to the wall of a cyst. Soft tissue gas may also be visible. Inflammation-induced changes in surrounding tissues, such as soft tissue edema and fat stranding, are often evident.

For an acute abscess, MRI provides high sensitivity and moderate to good specificity. The presence of the penumbra sign enhances specificity in subacute, chronic, or acute-on-chronic abscesses. T1WI reveals a low to intermediate signal intensity within the abscess cavity, while the capsular rim exhibits low to high signal intensity, particularly when the penumbra sign is present. In T2/T2FS/IMFS sequences, the abscess cavity demonstrates a high signal intensity, contrasting with the capsular rim, which displays a low signal intensity. DWI and ADC sequences depict a hyperintense diffusion restriction within the abscess cavity as well as a high b-value and a low signal on ADC. On T1 C+ (Gd), the abscess cavity exhibits low signal intensity with no enhancement, while the capsular rim shows avid enhancement.

Practical points in MRI include the use of post-contrast imaging to improve conspicuity and viewer confidence, while DWI/ADC enhances the detection of soft tissue abscesses, especially in cases where IV contrast agents are contraindicated [7, 8].

Differential Diagnosis

- Sterile fluid collections, post-operative seroma, and an evolving hematoma
- Early myositis ossificans
- Foreign body reaction
- Necrotic tumor
- Devitalized tissue
- Compartment syndrome with myonecrosis

Questions

1. **Spinal tuberculous abscesses, the FALSE answer is:**
   A. Chronic venous insufficiency is a risk factor for soft tissue abscesses.
   B. If the penumbra sign is present, it offers high specificity in subacute abscesses on MRI.
   C. US often shows a fluid accumulation that is regular and highly vascularized in the center.
   D. They contain pus with abundant organisms.
   E. If IV contrast agents are contraindicated, DWI can improve abscess detection.
   The answer is C.

   Ultrasonography typically reveals a fluid accumulation that is irregular and avascular in the center.

Case 90: Pyogenic Spondylitis

Case Scenario

A 37-year-old female with a history of cervical laminectomies at the level of C5/C6 20 days ago presented with neck pain and fever (Fig. 13.6).

Imaging Description

The MRI reveals osteomyelitis/discitis of C5/C6 with associated degradation of the disc space, loss of height, and phlegmon-induced compression of the nearby cord.

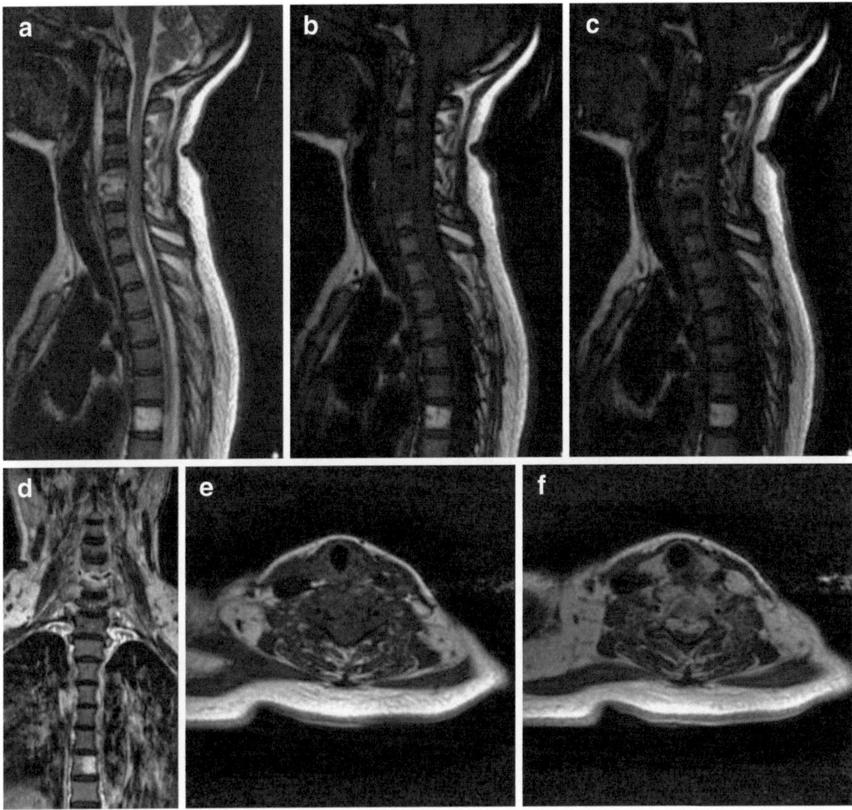

Fig. 13.6 (a) Sagittal T2, (b) Sagittal T1, (c) Sagittal T1 C+, (d) Coronal T1 C+, (e) Axial T1, (f) Axial T1 C+. (Figure courtesy of Dr. Samer Hoz)

Small epidural abscesses are also seen, showing focal enhancement after gadolinium administration, and are located behind T1–T2 in the anterior thecal sac. Additionally, the vertebral bodies exhibit diffuse hypo-intensity in T1 and slight hyper-intensity in T2. A hyper-intense signal within the intravertebral disc in T2 suggests medullary reconversion in response to the infectious process.

Pyogenic Spondylitis

Infections of the spine that affect the vertebrae, intervertebral disc, paraspinal soft tissue, or epidural space are referred to as pyogenic spondylitis. It is an umbrella term that includes disease entities such as epidural abscess, spondylodiscitis, and spinal osteomyelitis. Males are affected twice as often as females, and it commonly occurs in children and individuals over 50 years old. Risk factors include diabetes mellitus, sepsis, immunosuppression, and recent trauma or spinal surgery.

Unlike tuberculous spondylitis, symptoms of pyogenic spondylitis typically appear earlier. *Staphylococcus aureus* and *Streptococcus* spp. are the most common causative organisms. The term "discitis" is often avoided in adults when referring to an isolated spinal disc infection. This is because in most cases, infection starts with the vertebral endplate and involves the disc secondarily, and hence terms like "spondylodiscitis" or "vertebral osteomyelitis" are preferred for greater accuracy.

Plain radiography may not detect early changes due to spondylodiscitis, and a normal appearance can persist for 2–4 weeks. Irregular or poorly defined vertebral endplates and narrowing of the disc space become apparent later, with potential bony sclerosis in untreated cases after 10–12 weeks. CT results are similar to those observed on plain radiography, but CT is more sensitive to early changes.

MRI, due to its high sensitivity and specificity, is the recommended imaging modality. It helps differentiate between neoplastic processes and various infections (pyogenic, tuberculous, and fungal). T1 demonstrates a low signal in adjacent endplates (indicating bone marrow edema) and disc space (fluid). Signal characteristics on T2 include a high signal in disc space (fluid) and adjacent endplates (bone marrow edema), loss of low signal cortex at endplates, high signal in paravertebral soft tissues, and hyper-intensity within the psoas muscle (the imaging psoas sign). T2 fat-saturated and STIR sequences are especially useful. The sensitivity and specificity of MRI for spondylodiscitis are both approximately 92%. T1 post-contrast (Gd) shows peripheral enhancement around fluid collections, enhancement of spinal endplates and paravertebral soft tissues, as well as enhancement around a low-density center indicating abscess formation. The DWI sequence helps differentiate between the acute and chronic stages of the illness [9, 10].

Differential Diagnosis

- Tuberculous spondylitis
- Charcot joint
- Modic type I degenerative changes
- Schmorl nodes
- Langerhans cell histiocytosis

Questions

1. **Pyogenic spondylitis, the FALSE answer is:**
    A. Risk factors for pyogenic spondylitis include spinal surgery.
    B. Differential diagnosis for pyogenic spondylitis includes Schmorl nodes.
    C. The infection usually starts with the vertebral endplate.
    D. *Staphylococcus aureus* is one of the most common causative organisms.
    E. Females are more frequently affected than males.
     The answer is E.
     Pyogenic spondylitis affects males twice as frequently as females.

2. **Pyogenic spondylitis imaging, the FALSE answer is**
   A. The DWI sequence distinguishes between the acute and chronic stages of an illness.
   B. T2 shows a low signal in the disc space and adjacent endplates.
   C. Plain radiography is not sensitive to the early spondylodiscitis changes.
   D. MRI helps distinguish between pyogenic and neoplastic processes.
   E. T1 demonstrates a low signal in the disc space and adjacent endplates.
   The answer is B.
   T2 shows a high signal in the disc space and adjacent endplates.

References

1. Singh N, Anand T, Singh DK. Imaging diagnosis and management of primary spinal hydatid disease: a case series. Egypt J Radiol Nucl Med. 2022;53(1):1–7.
2. Naumov DG, Vishnevskiy AA, Tkach SG, Avetisyan AO. Spinal hydatid disease of cervico-thoracic in pregnant women: a case report and review. Traumatol Orthop Russia. 2021;27(4):102–10.
3. Prijambodo B, Pratama US. Clinical, laboratory and radiologic evaluations in patients with malignant tuberculous spondylitis. Indian J Forensic Med Toxicol. 2020;14(4):4476–81.
4. Yanardag H, Tetikkurt C, Bilir M, Demirci S, Canbaz B, Ozyazar M. Tuberculous spondylitis: clinical features of 36 patients. Case Rep Clin Med. 2016;5(10):411–7.
5. Sharfman ZT, Gelfand Y, Shah P, Holtzman AJ, Mendelis JR, Kinon MD, Krystal JD, Brook A, Yassari R, Kramer DC. Spinal epidural abscess: a review of presentation, management, and medicolegal implications. Asian Spine J. 2020;14(5):742.
6. Tetsuka S, Suzuki T, Ogawa T, Hashimoto R, Kato H. Spinal epidural abscess: a review highlighting early diagnosis and management. JMA J. 2020;3(1):29–40.
7. Agrawal V, Patgaonkar PR, Nagariya SP. Tuberculosis of spine. J Craniovertebr Junction Spine. 2010;1(2):74.
8. Nazirov P, Fakhridinova A, Makhmudova Z, Djuraev B. Differentiated approach to the diagnosis and treatment of tuberculous spondylitis in adults. Acta Med Iran. 2021;59:191–6.
9. Sato K, Yamada K, Yokosuka K, Yoshida T, Goto M, Matsubara T, Iwahashi S, Shimazaki T, Nagata K, Shiba N. Pyogenic spondylitis: clinical features, diagnosis and treatment. Kurume Med J. 2018;65(3):83–9.
10. Yoshimoto M, Takebayashi T, Kawaguchi S, Tsuda H, Ida K, Wada T, Yamashita T. Pyogenic spondylitis in the elderly: a report from Japan with the most aging society. Eur Spine J. 2011;20:649–54.

Sajjad G. Al-Badri, Mustafa Ismail, Fatimah O. Ahmed,
Ahmed Muthana, Haneen A. Salih, Awfa Aktham,
and Maliya Delawan

Case 91: Disc Protrusions

Case 91.1: Left Subarticular Disc Protrusion

Case Scenario

A 40-year-old female presented with complaints of low back pain and left lower limb radiculopathy (Fig. 14.1).

Imaging Description

The MRI reveals a left focal paracentral disc protrusion at the L5/S1 level. This protrusion is indenting the ventral theca sac, causing complete effacement of the epidural fat within the left neural exit foramen and compression of the cauda equina.

S. G. Al-Badri · M. Ismail · A. Muthana
College of Medicine, University of Baghdad, Baghdad, Iraq

F. O. Ahmed
College of Medicine, University of Mustansiriyah, Baghdad, Iraq

H. A. Salih
Department of Molecular and Medical Biotechnology, College of Biotechnology, Al-Nahrain University, Baghdad, Iraq

A. Aktham
Department of Neurosurgery, Tokyo General Hospital, Nakano, Japan

M. Delawan (✉)
College of Medicine, Gulf Medical University, Ajman, United Arab Emirates

S. Hoz et al. (eds.), *Neuroradiology Board's Favorites*,
https://doi.org/10.1007/978-3-031-64261-6_14

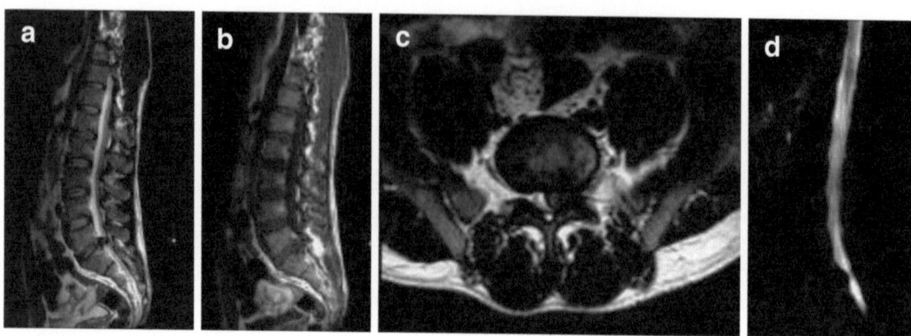

Fig. 14.1 Serial MRI images of the spine with the following: (**a**) Sagittal T2, (**b**) Sagittal T1, (**c**) Axial T2, (**d**) Myelography. (Figure courtesy of Dr. Samer Hoz)

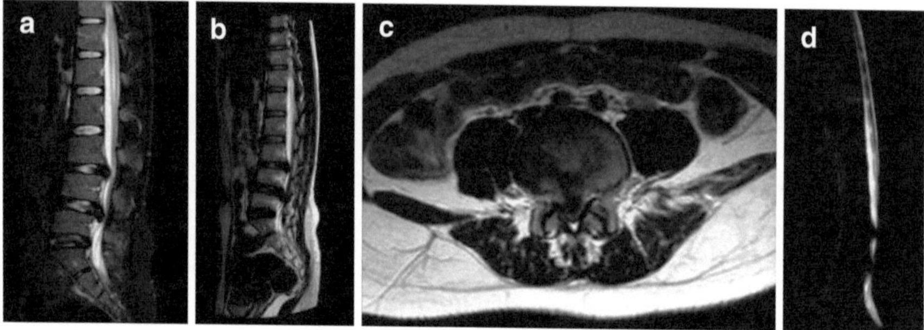

Fig. 14.2 Serial MRI images of the spine with the following: (**a**) Sagittal STIR, (**b**) Sagittal T2, (**c**) Axial T2, (**d**) Myelography. (Figure courtesy of Dr. Samer Hoz)

Case 91.2: L3/L4 and L4/L5 Central Disc Protrusion with Caudal Migration

Case Scenario

A 34-year-old male with a history of leg pain presented with complaints of numbness and foot drop upon awakening from sleep. Additionally, he reported experiencing low back pain (Fig. 14.2).

Imaging Description

The MRI reveals a large central disc protrusion at L4/L5, causing compression of the cauda equina. The epidural fat is completely effaced by caudal migration. Another small compressive central disc protrusion can be seen at L3/L4. The marrow signal appears normal.

Disc Protrusions

Disc protrusions refer to a type of disc herniation where the content of the disc extends beyond its normal confines in the intervertebral space. This protrusion occurs over a segment that is less than 25% of the disc's circumference, and the base of the protrusion is wider than the largest diameter of the disc material that extends beyond the normal margins. However, in protrusion, the disc does not extend above or below the relevant vertebral endplates.

It is essential to distinguish disc protrusions from disc bulges, where a bulge involves more than 25% of the disc's circumference. Additionally, disc extrusions differ from protrusions as the base of the extruded disc material is narrower than its dome shape and may extend above or below the disc level. Disc protrusions can be categorized based on their axial position, including central, subarticular, foraminal, extraforaminal, or anterior. The terms "contained" and "non-contained" describe whether the outer annulus fibrosus laminae are intact or deficient, respectively.

In CT imaging, disc protrusions may be indirectly inferred from signs such as obliteration of perineural fat, nerve root displacement, or changes in the size of the spinal canal or intervertebral foramina. However, CT is less sensitive and specific than MRI for disc pathology.

On MRI, the details of the anatomical structures, including the spinal cord, vertebrae, and surrounding soft tissues, are clearer in T1WI, while T2WI is particularly useful for identifying pathologies involving fluid. In T1WI, disc protrusions typically appear hypointense in comparison to the surrounding spinal fluid. In T2WI, disc protrusions appear hyperintense against the spinal cord and surrounding tissues. In the case of a herniated disc, the area of herniation often appears hyperintense due to the presence of fluid in the nucleus pulposus [1, 2].

Questions

1. **Disc protrusions, the FALSE answer is:**
   A. Disc protrusions involve less than 25% of the disc's circumference.
   B. Disc protrusions may extend above or below the relevant vertebral endplates.
   C. "Contained" disc protrusions have an intact outer annulus fibrosus lamina.
   D. "Non-contained" disc protrusions have a deficient lamina.
   E. A disc bulge involves more than 25% of the disc's circumference.
      The answer is B.
      Unlike disc extrusions, disc protrusions do not extend above or below the relevant vertebral endplates.

2. **Disc protrusions, the FALSE answer is:**
   A. Appear hypointense on T1WI and T2WI.
   B. Disc protrusions are more common in the lumbar spine.
   C. In CT scans, disc protrusions are typically seen as areas of soft tissue attenuation projecting into the spinal canal.
   D. MRI can show annular tears associated with disc protrusions.
   E. Disc protrusions can present with back pain, sciatica, and, in severe cases, cauda equina syndrome.
      The answer is A.
      Disc protrusions appear hyperintense on T2WI and hypointense on T1WI.

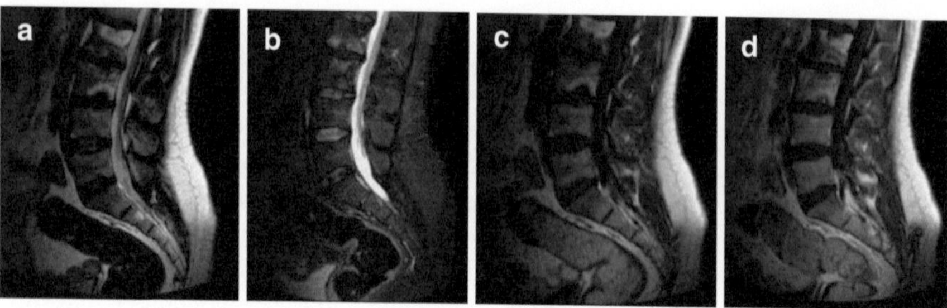

Fig. 14.3 Serial MRI images of the spine with the following: (**a**) Sagittal T2, (**b**) Sagittal STIR, (**c**) Sagittal T1, (**d**) Sagittal T1 C+. (Figure courtesy of Dr. Samer Hoz)

Case 92: Andersson Lesion

Case Scenario

A 25-year-old male presented with a history of progressively worsening low back pain without radiculopathy (Fig. 14.3).

Imaging Description

The MRI sequences reveal an abnormal signal intensity in a hemispherical shape, affecting the central regions of both vertebral halves and the discs at L3/L4 levels. These abnormalities exhibit low signal on T1, high signal on T2/STIR, and show enhancement on post-contrast sequences.

Andersson Lesion

Andersson lesion, also known as rheumatic spondylodiscitis, refers to the inflammatory involvement of intervertebral discs caused by spondyloarthritis. This non-infectious condition has been observed in approximately 8% of patients with ankylosing spondylitis, as identified through radiography. The exact pathophysiology of Andersson lesion is still a matter of debate, but proposed hypotheses include spinal stress fractures and a localized delay in the ankylosing process compared to adjacent levels (resulting in the segment becoming the last mobile region).

Radiographically, irregularities and erosions of the vertebral endplates are observed, primarily in the central portion rather than the anterior or posterior edges. These findings are now recognized as late features of spondyloarthritis. On MRI, Andersson lesions are visualized as abnormalities in the signal intensity of the affected disc and adjacent vertebral halves of a discovertebral unit. They appear hyperintense on STIR images and hypointense on T1W images, often exhibiting a hemispherical shape. In early disease, increased signal intensity lines may be observed at the interface between the annulus fibrosus and nucleus pulposus or within the nucleus pulposus itself.

Compared to conventional radiography, MRI is superior in depicting anterior spondylitis (or Romanus lesions) and spondylodiscitis, as it can detect edematous changes in early disease not visible on X-rays. There is no consensus regarding the management of Andersson lesions, especially regarding when to perform surgical intervention. Surgical treatment primarily involves instrumentation and fusion, aiming to correct any kyphotic deformity if it exists.

A question arises about whether localized lesions should be classified as Andersson lesions, as their radiological appearance, mechanical consequences, prognosis, and management differ from extensive lesions. Some suggest preserving the term "Andersson lesion" for extensive lesions, which represent spinal pseudarthrosis resulting from various underlying causes [3, 4].

Questions

1. **Andersson lesion, the FALSE answer is:**
   A. Andersson lesions are the inflammatory involvement of intervertebral discs caused by spondyloarthritis.
   B. Andersson lesions are hyperintense on T1W images.
   C. Andersson lesions often exhibit a hemispherical shape.
   D. They appear hyperintense on STIR and T2W images.
   E. They are most commonly located in the lumbar and thoracic regions.
   The answer is B.
   Andersson lesions appear hypointense on T1W MRI images, not hyperintense.

Case 93: Schmorl Nodes

Case 93.1: Acute Schmorl Nodes

Case Scenario

A 20-year-old female athlete presented with a 3-week history of back pain without radiculopathy. Her CRP and white cell count are within normal ranges with no signs of fever (Fig. 14.4).

Imaging Description

The MRI findings indicate a well-defined, rounded vertebral lesion at the lower endplate of L2 with a hypointense rim on T1 and mild perifocal edema, consistent with an acute Schmorl node. At the L4/L5 level, there is a mild circumferential disc bulge that is indenting the thecal sac, causing relative narrowing of the bilateral neural foramina and resulting in mild compromise at the nerve root.

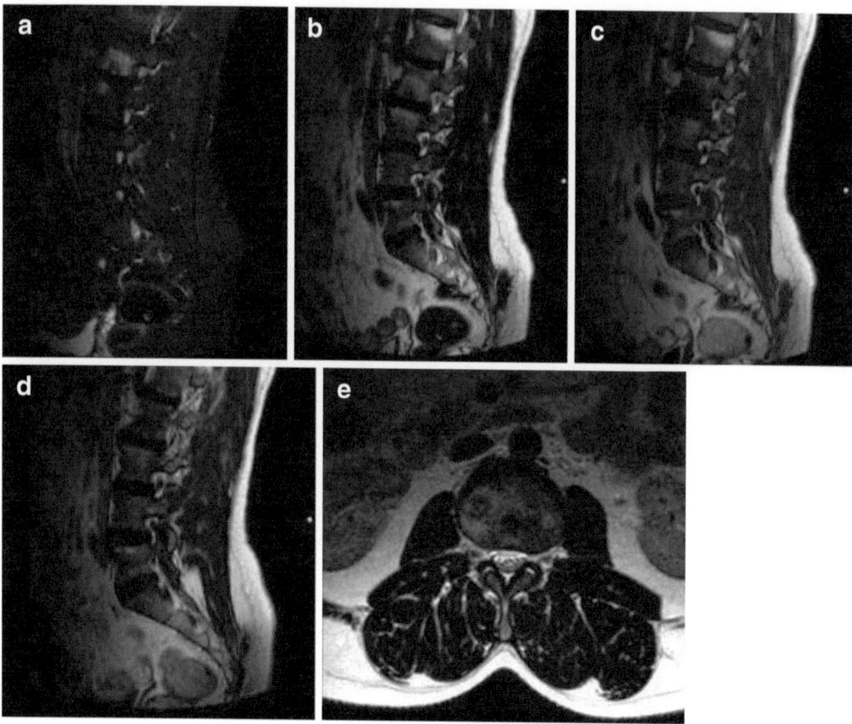

Fig. 14.4 Serial MRI images of the spine with the following: (**a**) Sagittal STIR, (**b**) Sagittal T2, (**c**) Sagittal T1, (**d**) Sagittal T1 C+, (**e**) Axial T2. (Figure courtesy of Dr. Samer Hoz)

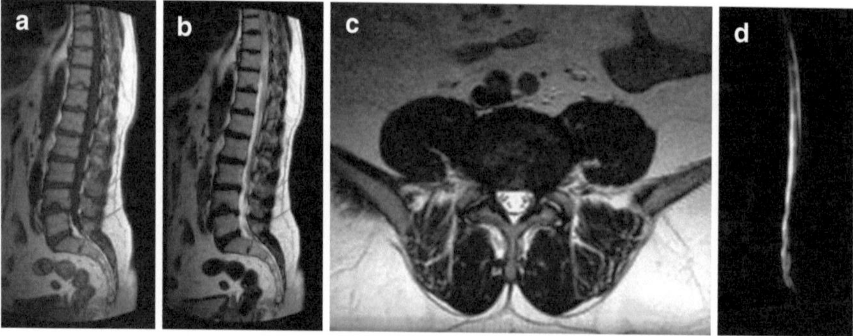

Fig. 14.5 Serial MRI images of the spine with the following: (**a**) Sagittal T1, (**b**) Sagittal T2, (**c**) Axial T2, (**d**) Myelography. (Figure courtesy of Dr. Samer Hoz)

Case 93.2: Chronic Schmorl Nodes

Case Scenario

A 50-year-old male presented with a 2-week history of low back pain without accompanying radiculopathy (Fig. 14.5).

Imaging Description

The MRI reveals an incidental finding of Schmorl nodes at multiple levels in the inferior endplates of lower dorsal vertebral bodies. At the L4/L5 level, there is a circumferential disc bulge indenting the thecal sac, causing relative obliteration of the bilateral neural foramina and resulting in compromise at the nerve roots with spinal canal stenosis.

Schmorl Nodes

Schmorl nodes, also known as intravertebral disc herniations, occur when the cartilage of the intervertebral disc protrudes through the endplate of the vertebral body into the adjacent vertebra. These nodes, prevalent in about 75% of autopsies, are more frequently observed in males and can affect individuals of all ages.

With regards to clinical presentation, chronic Schmorl nodes typically do not cause symptoms, and their role in causing back pain is debated. However, acute Schmorl nodes, which are less common, are associated with inflammation and may lead to symptomatic manifestations. They are considered one of the diagnostic criteria for Scheuermann disease and are closely related to the condition known as limbus vertebrae. The development of Schmorl nodes is believed to be associated with previous back trauma, although the precise mechanism remains incompletely understood. Recent research suggests that pressure from the nucleus pulposus on the weakest part of the endplate or disturbances during vertebral development in early life could contribute to their formation.

Radiographically, diagnosing Schmorl nodes in the acute stage using plain radiography can be challenging due to the absence of well-developed sclerosis around the herniation margins. However, small nodular lucent lesions are typically observed in the lower thoracic and lumbar vertebral bodies, primarily involving the inferior endplate. CT imaging can enhance identification, and MRI, especially sagittal sequences, provides clear visualization.

On MRI, Schmorl nodes exhibit signal characteristics similar to those of adjacent discs, with a thin rim of sclerosis at the margins. During the acute phase, Schmorl nodes may present more aggressively, showing signs of bone marrow edema and peripheral enhancement. The gas extrusion sign, characterized by hypointensity in the intervertebral disc space across all pulse sequences, may be observed. These acute features evolve gradually over several months, with some cases taking over a year to fully manifest.

When considering the differential diagnosis, it is limited when it comes to chronic Schmorl nodes, as they exhibit a distinct appearance. In the case of acute Schmorl nodes, differentials include malignancy and spondylodiscitis. However, acute Schmorl nodes can be differentiated from infection by the presence of more localized endplate changes, the absence of fever, and the absence of phlegmonous changes in the epidural or prevertebral region [5, 6].

Questions

1. **Schmorl nodes, the FALSE answer is:**
   A. Schmorl nodes can be easily diagnosed in the acute stage using plain radiography.
   B. The development of Schmorl nodes is believed to be associated with previous back trauma.
   C. The gas extrusion sign in Schmorl nodes appears as a hypointensity in the intervertebral disc space on all pulse sequences.
   D. Schmorl nodes are considered one of the diagnostic criteria for Scheuermann disease.
   E. Chronic Schmorl nodes typically do not cause symptoms.
   The answer is A.

   Acute Schmorl nodes are challenging to diagnose on plain radiography due to the absence of well-developed sclerosis around the herniation margins.

Case 94: Osteoporotic Fracture

Case Scenario

A 70-year-old female presented with acute-on-chronic back pain, specifically described as severe discomfort in the mid-back area (Fig. 14.6).

Imaging Description

On MRI, the bone marrow exhibits a heterogeneous signal appearance with rounded focal fatty lesions replacing normal marrow, coalescing to form a well-defined rounded lesion that appears hyperintense on both T1 and T2 sequences, and shows suppression in STIR at the L2 vertebral body (arrow). There are also osteoporotic wedge compression fractures noted at the D8 vertebral body. No paravertebral soft tissue edema or swelling is observed. These changes are more likely indicative of an old compression fracture and are unlikely to represent a malignant process. Additionally, Schmorl nodes are identified at L4, accompanied by a circumferential disc bulge at multiple levels, resulting in indentation of the thecal sac, relative obliteration of the bilateral neural foramina with compromise at the nerve roots, and spinal canal stenosis.

Osteoporosis

Osteoporosis is a metabolic bone disease characterized by a decrease in bone mass and increased vulnerability to fractures. According to the World Health Organization (WHO), osteoporosis is defined operationally as having a bone mineral density T-score lower than -2.5 standard deviations from the mean of young adults. This measurement is typically obtained using dual-energy X-ray absorptiometry and applies to postmenopausal women and men aged 50 and older. The femoral neck is the reference standard site for assessing bone mineral density, although other sites

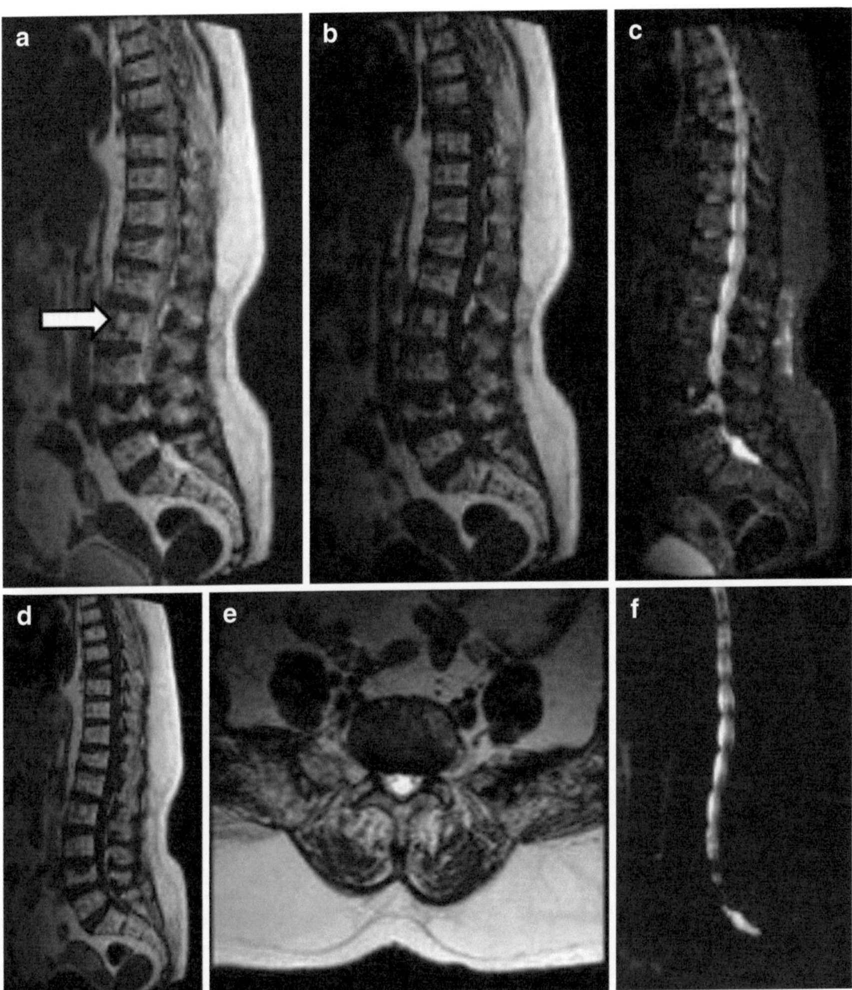

Fig. 14.6 Serial MRI images of the spine with the following: (**a**) Sagittal T2, (**b**) Sagittal T1, (**c**) Sagittal STIR, (**d**) Sagittal T1 C+, (**e**) Axial T2, (**f**) Myelography. (Figure courtesy of Dr. Samer Hoz)

such as the lumbar spine can also be used. In addition to measuring bone density, a clinical diagnosis of osteoporosis can be established without bone mineral density measurement based on the presence of fragility fractures, particularly at common locations such as the spine, hip, pelvis, wrist, humerus, or rib.

Risk factors play an important role in assessing the risk of fractures in individuals with osteoporosis. These include sex (females have a higher risk), age (older adults are more susceptible to fractures), body mass index (lower body mass carries higher risk), prior fragility fracture, parental history of hip fracture, current tobacco smoking, daily alcohol consumption of at least 3 units, long-term use of oral glucocorticoids (more than 3 months at a dose equivalent to at least 5 mg daily prednisolone), rheumatoid arthritis, and other causes of secondary osteoporosis. Secondary osteoporosis risk factors encompass conditions such as type I (insulin-dependent) diabetes, osteogenesis imperfecta in adults, untreated long-standing hyperthyroidism, hypogonadism, premature menopause (<45 years), chronic malnutrition, malabsorption, chronic liver disease, and HIV/AIDS, especially with certain antiretroviral therapies (e.g., tenofovir disoproxil fumarate).

Radiographic features reveal osteoporosis in its early stages, with reduced bone density and loss of bony trabeculae. Key bones like vertebrae, proximal femur, calcaneum, and tubular bones are examined for signs. Dual-energy X-ray absorptiometry is considered the gold standard for diagnosing osteoporosis. Vertebral osteoporosis may show narrowing of vertebrae, loss of cortical and trabecular bone, compression fractures, and vertebral plana. Bone mineral density measurement is used to estimate calcium hydroxyapatite levels in bones.

In CT imaging, quantitative CT attenuation values can provide information about bone mineral density, which correlates with traditional measurements. When performing routine CT scans at 120 kV, a trabecular measurement of less than 90–135 HU in the L1 vertebral body indicates osteoporosis. Higher values within this range are more sensitive, while lower values are more specific. Recently, quantitative ultrasound of the calcaneal bone has become a cost-effective screening method for assessing bone quality in osteoporosis.

In MRI, the bone marrow signal in osteoporosis shows a heterogeneous appearance, with rounded focal fatty lesions replacing normal marrow. Coalescence of these lesions can also occur. On T1W images, the signal is heterogeneously hyperintense, while on T2W images, the signal is variable. The signal characteristics of osteoporotic wedge compression fractures may vary depending on the age of the patient.

The use of bisphosphonates and denosumab, although rare, has been linked to serious side effects such as bisphosphonate-related atypical femoral fractures and bisphosphonate-related osteonecrosis of the jaw. In the differential diagnosis, ochronosis should be considered, which is characterized by severe osteoporosis accompanied by intervertebral disc calcifications [7, 8].

Questions

1. **Osteoporosis, the FALSE answer is:**
   A. Dual-energy X-ray absorptiometry is typically used to measure bone density.
   B. The femoral neck is the reference standard site for assessing bone mineral density.
   C. The diagnosis of osteoporosis cannot be established without measuring bone mineral density.
   D. Dual-energy X-ray absorptiometry is considered the gold standard for diagnosing osteoporosis.
   E. Plain radiography requires a significant bone loss of 30–50% to be evident.
   The answer is C.

The diagnosis of osteoporosis can be established without measuring bone mineral density.

2. **Osteoporosis, the FALSE answer is:**
   A. CT scan may show trabecular bone loss and cortical thinning in osteoporosis.
   B. MRI is more sensitive than CT in detecting early changes of osteoporosis.
   C. Quantitative CT can provide an accurate assessment of bone mineral density.
   D. On T1W MRI, osteoporotic bone appears as high signal intensity.
   E. On MRI, osteoporosis is characterized by increased signal intensity on T1W images and decreased signal intensity on T2W images.

   The answer is D.

   Osteoporotic bone typically appears as lower signal intensity on T1W MRI compared to normal bone due to the conversion of red marrow to fatty yellow marrow.

3. **Osteoporosis, the FALSE answer is:**
   A. Osteoporosis is defined as having a bone mineral density T-score lower than −2.5 SD.
   B. Fragility fractures can contribute to the clinical diagnosis of osteoporosis.
   C. Vertebral osteoporosis can be identified by the narrowing of vertebrae (pencilling) on radiographs.
   D. Thinning of the cortex in metacarpals indicates osteoporosis.
   E. Loss of trabeculae can be observed in the metaphysis of the femur.

   The answer is E.

   Loss of trabeculae can be observed in the proximal femur.

Case 95: Modic Type II Endplate Changes at L4–L5

Case Scenario

A 56-year-old male presented with a history of low back pain, predominantly worse in the morning, with no documented radiculopathy (Fig. 14.7).

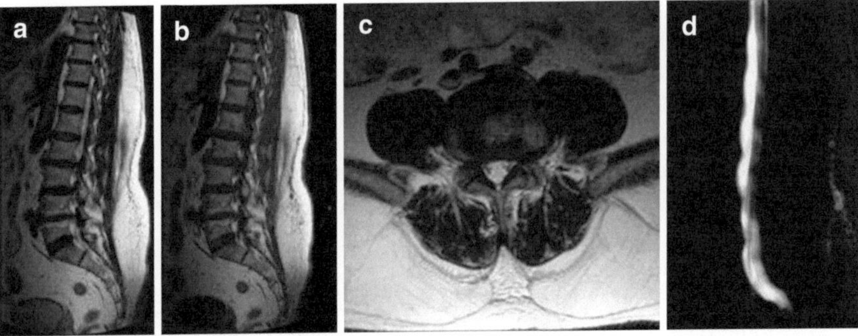

Fig. 14.7 Serial MRI images of the spine with the following: (**a**) Sagittal T2, (**b**) Sagittal T1, (**c**) Axial T2, (**d**) Myelography. (Figure courtesy of Dr. Samer Hoz)

Imaging Description

High T1 and T2 endplate changes at L4/L5 are indicative of Modic type II changes. Additionally, there is a circumferential disc bulge at the L4/L5 level, causing indentation of the thecal sac, relative obliteration of the bilateral neural foramina with compromise at the nerve roots, and spinal canal stenosis.

Modic-Type Endplate Changes

Modic-type endplate changes serve as a classification system for MRI signal changes in the vertebral body endplates. Named after the American neuroradiologist Michael T. Modic, this classification is widely recognized and utilized by radiologists and clinicians as a convenient way to report spine MRIs.

Modic type I changes are characterized by bone marrow edema and inflammation, appearing as low signal on T1W images, high signal on T2W images, and showing enhancement with gadolinium contrast. On the other hand, Modic type II changes represent the conversion of normal red hemopoietic bone marrow into yellow fatty marrow as a result of marrow ischemia. They appear as high signal on T1W images and iso to high signal on T2W images. Lastly, Modic type III changes indicate subchondral bony sclerosis, appearing as low signal on both T1W and T2W. In recent times, Modic type I changes have gained renewed attention due to the possibility of indicating low-grade indolent infection [9, 10].

Questions

1. **Modic-type endplate changes, the FALSE answer is:**
   A. Modic-type endplate changes serve as a classification system for MRI signal changes in the vertebral body endplates.
   B. Modic type I may indicate low-grade indolent infection.
   C. Modic type I changes appear as high signal on T1W images.
   D. Modic type II changes appear as high signal on T1W images.
   E. Modic type III changes indicate subchondral bony sclerosis with low signal intensity in both T1 and T2.
   The answer is C.
   Modic type I changes are characterized by bone marrow edema and inflammation, which appear as low signal on T1W images.
2. **Modic-type endplate changes, the FALSE answer is:**
   A. Modic type I changes typically represent bone edema and inflammation.
   B. Modic type III changes represent fatty degeneration and are often found in the context of degenerative disc disease.
   C. Modic type II changes are the most common type of Modic changes.
   D. Modic type III changes are associated with reduced disc height.
   E. Modic type I changes may be associated with discogenic low back pain.
   The answer is B.

Modic type III changes are characterized by subchondral bone sclerosis, not fatty degeneration. Modic type II changes represent fatty degeneration.

Case 96: Aggressive Vertebral Hemangioma

Case Scenario

A 30-year-old male presented with a history of back pain persisting for several weeks (Fig. 14.8).

Imaging Description

The MRI findings indicate the presence of an aggressive spinal hemangioma causing compression of the spinal canal. This is evident as a thickening that is hypointense in T1WI and hyperintense in STIR and T2WI. There is an epidural extension (arrow) observed as an extraosseous component. These features are consistent with

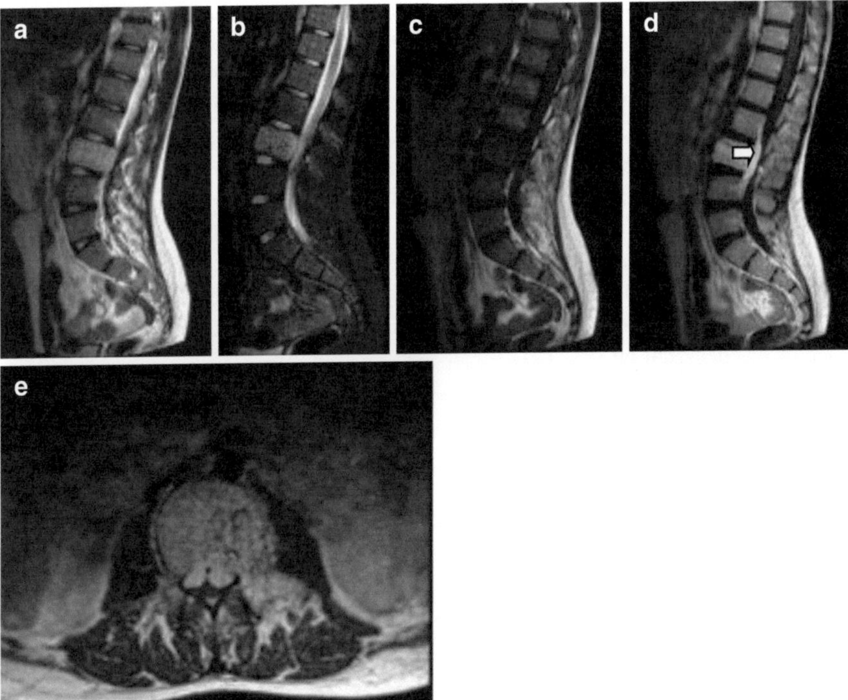

Fig. 14.8 Serial MRI images of the spine with the following: (**a**) Sagittal T2, (**b**) Sagittal STIR, (**c**) Sagittal T, (**d**) Sagittal T1 C+, (**e**) Axial T1 C+. (Figure courtesy of Dr. Samer Hoz)

the usual features of a hemangioma across all pulse sequence signals, and there is uniform enhancement post-contrast.

Aggressive Vertebral Hemangiomas

Aggressive vertebral hemangiomas are a rare variant of vertebral hemangiomas, characterized by notable vertebral expansion, the presence of an extra-osseous component with epidural extension, disrupted blood flow, and occasional compression fractures that lead to the compression of the spinal cord and/or nerve roots. While this condition can manifest at any age, it is most commonly observed in young adults, reaching peak prevalence during that period. They account for approximately 1% of spinal hemangiomas.

Around 75% of these lesions are found in the thoracic spine, specifically between vertebral segments D3 and D9. Unlike typical vertebral hemangiomas, which are often asymptomatic, the aggressive form almost always presents with neurological symptoms. This is primarily attributed to the epidural component exerting pressure on the spinal cord, nerve roots, or both, resulting in compressive myelopathy and/or radiculopathy.

Radiographically, CT scans reveal hypodense, expansive masses within the vertebrae, accompanied by cortical defects, soft tissue extension, and compression of the spinal cord or nerve roots. The presence of thickened vertebral trabeculae gives rise to the classic "polka dot" and "corduroy" signs in the vertebral body. These lesions generally occupy the entire vertebral body, extend into the neural arch, expand the osseous margins, and contain a soft tissue component.

MRI imaging reveals low signal areas corresponding to the thickened trabeculae on both T1 and T2 images. The extra-osseous component exhibits typical hemangioma characteristics in all pulse sequences, including high T1 and T2 signals, as well as uniform enhancement after contrast administration. MRI is particularly valuable for evaluating compression of the spinal cord or nerve roots. Chemical shift imaging can also be employed to identify signal dropout, indicating the presence of intralesional fat [11, 12].

Differential Diagnosis

When considering a differential diagnosis, other conditions such as plasmacytoma, metastases, lymphoma, and chordoma should be taken into account. Plasmacytomas exhibit a characteristic "mini-brain" appearance. Metastases typically display decreased signal intensity on T1 and increased signal intensity on T2 images. Lymphoma may show a hypointense epidural component on T1 and less hyperintensity on T2. Chordomas can resemble aggressive vertebral hemangiomas on T2 and post-contrast imaging, but they generally lack a high T1 signal. Additionally, the presence of frequent peripheral calcifications in chordomas can mimic the thickened trabeculae seen in vertebral hemangiomas.

Questions

1. **Aggressive vertebral hemangiomas, the FALSE answer is:**
   A. The presence of thickened vertebral trabeculae gives rise to the classic "polka dot" and "corduroy" signs in the vertebral body.
   B. CT scans show hyperdense, expansive masses within the vertebrae in hemangiomas.
   C. MRI imaging reveals low signal areas corresponding to the thickened trabeculae on both T1 and T2 images.
   D. MRI is particularly useful for evaluating compression of the spinal cord or nerve roots in hemangiomas.
   E. The most important differential diagnoses are plasmacytoma, metastases, lymphoma, and chordoma.
   The answer is B.
   CT scans typically show hypodense (not hyperdense) expansive masses within the vertebrae in hemangiomas.
2. **Differential diagnosis of aggressive vertebral hemangiomas, the FALSE answer is:**
   A. Plasmacytoma is a differential diagnosis that exhibits a characteristic "mini-brain" appearance.
   B. Metastases typically display decreased signal intensity on T1 and increased signal intensity on T2 images.
   C. Lymphoma may show a hypointense epidural component on T1 and less hyperintensity on T2.
   D. Chordomas can resemble aggressive vertebral hemangiomas on T2.
   E. Chordomas show increased signal intensity on T1.
   The answer is E.
   Generally, chordomas do not exhibit increased signal intensity on T1.

References

1. Fenn J, Olby NJ, Canine Spinal Cord Injury Consortium (CANSORT-SCI). Classification of intervertebral disc disease. Front Vet Sci. 2020;7:579025.
2. Ma XL. A new pathological classification of lumbar disc protrusion and its clinical significance. Orthop Surg. 2015;7(1):1–2.
3. Kim SK, Shin K, Song Y, Lee S, Kim TH. Andersson lesions of whole spine magnetic resonance imaging compared with plain radiography in ankylosing spondylitis. Rheumatol Int. 2016;36:1663–70.
4. Bron JL, de Vries MK, Snieders MN, van der Horst-Bruinsma IE, Van Royen BJ. Discovertebral (Andersson) lesions of the spine in ankylosing spondylitis revisited. Clin Rheumatol. 2009;28:883–92.
5. Dar G, Peleg S, Masharawi Y, Steinberg N, May H, Hershkovitz I. Demographical aspects of Schmorl nodes: a skeletal study. Spine. 2009;34(9):E312–5.
6. Mattei TA, Rehman AA. Schmorl's nodes: current pathophysiological, diagnostic, and therapeutic paradigms. Neurosurg Rev. 2014;37:39–46.

7. Link TM. Osteoporosis imaging: state of the art and advanced imaging. Radiology. 2012;263(1):3–17.
8. Martel D, Monga A, Chang G. Osteoporosis imaging. Radiol Clin. 2022;60(4):537–45.
9. Udby PM, Samartzis D, Carreon LY, Andersen MØ, Karppinen J, Modic M. A definition and clinical grading of Modic changes. J Orthop Res. 2022;40(2):301–7.
10. Zhang YH, Zhao CQ, Jiang LS, Chen XD, Dai LY. Modic changes: a systematic review of the literature. Eur Spine J. 2008;17(10):1289–99.
11. Vasudeva VS, Chi JH, Groff MW. Surgical treatment of aggressive vertebral hemangiomas. Neurosurg Focus. 2016;41(2):E7.
12. Gaudino S, Martucci M, Colantonio R, Lozupone E, Visconti E, Leone A, Colosimo C. A systematic approach to vertebral hemangioma. Skeletal Radiol. 2015;44:25–36.

Spinal Miscellaneous Lesions

15

Hasan M. Jabbar, Mohammed E. Al-Hamadani,
Teeba A. Al-Ageely, Hagar A. Algburi, Abbas F. A. Hussein,
Ahmed Muthana, and Asmaa H. AL-Sharee

Case 97: FASI

Case Scenario

A 30-year-old male with a known case of NF1 presented with cognitive impairment (Fig. 15.1).

Imaging Description

The MRI of a known case of NF1 reveals multiple FASI in the brain and spine. Abnormally hyper-intense areas are observed in T2 and SPIR sequences in the spinal cord at the C2 and C3 levels. These areas appear iso-intense in T1 without causing any mass effect, and no enhancement is observed.

FASI

FASI, also known as unidentified bright objects or focal abnormal signal intensity, are hyper-intense bright spots seen on T2W MRI, appearing more prominently on

H. M. Jabbar
College of Medicine, University of Misan, Misan, Iraq

M. E. Al-Hamadani · T. A. Al-Ageely · H. A. Algburi · A. Muthana
College of Medicine, University of Baghdad, Baghdad, Iraq

A. F. A. Hussein
College of Medicine, Babylon University, Babylon, Iraq

A. H. AL-Sharee (✉)
Department of Neuroradiology, Neurosurgery Teaching Hospital, Baghdad, Iraq

© The Author(s), under exclusive license to Springer Nature Switzerland AG 2024
S. Hoz et al. (eds.), *Neuroradiology Board's Favorites*,
https://doi.org/10.1007/978-3-031-64261-6_15

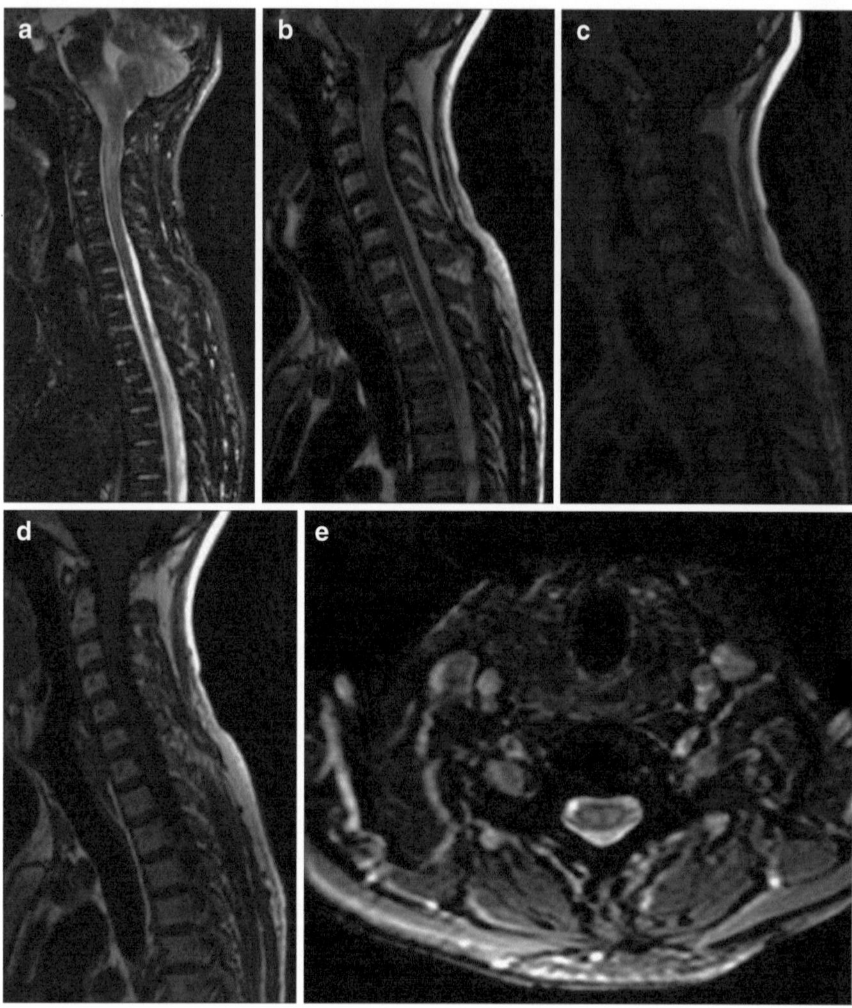

Fig. 15.1 Serial MRI images of the spine with the following: (**a**) Sagittal STIR, (**b**) Sagittal T2, (**c**) Sagittal T1, (**d**) Sagittal T1 C+, (**e**) Axial T2. (Figure courtesy of Dr. Samer Hoz)

FLAIR images. FASI are considered a benign process caused by increased fluid accumulation in the intra-myelinic vacuoles. However, they may be associated with hyperplastic or dysplastic glial proliferation. They are usually located in the basal ganglia, thalamus, brainstem, optic tracts, cerebellum, and cerebellar white matter.

FASI localization within the spine has recently been reported, as observed in this case. In NF1, the intramedullary lesions are typically low-grade astrocytomas (15% of patients). Rarely, cases of ependymomas and gangliogliomas have also been reported. FASI are the most frequent neuroimaging feature encountered in NF1 patients and is well documented in the literature, with approximately 86% of children with NF1 exhibiting one or more FASI. However, FASI are not included in the

National Institutes of Health diagnostic criteria for NF1. An increase in either size or number of FASI is common in patients younger than 10, but the occurrence of such an increase beyond 10 years of age raises concern for a neoplasm.

On MRI, FASI have no mass effect and appear as iso-intense to hyper-intense on T1W images and hyper-intense on T2W and FLAIR. Additionally, T1 C+ (Gd) shows a lack of contrast enhancement, although enhancing FASI have the potential to regress gradually. MRS, which can be useful to distinguish from tumors, is normal [1, 2].

Questions

1. **FASI, the FALSE answer is:**
   A. FASI are caused by increased fluid accumulation in the intra-myelinic vacuoles.
   B. Intramedullary FASI can be seen in patients with NF1.
   C. FASI may be associated with increased glial cell proliferation.
   D. An increase in the size or number of FASI after the age of 10 raises concern for a neoplasm.
   E. Ependymomas are the most common intramedullary spinal cord tumors associated with NF1.
   The answer is E.
   Astrocytomas are the most common intramedullary spinal cord tumors in NF1.

2. **FASI, the FALSE answer is:**
   A. FASI are iso-intense to hyper-intense on T1WI.
   B. FASI are hypo-intense on T2WI/FLAIR.
   C. FASI have no mass effect on MRI.
   D. Enhancing FASI may have the potential to regress gradually.
   E. MRS can distinguish FASI from tumors.
   The answer is B.
   FASI are hyper-intense on T2WI/FLAIR.

Case 98: Dorsal Thoracic Arachnoid Web

Case Scenario

A 60-year-old female with a history of 3 weeks of worsening back pain, ataxia, and lower limb sensory changes (Fig. 15.2).

Imaging Description

A T2W MRI reveals a prominence of the dorsal arachnoid space at the D5–D6 disc space level, with consequent ventral displacement of the spinal cord denoting a possible "scalpel" sign. Nonetheless, the absence of enhancement in the post-contrast

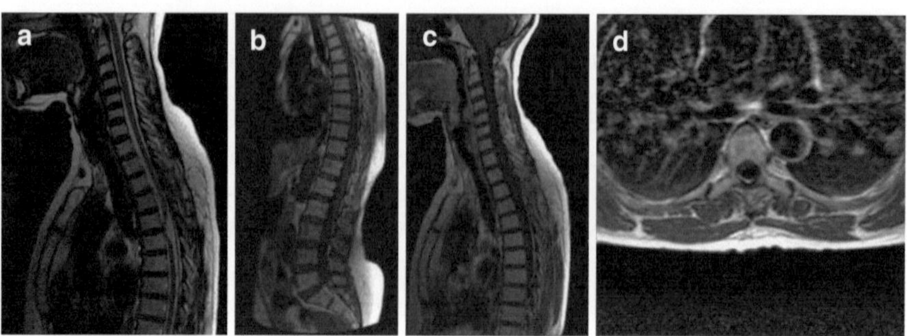

Fig. 15.2 MRI images of the spine with the following: (**a**) Sagittal T2, (**b**) Sagittal T1, (**c**) Sagittal T1 C+, (**d**) Axial T1 C+. (Figure courtesy of Dr. Samer Hoz)

sagittal and axial images excludes tumor. Differential diagnosis includes a dorsal arachnoid web and ventral cord herniation.

Dorsal Thoracic Arachnoid Web

A dorsal thoracic arachnoid web is an exceedingly rare condition characterized by a band-like intradural extramedullary thickening of arachnoid tissue in the dorsal aspect of the spinal cord, often causing a focal deformity by compressing the thoracic cord with consequent neurological dysfunction. As it is almost exclusively found in the thoracic region, it is referred to as the thoracic arachnoid web rather than the arachnoid web of the spine. Clinical presentation typically involves the lower extremities, presenting with myelopathy, progressive localized back pain, and most commonly, motor and sensory weakness, according to the level affected.

The characteristic imaging feature, known as the "scalpel sign," is identified by a focal dorsal indentation and ventral shift of the thoracic spinal cord, with concomitant widening of the dorsal CSF space. The expanded dorsal CSF space resembles the silhouette of a surgical scalpel on sagittal imaging, hence the origin of the term "scalpel sign." While MRI is the preferred modality for visualizing this deformity, a CT myelogram can be considered as an alternative. The thoracic cord above or below the thickened band usually shows a high-intensity T2 signal, and in approximately two-thirds of the cases, it appears with a defined fluid-filled cyst (syrinx). Direct visualization of the web is typically unachievable by means of routine imaging; the diagnosis is based on the key finding of an indentation and displacement of the thoracic cord [3, 4].

Differential Diagnosis

- Ventral cord herniation: There is a focal distortion at the point of herniation, as the lesion is due to the pulling of the cord rather than compression. Due to herniation, the space between the cord and theca will be closed ventrally.
- Dorsal spinal arachnoid cyst: Looks very similar to the arachnoid web; hence, CT myelography is used to demonstrate the cyst, which fills with contrast at a slower rate. Unlike the dorsal thoracic arachnoid web, the distortion pattern of the spinal cord is more widespread, lacking the characteristic scalpel sign.

Questions

1. **Dorsal thoracic arachnoid web, the FALSE answer is:**
   A. The arachnoid web is an intradural extramedullary thickening.
   B. The web causes dorsal displacement of the spinal cord.
   C. Males are more affected than females.
   D. The web is associated with the presence of a syrinx.
   E. Almost always occurs in the thoracic region.
   The answer is B.
   The web causes ventral displacement of the spinal cord.
2. **Dorsal thoracic arachnoid web, the FALSE answer is:**
   A. The scalpel sign reflects the shape of the expanded dorsal CSF space on sagittal imaging.
   B. The cord adjacent to the web shows a hyper-intense T2 signal.
   C. The scalpel sign is best seen by MRI.
   D. Diagnosis can be made without direct visualization of the web.
   E. MRI is the modality of choice to differentiate arachnoid webs from arachnoid cysts.
   The answer is E.
   CT myelography is the modality of choice to differentiate arachnoid webs from arachnoid cysts.

Case 99: Spinal Synovial Cyst

Case Scenario

A 30-year-old male presented with neck pain and radiculopathy predominantly affecting the left side (Fig. 15.3).

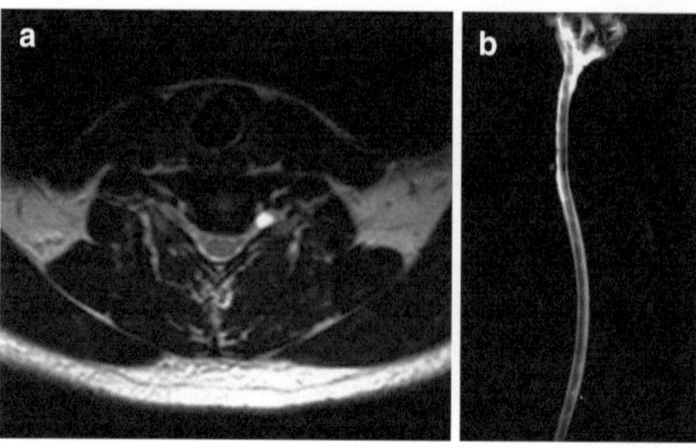

Fig. 15.3 MRI images of the spine with the following: (**a**) Axial T2, (**b**) Myelography. (Figure courtesy of Dr. Samer Hoz)

Imaging Description

The MRI reveals a posterior diffuse disc bulge at the C6–C7 level, causing indentation of the thecal sac and encroaching on the exiting foramina bilaterally as well as both lateral recesses, with more pronounced involvement on the left side. There is also evidence of a left facet joint synovial cyst.

Spinal Synovial Cyst

Spinal synovial cysts, also known as facet joint cysts, are benign cystic formations that develop within a synovium-lined cavity of a facet joint, which is lined with cuboid or pseudostratified columnar epithelium and enclosing synovial fluid. Synovial cysts are typically associated with adjacent facet joint degenerative or traumatic arthropathy. Additionally, they may be asymptomatic and found incidentally during MRI, or they may result in radiculopathy in a significant number of cases.

Compared to the cervical and thoracic regions, the lumbar spine is the most common site of synovial cysts, with an increased tendency toward the L4–L5 level. Indeed, spinal synovial cysts are a cause of peripherally enhancing masses in the extrathecal space anywhere throughout the spinal canal. On CT scans of the spine, a synovial cyst appears as a calcified cystic lesion next to a facet joint, which is often affected by arthropathy. Moreover, gas within the cyst can often be seen.

MRI is the modality of choice for the diagnosis of a synovial cyst. On MRI, synovial cysts typically appear hypo-intense to iso-intense on T1W images and hyper-intense on T2W, but the signal varies depending on several factors, including the proteinaceous content and blood products within the cyst. Furthermore, calcification within the cyst wall shows low signal intensity on T1W and T2W, whereas hemorrhagic cysts display increased intensity compared to CSF, probably due to T1 shortening from methemoglobin. In addition, the demonstration of gas within the cyst is pathognomonic for a synovial cyst [5–7].

Questions

1. **Spinal synovial cyst, the FALSE answer is:**
   A. They are lined with cuboid or pseudostratified columnar epithelium.
   B. The most common site of spinal synovial cysts is the L4–L5 level.
   C. Myelopathy is the most common presenting symptom.
   D. They are typically associated with facet joint arthropathy.
   E. Trauma plays a role in the development of spinal synovial cysts.
   The answer is C.
   Radiculopathy is the most common presenting symptom.
2. **Spinal synovial cyst, the FALSE answer is:**
   A. On CT scan, they appear as calcified cystic lesions adjacent to facet joints.
   B. Spinal synovial cysts appear hyper-intense on T2WI.
   C. Spinal synovial cysts appear hypo-intense on T1WI.
   D. Non-hemorrhagic spinal synovial cysts typically appear hypo-intense on T2WI.
   E. The demonstration of gas within the cyst is pathognomonic for spinal synovial cyst.
   The answer is D.
   Non-hemorrhagic spinal synovial cysts typically appear hyper-intense on T2WI.

Case 100: Bilateral Acoustic Schwannomas: NF2

Case Scenario

A 21-year-old female with a history of lower limb weakness presented with bilateral hearing loss (Fig. 15.4).

Imaging Description

On the MRI, the post-contrast T1 axial imaging reveals bilateral enhancing masses in the cerebellopontine angle. These masses extend into the internal acoustic meatuses, consistent with acoustic schwannomas. The bilateral nature of these masses almost certainly indicates the presence of NF2. Additionally, there are prominent enhancing lesions spread throughout the spinal canal at the lumbar level from D12 to L4, protruding through some vertebral foramina. There is bilateral lateral expansion of a meningocele through multilevel lumbar foramina with a circumferential dural envelope. There is also scalloping observed on the posterior aspect of the lower lumbar vertebrae.

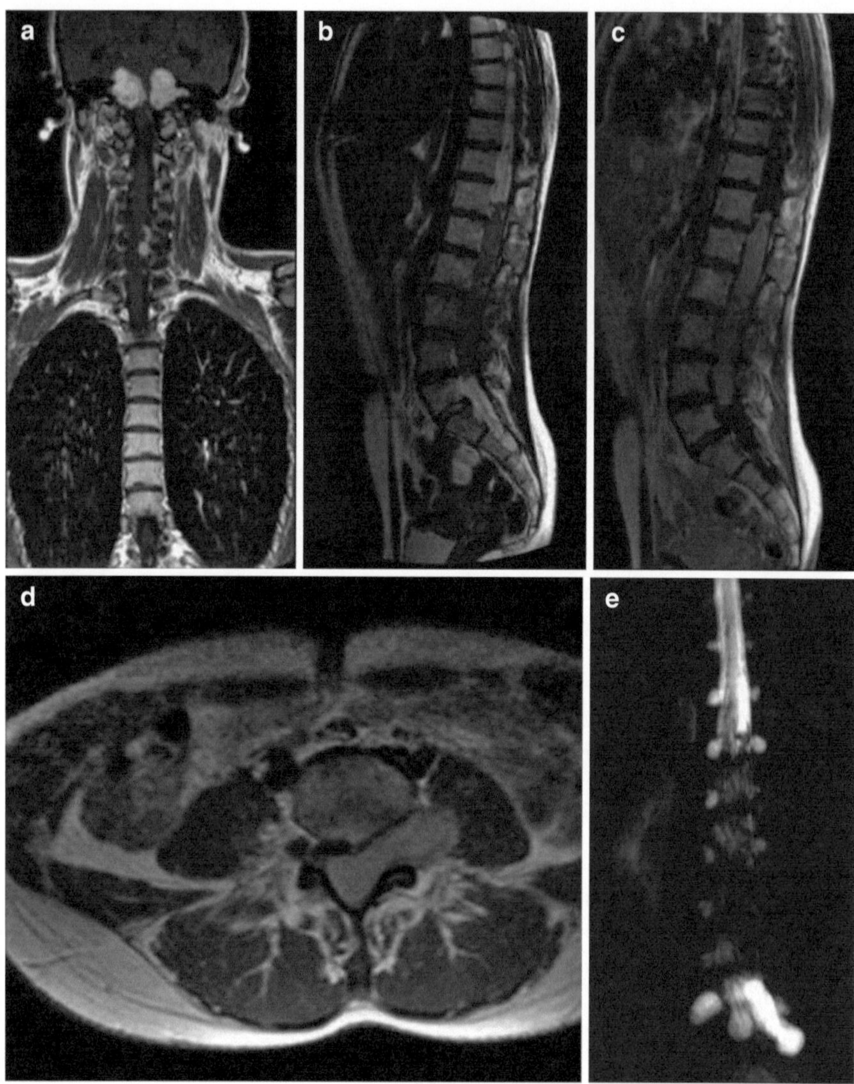

Fig. 15.4 Serial MRI images of the spine with the following: (**a**) Coronal T1 C+, (**b**) Sagittal T2, (**c**) Sagittal T1 C+, (**d**) Axial T1 C+, (**e**) Myelography. (Figure courtesy of Dr. Samer Hoz)

Neurofibromatosis 2 (NF2)

NF2 is a rare autosomal dominant neurocutaneous disorder (phakomatosis), with an estimated prevalence of 1 in every 50,000 individuals. Typically manifesting in young adults aged 18–24, NF2 is distinguished by the growth of multiple CNS tumors, unlike NF1, which involves peripheral nerves. CNS lesions in NF2 include intracranial schwannomas, intracranial and spinal meningiomas, and intraspinal-intramedullary ependymomas. Intracranial schwannomas typically present as vestibular schwannomas; bilateral vestibular schwannomas are a distinctive feature of NF2. Following this, spinal schwannomas are the second most common intracranial schwannoma. The primary manifestations of NF2 can be summarized with the mnemonic MISME (multiple inherited schwannomas, meningioma, and ependymomas). Notably, neurofibromas are not a typical manifestation of NF2, making the term a misnomer.

The pattern of NF2 occurrence appears to be inherited from an affected parent in 50% of cases (autosomal dominant), while the other 50% occur spontaneously without family history (de novo mutation). Meningiomas in adults are usually isolated findings and do not strongly suggest NF2. However, in children, meningiomas, especially when multiple or involving the spine, raise suspicion for NF2. NF2 can also be associated with spinal lesions such as syringohydromyelia and ocular manifestations like cataracts.

MRI is the preferred modality for both diagnosis and characterization. Meningiomas typically exhibit intense and homogeneous enhancement in T1 post-contrast images. For schwannomas, which can arise from the facial nerve or the inferior vestibular division of the vestibulochochlear nerve, MRI shows iso-intense to hypo-intense lesions in T1 and intense enhancement in T1 post-contrast. Additional signs like the split-fat sign, target sign, and fascicular sign can aid in diagnosis. For spinal ependymomas, which commonly affect the cervical cord, plain films may show features such as scoliosis, spinal canal widening, vertebral body scalloping, pedicle erosion, and laminar thinning. On MRI T1, ependymomas often appear iso-intense to hypo-intense [8–10].

Questions

1. **NF2, the FALSE answer is:**
   A. NF2 affects the CNS rather than peripheral nerves.
   B. Neurofibroma is not a feature of NF2.
   C. Multiple spinal tumors suggest NF2.
   D. Meningiomas in adults are commonly associated with NF2.
   E. Bilateral vestibular schwannomas are a distinctive feature of NF2.
      The answer is D.
      Meningiomas in adults are usually isolated findings and do not strongly suggest NF2.
2. **NF2 imaging, the FALSE answer is:**
   A. Intense enhancement is seen in schwannoma on post-contrast T1 MRI.
   B. Meningiomas exhibit an intense homogenous enhancement on T1 post-contrast.
   C. The split-fat sign, target sign, and fascicular sign are seen in meningiomas.
   D. Ependymomas are iso- to hypo-intense on T1WI.
   E. Plain radiographs of ependymoma show structural changes in the spinal column.
      The answer is C.
      The split-fat sign, target sign, and fascicular sign are seen in schwannomas.

References

1. Calvez S, Levy R, Calvez R, Roux CJ, Grévent D, Purcell Y, Beccaria K, Blauwblomme T, Grill J, Dufour C, Bourdeaut F. Focal areas of high signal intensity in children with neurofibromatosis type 1: expected evolution on MRI. Am J Neuroradiol. 2020;41(9):1733–9.
2. D'Amico A, Mazio F, Ugga L, Cuocolo R, Cirillo M, Santoro C, Perrotta S, Melis D, Brunetti A. Medullary unidentified bright objects in neurofibromatosis type 1: a case series. BMC Pediatr. 2018;18(1):1–5.
3. Laxpati N, Malcolm JG, Tsemo GB, Mustroph C, Saindane AM, Ahmad F, Refai D, Gary MF. Spinal arachnoid webs: presentation, natural history, and outcomes in 38 patients. Neurosurgery. 2021;89(5):917–27.
4. Reardon MA, Raghavan P, Carpenter-Bailey K, Mukherjee S, Smith JS, Matsumoto JA, Yen CP, Shaffrey ME, Lee RR, Shaffrey CI, Wintermark M. Dorsal thoracic arachnoid web and the "scalpel sign": a distinct clinical-radiologic entity. Am J Neuroradiol. 2013;34(5):1104–10.
5. Mak D, Vidoni A, James S, Choksey M, Beale D, Botchu R. Magnetic resonance imaging features of cervical spine intraspinal extradural synovial cysts. Can Assoc Radiol J. 2019;70(4):403–7.
6. Bruder M, Cattani A, Gessler F, Droste C, Setzer M, Seifert V, Marquardt G. Synovial cysts of the spine: long-term follow-up after surgical treatment of 141 cases in a single-center series and comprehensive literature review of 2900 degenerative spinal cysts. J Neurosurg Spine. 2017;27(3):256–67.
7. Cannarsa G, Clark SW, Chalouhi N, Zanaty M, Heller J. Hemorrhagic lumbar synovial cyst: case report and literature review. Nagoya J Med Sci. 2015;77(3):481.
8. Ahlawat S, Blakeley JO, Langmead S, Belzberg AJ, Fayad LM. Current status and recommendations for imaging in neurofibromatosis type 1, neurofibromatosis type 2, and schwannomatosis. Skeletal Radiol. 2020;49:199–219.
9. Alkadeem RM, El-Shafey MH, Eldein AE, Nagy HA. Magnetic resonance diffusion tensor imaging of acute spinal cord injury in spinal trauma. Egypt J Radiol Nucl Med. 2021;52(1):1–3.
10. Coy S, Rashid R, Stemmer-Rachamimov A, Santagata S. An update on the CNS manifestations of neurofibromatosis type 2. Acta Neuropathol. 2020;139:643–65.